D1000382

OXYGEN-ENHANCED COMBUSTION

OXYGEN-ENHANCED COMBUSTION

Edited by

CHARLES E. BAUKAL, Jr., Ph.D.

Air Products and Chemicals, Inc.
Allentown, Pennsylvania

CRC Press
Boca Raton London New York Washington, D.C.

Acquiring Editor: Robert Stern
Project Editor: Andrea Demby
Marketing Manager: Jane Stark
Cover design: Dawn Boyd
PrePress: Greg Cuciak
Manufacturing: Carol Royal

Library of Congress Cataloging-in-Publication Data

Oxygen-enhanced combustion / edited by Charles E. Baukal, Jr.
p. cm.
Includes bibliographical references and index.
ISBN 0-8493-1695-2 (alk. paper)
1. Combustion engineering. 2. Combustion gases--Environmental aspects.. I. Baukal, Charles E..
TJ254.5.099 1998
621.402'3--dc21 97-46803
CIP

No claim to original U.S. Government works
International Standard Book Number 0-8493-1695-2
Library of Congress Card Number 97-46803
Printed in the United States of America 2 3 4 5 6 7 8 9 0
Printed on acid-free paper

Preface

This book is a compilation of information about using oxygen to enhance industrial combustion processes. It is important to state what is not covered in this book. It does not include the use of oxygen to enhance internal combustion (IC) engines, gas turbines, or other pressurized combustors including those used in aerospace propulsion. In the case of the use of oxygen in IC engines, this is an emerging area which is currently receiving a significant amount of attention for reducing carbon monoxide and hydrocarbon emissions.[1] Very little work has been done on using oxygen in turbines, primarily because of the temperature limitations for the materials of construction. One of the challenges of turbine design is adequate cooling for the metal parts. As will be shown in the book, using oxygen in combustion normally increases the flame temperature which would only exacerbate the problem in the turbine. Much work has been done by NASA in studying the use of liquid oxygen in rocket propulsion. The interested reader is advised to consult relevant works in the aerospace industry. In this book, there is also very little discussion of the use of oxygen in the power generation industry because of the limited use in that application at the present time. As the cost of oxygen continues to decline, this may become an important application in the future. There are some specialty uses for oxygen in combustion, such as oxygen/acetylene torches, that are not included here since they are usually adequately covered in detail in other publications. In many molten metal processes in the steel industry, oxygen is injected into and reacts with the liquid metal. For example, oxygen is injected into molten steel to react with and reduce the carbon in the bath to the design level for the specific grade of steel. Although this could be considered to be a combustion application because of the high-temperature reaction of oxygen with some of the chemicals in the molten metal which could be considered to be fuels, it is not considered here as oxygen injection into molten metals is not a traditional type of combustion and it is fairly specialized in nature.

This book addresses high-temperature, atmospheric-pressure heating and melting processes. Currently, there is very little information on this subject in standard combustion textbooks. The main area discussed in those books related to the use of oxygen is the combustion of hydrogen and oxygen because of its relatively simple chemistry. However, this type of combustion process is rarely used in industry and therefore has little relevance for most users. This book has a decided bias toward the use of natural gas as the fuel which is by far the most predominant choice in the U.S. However, there is a chapter which does discuss the use of other fuels, including both solid and liquid fuels.

This book is targeted primarily toward that end user, the one who is responsible for implementing the use of oxygen in a combustion system. However, others should find the book of interest as well, including combustion equipment and industrial gas suppliers, industrial and academic researchers, funding agencies who sponsor research on advanced combustion technologies, and government agencies that are responsible for modifying existing and developing new regulations related to the safe and efficient use of combustion systems and to the minimization of pollutant emissions.

As with any book of this type, it is necessarily limited in scope. Every attempt has been made to include the important commercialized uses of oxygen-enhanced combustion (OEC) in a wide range of industries. Since this is a dynamic and constantly changing area of technology, new uses of OEC are continuously being developed. This book provides the basic information on OEC which can be used to evaluate those emerging developments.

The book is divided into three parts. The first part concerns the fundamentals of using oxygen in combustion and consists of the first four chapters on basic principles, pollutant emissions, oxygen production, and heat transfer. The middle part of the book concerns the application of oxygen in specific industries and consists of Chapters 5 through 8 on ferrous metals, nonferrous metals, glass, and incineration. The last part of the book concerns equipment considerations for the use of oxygen in combustion and consists of Chapters 9 through 11 on safety, equipment design, and fuels. The book contains almost 150 figures, over 50 tables, and over 250 references.

REFERENCES

1. Poola, R. B., Ng, H. K., Sekar, R. R., Baudino, J. H., and Colucci, C. P., Utilizing Intake-Air Oxygen-Enrichment Technology to Reduce Cold-Phase Emissions, SAE Technical Paper 952420, Warrendale, PA, 1995.

Acknowledgments

All of the authors owe a deep debt of gratitude to Air Products and Chemicals, Inc. for its support of this effort and for allowing us to publish this information. In addition, Chuck Baukal would like to thank his wife Beth for her support and encouragement during the preparation of this book. Roger McGuinness (Chapter 3) would like to thank George Dous and Cindy Riedy for drafting and editing; Rakesh Agrawal, Richard O'Reilly, and Tom Copeman for their advice and review; and Bill Kotke for providing a mechanical engineering perspective of the section on turbo-expanders. Marie DeGregorio Kistler and Scott Becker would like to thank Tom Ward and Debabrata Saha for their help with Chapter 5. Debabrata Saha gratefully acknowledges the assistance of Jane McQuillan and Lori DaCosta in typing the manuscript and preparing figures. Buddy Eleazer would like to thank his children for their patience while he spent evenings working on Chapter 7 and the glass industry for allowing us the chance to explore their heating processes. Mark Niemkiewicz and Scott Becker would like to thank James G. Hansel for his expertise in oxygen safety and flammability and Barry L. Werley, who is a world-renowned expert in oxygen safety and compatibility. Mark Niemkiewicz would also like to thank his wife, Marie, and father, John, who greatly assisted in editing Chapters 9 and 10. Yanping Zhang would like to thank his wife, Shuxiang, and his father for taking care of his newborn son and his two-year-old daughter when he wrote Chapter 11 at nights and on weekends at home. He would also like to thank Leon Chang of Air Products China, Thomas Niehoff of Air Products GmbH, and Richard Wilson of Air Products South Africa for providing data on fuels commonly used in China, Germany, and South Africa.

Table of Contents

1 Basic Principles

Charles E. Baukal, Jr.

CONTENTS

0-8493-1695-2/98/$0.00+$.50

1.1 INTRODUCTION

Most industrial heating processes require substantial amounts of energy, which are commonly generated by combusting such hydrocarbon fuels as natural gas or oil. Most combustion processes use air as the oxidant. In many cases, these processes can be enhanced by using an oxidant that contains a higher proportion of O_2 than that in air. This is known as *oxygen-enhanced combustion* or OEC. Air consists of approximately 21% O_2 and 79% N_2, by volume. One example of OEC uses an oxidant consisting of air blended with pure O_2. Another example uses high-purity O_2 as the oxidant, instead of air. This is usually referred to as *oxy/fuel* combustion.

New developments have made oxy/fuel combustion technology more amenable to a wide range of applications. In the past, the benefits of using oxygen could not always offset the added costs. New oxygen generation technologies, such as pressure and vacuum swing adsorption (see Chapter 3), have substantially reduced the cost of separating O_2 from air. This has increased the number of applications in which using oxygen to enhance performance is cost-justified. Another important development is the increased emphasis on the environment. In many cases, OEC can substantially reduce pollutant emissions (see Chapter 2). This has also increased the number of cost-effective applications. The Gas Research Institute in Chicago[1] and the U.S. Department of Energy[2] have sponsored independent studies that predict that OEC will be a critical combustion technology in the very near future.

Historically, air/fuel combustion has been the conventional technology used in nearly all industrial heating processes. OEC systems are becoming more common in a variety of industries. Where traditional air/fuel combustion systems have been modified for OEC, many benefits have been demonstrated. Typical improvements include increased thermal efficiency, increased processing rates, reduced flue gas volumes, and reduced pollutant emissions.

The use of oxygen in combustion has received relatively little attention from the academic combustion community. This may be for several reasons. Probably the most basic reason is the lack of research interest and funding to study OEC. The industrial gas companies that produce oxygen have been conducting research into OEC for many years which has been mostly applied R&D. Very little basic research has been done, compared with air/fuel combustion, to study the fundamental processes in atmospheric flames utilizing OEC. The aerospace industry has done a considerable amount of work, for example, to study the high-pressure combustion of liquid oxygen and liquid hydrogen used to propel space vehicles. However, that work has little relevance to the low-pressure combustion of fuels other than hydrogen in industrial furnace applications. Another reason little research has been done may be due to concerns about the safety issues of using oxygen, as well as the very high temperature flames that may be encountered using OEC. Yet another reason may be a cost issue since the small quantities of oxygen that might be used can be relatively expensive. Handling oxygen cylinders takes more effort than using either a houseline source of air or a small blower for the air used in small-scale flames.

1.2 OXYGEN

Oxygen is a colorless, odorless, tasteless gas at standard temperature and pressure (STP). In its normal uncombined form, it is a diatomic molecule, designated as O_2, with a molecular weight of 32.00. Gaseous oxygen, sometimes referred to as GOX, is slightly heavier than air. At atmospheric pressure, oxygen is a liquid below –297.3°F (90 K). A number of references are available for the thermodynamic and transport properties of both gaseous and liquid oxygen.[3-7] Liquid oxygen, sometimes referred to as LOX, is light blue in color, transparent, odorless, and slightly heavier than water. Oxygen is a strong oxidant, which means that it is nonflammable but that it can greatly accelerate the rate of combustion. Pure oxygen is very chemically reactive. Oxygen is required to support human life. However, breathing pure O_2 can produce coughing and chest pains. For general information on oxygen, the reader is referred to the following CGA (Compressed Gas Association) publications:

CGA G-4 Oxygen
CGA G-4.3 Commodity Specification for Oxygen

and to other general reference publications.[3,7,8-11]

The name *oxygen* means "acid-former" because of its ability to combine with other elements to form acids.[12] Oxygen is Earth's most abundant element and the fourth most abundant element in the universe, after hydrogen, helium, and neon.[13] About one fifth by volume of the air we breathe is O_2. Water (H_2O) is almost 90% oxygen by weight, and most living things need both air and water to survive. Oxygen is used in a wide variety of industrial applications. In the metals industry, it is used for cutting, welding, heating, melting, and in the manufacture of many types of metals. In the minerals industry, oxygen is used in the manufacturing of glass, ceramics, bricks, limestone, and cement. In the chemicals industry, oxygen is used in both heating applications and in chemical synthesis. In the environmental industry, it is used in wastewater treatment and in waste incineration. Most of these applications are discussed in some detail elsewhere in this book.

1.3 OXYGEN IN COMBUSTION

Many industrial heating processes may be enhanced by replacing some or all of the air with high-purity oxygen.[1,14] Typical applications include metal heating and melting, glass melting, and calcining. In a report done for the Gas Research Institute, the following applications were identified as possible candidates for OEC:

- Processes that have high flue gas temperatures, typically in excess of 2000°F (1400 K),
- Processes that have low thermal efficiencies, typically due to heat transfer limitations,
- Processes that have throughput limitations that could benefit from additional heat transfer without adversely affecting product quality, and
- Processes that have dirty flue gases, high NOx emissions, or flue gas volume limitations.[15]

When air is used as the oxidizer, only the O_2 is needed in the combustion process. By eliminating N_2 from the oxidizer, many benefits may be realized.

1.3.1 Typical Use Methods

Oxygen has been commonly used to enhance combustion processes in four primary ways: (1) adding O_2 into the incoming combustion airstream, (2) injecting O_2 into an air/fuel flame, (3) replacing the combustion air with high-purity O_2, and (4) separately providing combustion air and O_2 to the burner. These methods are discussed next.

1.3.1.1 Air Enrichment

Figure 1.1 shows an air/fuel process where the air is enriched with O_2. This may be referred to as low-level O_2 enrichment or premix enrichment. Many conventional air/fuel burners can be adapted for this technology.[16] The O_2 is injected into the incoming combustion air supply, usually through a diffuser to ensure adequate mixing. This is usually an inexpensive retrofit that can provide substantial benefits. Typically, the added O_2 will shorten and intensify the flame. However, there may be some concerns if too much O_2 is added to a burner designed for air/fuel. The flame shape may become unacceptably short. The higher flame temperature may damage the burner or burner block. The air piping may need to be modified for safety reasons to handle higher levels of O_2.

1.3.1.2 O_2 Lancing

Figure 1.2 shows another method for enriching an air/fuel process with O_2. As in the first method, this is also generally used for lower levels of O_2 enrichment.

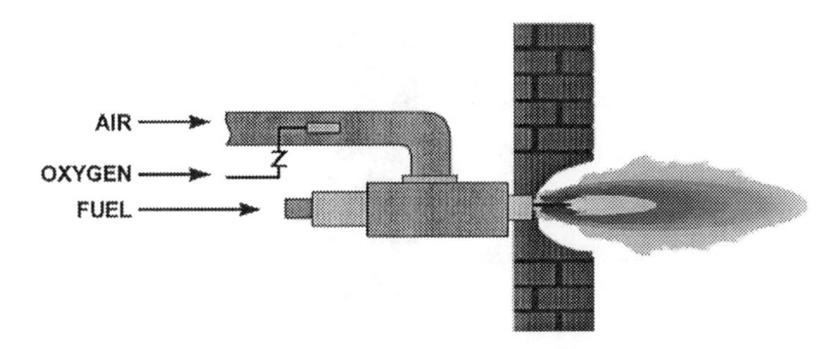

FIGURE 1.1 Schematic of premixing O_2 with air.

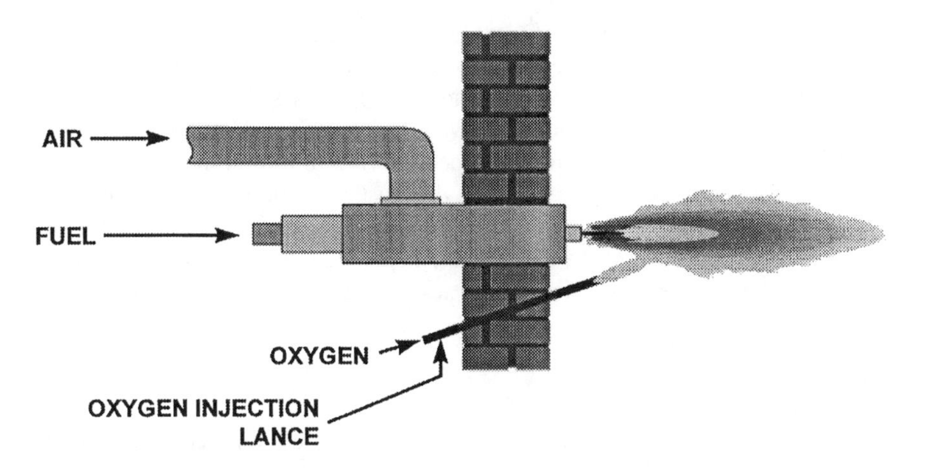

FIGURE 1.2 Schematic of O_2 lancing.

However, oxygen lancing may have several advantages over air enrichment. No modifications to the existing air/fuel burner need to be made. Typically, the NOx emissions are lower using O_2 lancing compared with premixing since this is a form of staging, which is a well-accepted technique for reducing NOx.[17] Depending on the injection location, the flame shape may be lengthened by staging the combustion reactions. The flame heat release is generally more evenly distributed than with premix O_2 enrichment. Under certain conditions, O_2 lancing between the flame and the load causes the flame to be pulled toward the material. This improves the heat transfer efficiency. Therefore, there is less likelihood of overheating the air/fuel burner, the burner block, and the refractory in the combustion chamber. Another variant of this staging method involves lancing O_2 not into the flame but somewhere else in the combustion chamber. One example of this technique is known as oxygen-enriched air staging, or OEAS, which is discussed in Chapter 7. That technology for O_2 lancing is an inexpensive retrofit for existing processes. One potential disadvantage is the cost of adding another hole in the combustion chamber for the lance. This includes both the installation costs and the lost productivity. However, the hole is typically very small.

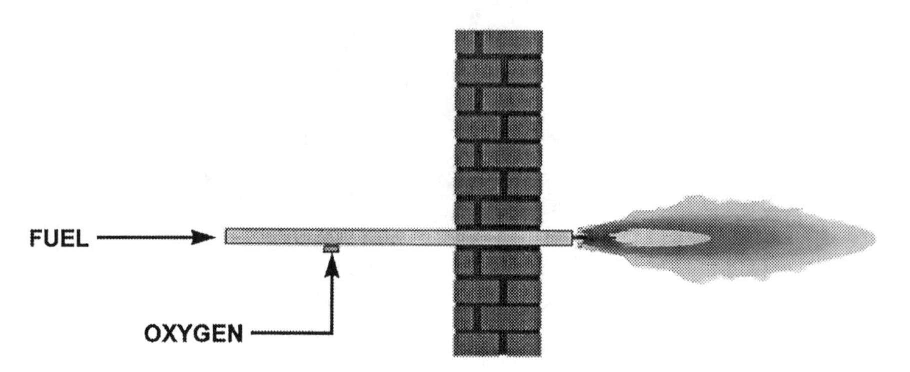

FIGURE 1.3 Schematic of oxy/fuel.

In one specific embodiment of O_2 lancing, known as undershot enrichment, O_2 is lanced into the flame from below. The lance is located between the burner and the material being heated. While air enrichment increases the flame temperature uniformly, the undershot technique selectively enriches the underside of the conventional flame, thereby concentrating extra heat downward toward the material being heated. While the mixing of oxygen and combustion air is not as complete with undershot oxygen as with premixing, this disadvantage is often outweighed by the more effective placement of the extra heat. Another benefit is that the refractory in the roof of the furnace generally receives less heat compared with air enrichment. This usually increases the life of the roof.

1.3.1.3 Oxy/Fuel

Figure 1.3 shows a third method of using OEC, commonly referred to as oxy/fuel combustion. In nearly all cases, the fuel and the oxygen remain separated inside the burner and do not mix until reaching the outlet of the burner. This is commonly referred to as a nozzle-mix burner, which produces a diffusion flame. There is no premixing of the gases for safety reasons. Because of the extremely high reactivity of pure O_2, there is the potential for an explosion if the gases are premixed. In this method, high-purity oxygen (>90% O_2 by volume) is used to combust the fuel. As will be discussed later, there are several ways of generating the O_2. In an oxy/fuel system, the actual purity of the oxidizer will depend on which method has been chosen to generate the O_2. As will be shown later, oxy/fuel combustion has the greatest potential for improving a process, but it also may have the highest operating cost.

One specific variation of oxy/fuel combustion, known as dilute oxygen combustion, is where fuel and oxygen are separately injected into the combustion chamber.[18] In order to ensure ignition, the chamber temperature must be above the autoignition temperature of the fuel. Depending on the exact geometry, this can produce an almost invisible flame, sometimes referred to as flameless oxidation. The advantage of this technique is very low NOx emissions because hot spots in the "flame" are minimized, which generally reduces NOx (see Chapter 2). A potential disadvantage, besides the

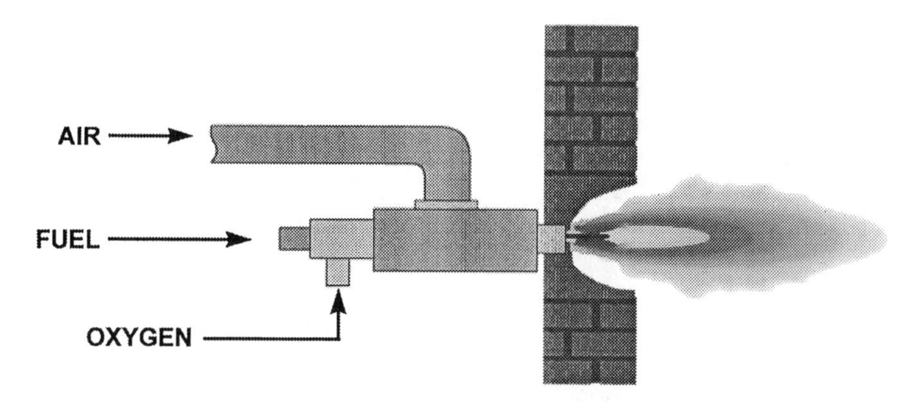

FIGURE 1.4 Schematic of air–oxy/fuel burner.

safety concern, is a reduction in heat transfer as both the temperature and effective emissivity of the flame may be reduced.

1.3.1.4 Air–Oxy/Fuel

The fourth common method of using OEC involves separately injecting air and O_2 through a burner, as shown in Figure 1.4. It is sometimes referred to as an air–oxy/fuel burner. This is a variation of the first three methods discussed above. In some cases, an existing air/fuel burner may be easily retrofitted by inserting an oxy/fuel burner through it.[19] In other cases, a specially designed burner may be used.[20] This method of OEC can have several advantages. It can typically use higher levels of O_2 than air enrichment or O_2 lancing, which yields higher benefits. Furthermore, the operating costs are less than for oxy/fuel, which uses very high levels of O_2. The flame shape and heat release pattern may be adjusted by controlling the amount of O_2 used in the process. It is also generally an inexpensive retrofit. Many air/fuel burners are designed for dual fuels, usually a liquid fuel like oil and a gaseous fuel like natural gas. The oil gun in the center of the dual fuel burner can usually be easily removed and replaced by either an O_2 lance or an oxy/fuel burner.

With this method of using OEC, the oxidizer composition may be specified in an alternative way. Instead of giving the overall O_2 concentration in the oxidizer, the oxidizer may be given as the fraction of the total oxidizer that is air and the fraction of the total oxidizer that is pure O_2. The equivalent overall O_2 in the oxidizer can be calculated as follows:

$$\Omega = \frac{20.9}{0.209\left(\text{vol\% O}_2\right) + \left(\text{vol\% air}\right)} \tag{1.1}$$

This conversion in Equation 1.1 is graphically shown in Figure 1.5. For example, the oxidizer may be specified as a blend of 60% O_2 and 40% air. That ratio of O_2 to air produces an equivalent of 39.8% overall O_2 in the oxidizer.

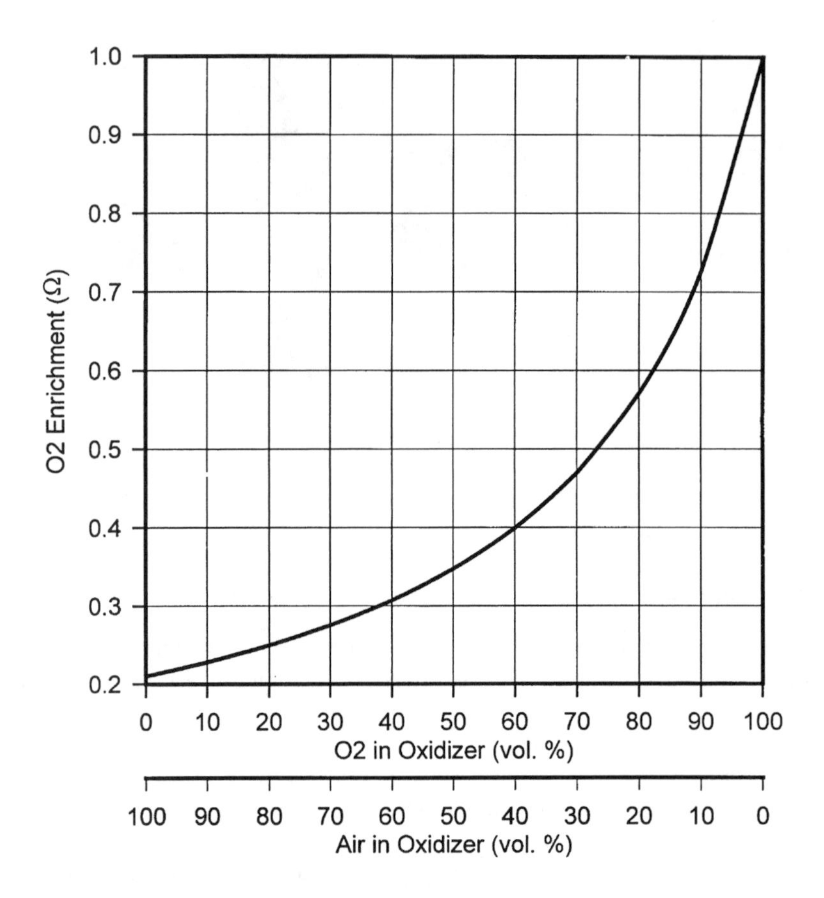

FIGURE 1.5 Oxidizer compositions for blends of air and pure O_2.

1.3.2 Basics

In this section, methane will be used to represent a typical fuel in a heating process. The analysis would be similar for other fuels.

1.3.2.1 Definitions

A generalized CH_4 combustion reaction may be written as

$$CH_4 + (xO_2 + yN_2) \rightarrow CO, CO_2, H_2, H_2O, N_2, NOx, O_2, \text{trace species} \quad (1.2)$$

The stoichiometry may be defined as

$$S = \frac{\text{volume flow rate of } O_2 \text{ in the oxidizer}}{\text{volume flow rate of } CH_4} \quad (1.3)$$

Note that this definition differs slightly from the one commonly used in industry in which the stoichiometry is usually defined as the total oxidizer flow divided by the fuel flow. The problem with the definition commonly used in industry is that the stoichiometry must be recalculated whenever the oxidizer composition changes and that stoichiometric conditions change for each oxidizer composition. This is not a concern if air is always used as the oxidizer, which is the case for the vast majority of combustion processes. The benefit of the definition used here is that stoichiometry is independent of the oxidizer composition, so stoichiometric conditions are the same for any oxidizer composition. In Equation 1.2, $S = x/1 = x$. Theoretically, for the complete combustion of CH_4, $S = 2.0$. Actual flames generally require some excess O_2 for complete combustion of the fuel. This is due to incomplete mixing between the fuel and oxidant. For the fuel rich combustion of CH_4, $S < 2.0$. For the fuel lean combustion of CH_4, $S > 2.0$.

The O_2 mole fraction in the oxidizer may be defined as

$$\Omega = \frac{\text{volume flow rate of } O_2 \text{in the oxidizer}}{\text{total volume flow rate of oxidizer}} \tag{1.4}$$

By using Equation 1.2, $\Omega = x/(x + y)$. If the oxidizer is air, $\Omega = 0.21$. If the oxidizer is pure O_2, $\Omega = 1.0$. The O_2 enrichment level is sometimes used. This refers to the incremental O_2 volume above that found in air. For example, if $\Omega = 0.35$, then the O_2 enrichment would be 14% (35% – 21%).

1.3.2.2 Operating Regimes

There are two common operating regimes for OEC. The first, or lower, regime is usually referred to as *low-level enrichment* ($\Omega < 0.30$). This is commonly used in retrofit applications where only a few modifications need to be made to the existing combustion equipment. It is used when only incremental benefits are required. For example, in many cases the production rate in a heating process can be significantly increased even with only relatively small amounts of oxygen enrichment. In most cases, air/fuel burners can successfully operate up to about $\Omega = 0.28$ with no modifications.[21] For $\Omega > 0.28$, the flame may become unstable or the flame temperature may become too high for a burner designed to operate under air/fuel conditions. In some cases, it may be possible to make minor burner modifications to permit operation at slightly higher O_2 concentrations.

In the other common operating regime, usually referred to as *high-level enrichment,* high-purity oxygen ($\Omega > 0.90$) is used. This is used in higher-temperature applications where the benefits of higher-purity oxygen justify the added costs. The heating process is greatly intensified by the high-purity oxygen. In a retrofit situation, existing air/fuel burners are replaced by burners specifically designed to use the higher levels of O_2.

It has only been in the last decade that a significant number of combustion systems have been operated in the intermediate oxygen regime or *medium-level enrichment* ($0.30 < \Omega < 0.90$). Again, these usually require specially designed burners or retrofits of existing burners.

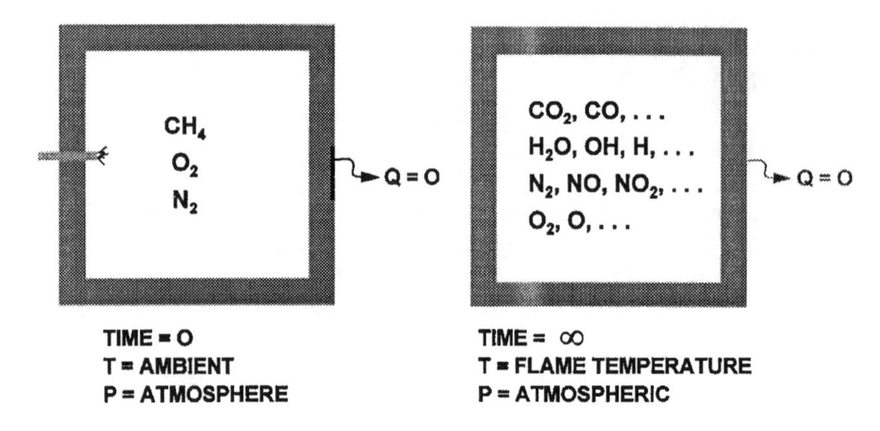

FIGURE 1.6 Adiabatic equilibrium reaction process.

1.3.2.3 Combustion Products

The stoichiometric combustion of CH_4 with air may be represented by the following global equation:

$$CH_4 + 2O_2 + 7.52N_2 \rightarrow CO_2, 2H_2O, 7.52N_2, \text{trace species} \quad (1.5)$$

It may be seen that over 70 vol% of the exhaust gases is N_2. Similarly, a stoichiometric O_2/CH_4 combustion process may be represented by

$$CH_4 + 2O_2 \rightarrow CO_2, 2H_2O, \text{trace species} \quad (1.6)$$

The volume of exhaust gases is significantly reduced by the elimination of N_2. In general, a stoichiometric oxygen-enhanced methane combustion process may be represented by:

$$CH_4 + 2O_2 + xN_2 \rightarrow CO_2 + 2H_2O + xN_2 + \text{trace species} \quad (1.7)$$

where $0 \leq x \leq 7.52$, depending on the oxidizer.

The actual composition of the exhaust products from the combustion reaction depends on several factors including the oxidizer composition Ω, the temperature of the gases, and the stoichiometry S. A cartoon showing an adiabatic equilibrium combustion reaction is shown in Figure 1.6. This is not the case in an actual combustion process where heat is lost from the flame by radiation. Figure 1.7 shows the predicted major species for the adiabatic equilibrium combustion of CH_4 as a function of the oxidizer composition. The calculations were made using a NASA computer program that minimizes the Gibbs Free Energy of a gaseous system.[22] An equilibrium process means that there is an infinite amount of time for the chemical reactions to take place or that the reaction products are not limited by chemical kinetics. In actuality, the combustion reactions are completed in fractions of a second.

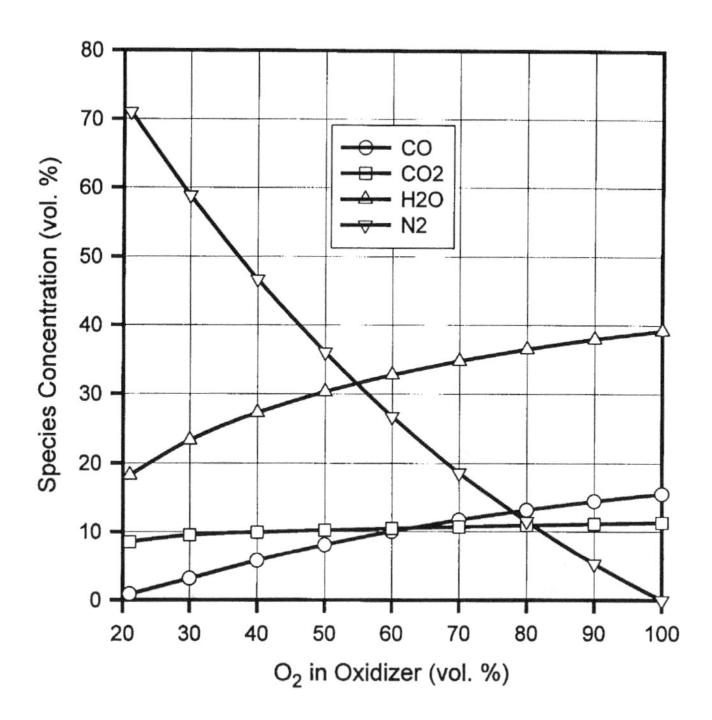

FIGURE 1.7 Major species concentrations vs. oxidizer composition, for an adiabatic equilibrium stoichiometric CH_4 flame.

An adiabatic process means that no heat is lost during the reaction or that the reaction occurs in a perfectly insulated chamber. As expected, Figure 1.7 shows that as N_2 is removed from the oxidizer, the concentration of N_2 in the exhaust products decreases correspondingly. Likewise, there is an increase in the concentrations of CO_2 and H_2O. For this adiabatic process, there is a significant amount of CO at higher levels of O_2 in the oxidizer. Figure 1.8 shows the predicted minor species for the same conditions as Figure 1.7. Note that trace species have been excluded from this figure. One of those trace species, NO, is discussed in detail in Chapter 2. The radical species H, O, and OH all increase with the O_2 in the oxidizer. Unburned fuel in the form of H_2 and unreacted oxidizer in the form of O_2 also increase with the O_2 concentration in the oxidizer. These increases in radical concentrations, unburned fuel in the form of CO and H_2, and unreacted O_2 are all due to chemical dissociation, which occurs at high temperatures.

The actual flame temperature is lower than the adiabatic equilibrium flame temperature because of heat loss from the flame. The actual flame temperature is determined by how well the flame radiates its heat and how well the combustion system, including the load and the refractory walls, absorbs that radiation. A highly luminous flame generally has a lower flame temperature than a highly nonluminous flame. The actual flame temperature will also be lower when the load and the walls are more radiatively absorptive. This occurs when the load and walls are at lower temperatures and have high radiant absorptivities. These effects are discussed in more detail in Chapter 4. As the gaseous combustion products exit the flame, they

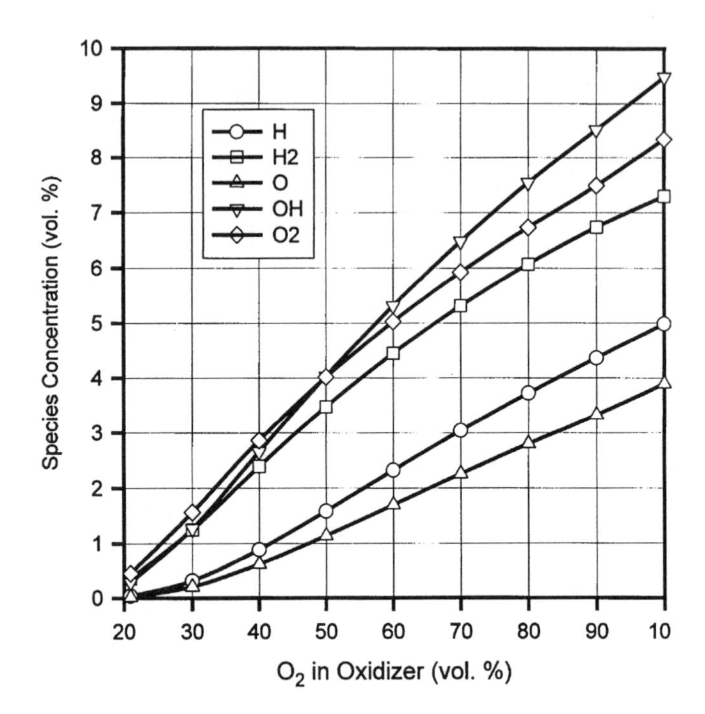

FIGURE 1.8 Minor species concentrations vs. oxidizer composition, for an adiabatic equilibrium stoichiometric CH_4 flame.

typically lose more heat by convection and radiation as they travel through the combustion chamber. The objective of a combustion process is to transfer the chemical energy contained in the fuel to the load, or in some cases to the combustion chamber. The more thermally efficient the combustion process, the more heat that is transferred from the combustion products to the load and to the combustion chamber. Therefore, the gas temperature in the exhaust stack is desirably much lower than in the flame in a thermally efficient heating process. The composition of the combustion products then changes with gas temperature.

Figure 1.9 shows the predicted major species for the equilibrium combustion of CH_4 with "air" (21% O_2, 79% N_2) and with pure O_2, as a function of the gas temperature. The highest possible temperature for the air/CH_4 and the O_2/CH_4 reaction is the adiabatic equilibrium temperature of 3537°F (2220 K) and 5038°F (3054 K), respectively. For the air/CH_4 reaction, there is very little change in the predicted gas composition as a function of temperature. For the O_2/CH_4 reaction, there is a significant change in the composition as the gas temperature increases above about 3000°F (1900 K). Figure 1.10 shows the predicted minor species for the same conditions as in Figure 1.9. Again, NO has been specifically excluded. For the air/CH_4, none of the minor species exceeds 1% by volume. As the gas temperature increases, chemical dissociation increases. For the O_2/CH_4 flame, significant levels of unreacted fuel (CO and H_2), radical species (O, H, and OH), and unreacted O_2 are present at high gas temperatures.

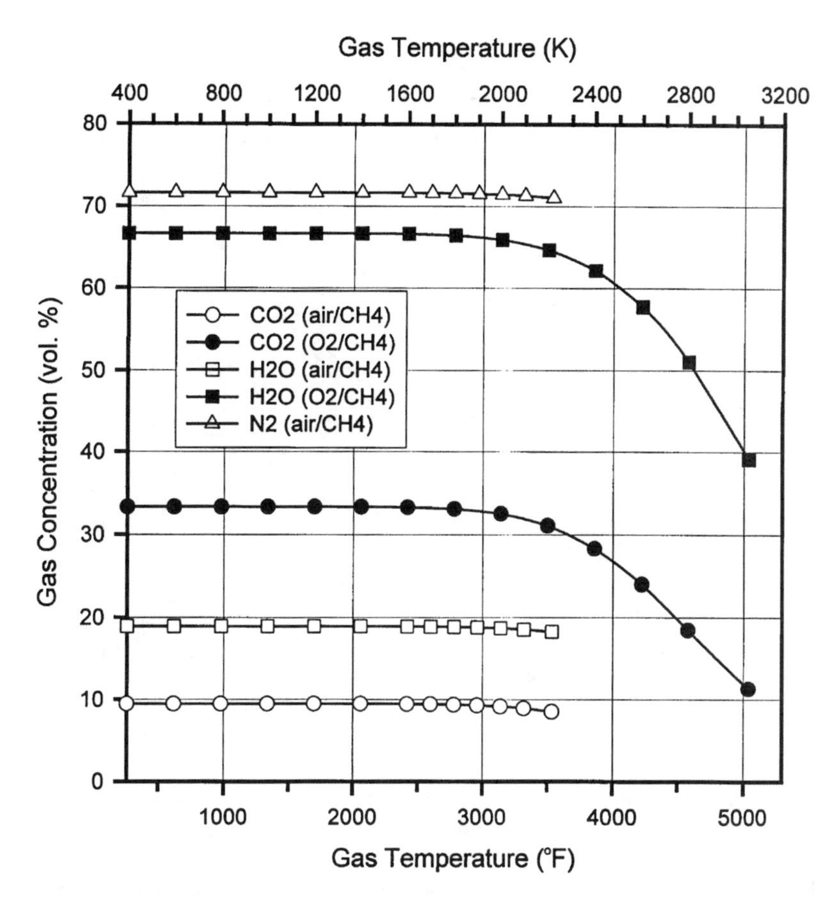

FIGURE 1.9 Equilibrium calculations for the predicted gas composition of the major species as a function of the gas temperature for air/CH_4 and O_2/CH_4 flames.

Figures 1.11 and 1.12 show the predicted major and minor species, respectively, for the adiabatic equilibrium combustion of CH_4 with "air" (21% O_2, 79% N_2) and with pure O_2, as a function of the stoichiometry. For the air/CH_4 flames, the N_2 concentration in the exhaust gases strictly increases with the stoichiometry. The H_2O and the CO_2 concentrations peak at stoichiometric conditions ($S = 2.0$). For the O_2/CH_4 flames, the peak H_2O concentration occurs at slightly fuel-rich conditions ($S < 2.0$). The predicted CO_2 concentration strictly increases with stoichiometry for the range of stoichiometries shown.

1.3.2.4 Flame Temperature

The flame temperature increases significantly when air is replaced with oxygen because N_2 acts as a diluent that reduces the flame temperature. Figure 1.13 is a plot of the adiabatic equilibrium flame temperature for CH_4 combustion, as a function of the oxidizer composition, for a stoichiometric methane combustion process. The flame temperature varies from 3600 to 5000°F (2300 to 3000 K) for air and pure

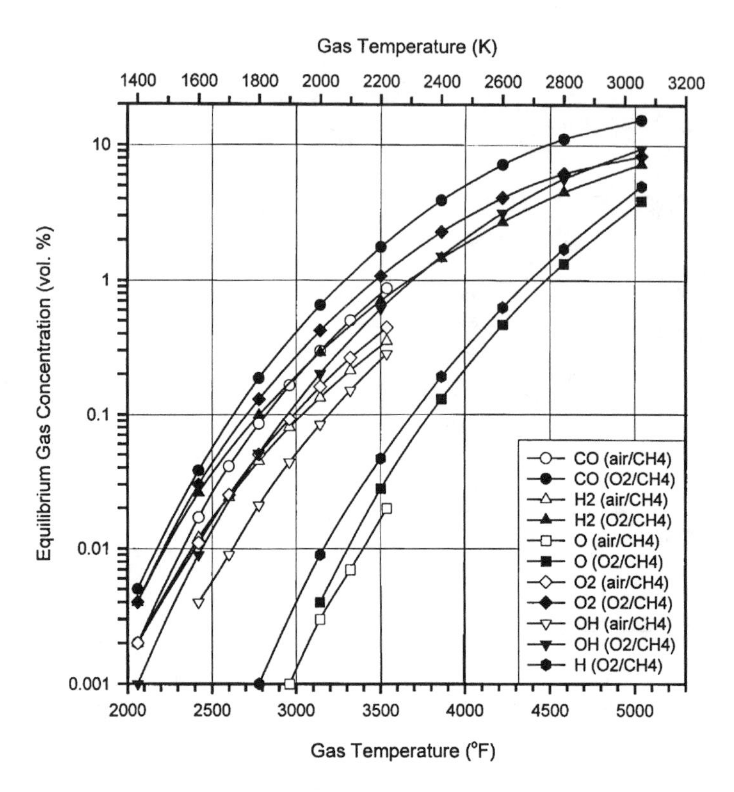

FIGURE 1.10 Equilibrium calculations for the predicted gas composition of the minor species as a function of the gas temperature for air/CH_4 and O_2/CH_4 flames.

oxygen, respectively. The graph shows a rapid rise in the flame temperature from air up to about 60% O_2 in the oxidizer. The flame temperature increases at a slower rate at higher O_2 concentrations. Table 1.1 shows the adiabatic flame temperatures for several fuels.

Figure 1.14 is a similar plot of the adiabatic equilibrium flame temperature for CH_4 flames as a function of the stoichiometry, for four different oxidizer compositions ranging from air to pure O_2. The peak flame temperatures occur at stoichiometric conditions. The lower the O_2 concentration in the oxidizer, the more the flame temperature is reduced by operating at nonstoichiometric conditions (either fuel rich or fuel lean). This is due to the higher concentration of N_2, which absorbs heat and lowers the overall temperature. Actual flame temperatures will be less than those given in Figures 1.13 and 1.14 because of heat losses from the flame, which is not an adiabatic process.

Figure 1.15 shows how the adiabatic flame temperature varies as a function of the oxidizer preheat temperature. The increase in flame temperature is relatively small for the O_2/CH_4 flame because the increased sensible heat of the O_2 is only a fraction of the chemical energy contained in the fuel. For the air/CH_4 flames, preheating the air has a more dramatic impact because the increase in sensible heat is very significant due to the large mass of air in the combustion reaction.

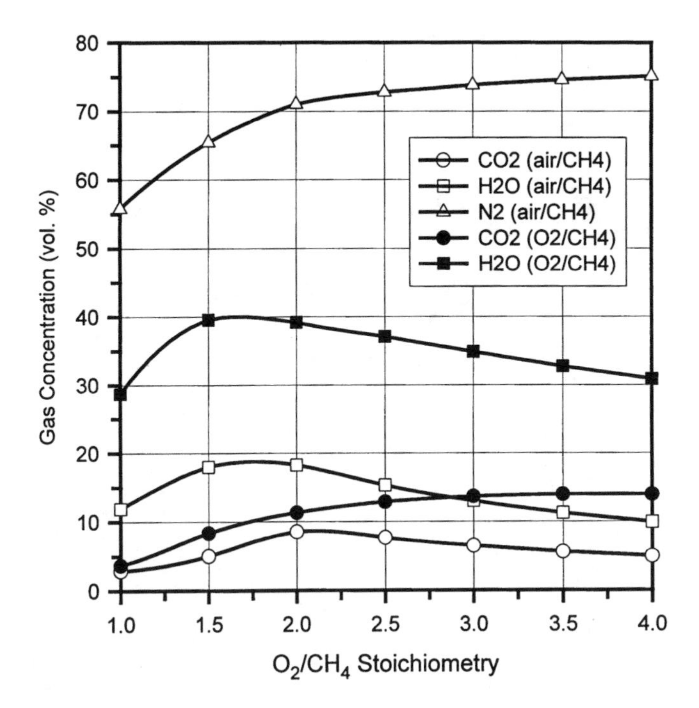

FIGURE 1.11 Adiabatic equilibrium calculations for the predicted gas composition of the major species as a function of the stoichiometry for air/CH_4 and O_2/CH_4 flames.

1.3.2.5 Available Heat

Available heat is defined as the gross heating value of the fuel less the energy carried out of the combustion process by the hot exhaust gases. N_2 in air acts as a ballast that carries energy out with the exhaust. Figure 1.16 is a graph of the available heat for the combustion of CH_4 as a function of the O_2 concentration in the oxidizer, for three different exhaust gas temperatures. As the exhaust gas temperature increases, the available heat decreases because more energy is carried out the exhaust stack. There is an initial rapid increase in available heat as the O_2 concentration in the oxidizer increases from the 21% found in air. That is one reason O_2 enrichment has been a popular technique for using OEC, as the incremental increase in efficiency is very significant.

Figure 1.17 shows how the available heat, for stoichiometric air/CH_4 and O_2/CH_4 flames, varies as a function of the exhaust gas temperature. As the exhaust temperature increases, more energy is carried out of the combustion system and less remains in the system. The available heat decreases to zero at the adiabatic equilibrium flame temperature where no heat is lost from the gases. The figure shows that even at gas temperatures as high as 3500°F (2200 K), the available heat of an O_2/CH_4 system is still as high as 60%. The figure also shows that it is usually not very economical to use air/CH_4 systems for high-temperature heating and melting processes. At an

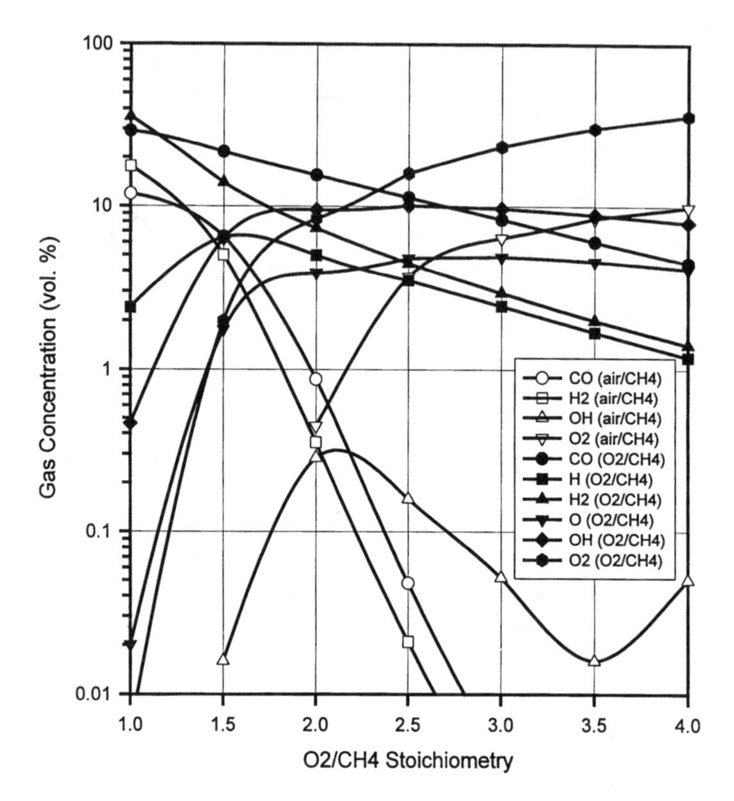

FIGURE 1.12 Adiabatic equilibrium calculations for the predicted gas composition of the minor species as a function of the stoichiometry for air/CH_4 and O_2/CH_4 flames.

exhaust temperature of 2500°F (1600 K), the available heat for the air/CH_4 system is only a little over 30%. Heat recovery in the form of preheated air is commonly used for higher-temperature heating processes to increase the thermal efficiencies.

Figure 1.18 shows how the available heat increases with the oxidizer preheat temperature. The thermal efficiency of the air/CH_4 doubles by preheating the air to 2000°F (1400 K). For the O_2/CH_4 flames, the increase in efficiency is much less dramatic by preheating the O_2. This is because the initial efficiency with no preheat is already 70% and because the mass of the O_2 is not nearly as significant in the combustion reaction as compared with the mass of air in an air/fuel flame. There are also safety concerns when flowing hot O_2 through piping, heat recuperation equipment, and a burner.

The fuel savings for a given technology can be calculated using the available heat curves:

$$\text{Fuel Savings }(\%) = \left(1 - \frac{AH_2}{AH_1}\right) \times 100 \tag{1.8}$$

where AH_1 is the available heat of the base case process and AH_2 is the available heat using a new technology. For example, if the base case process has an available

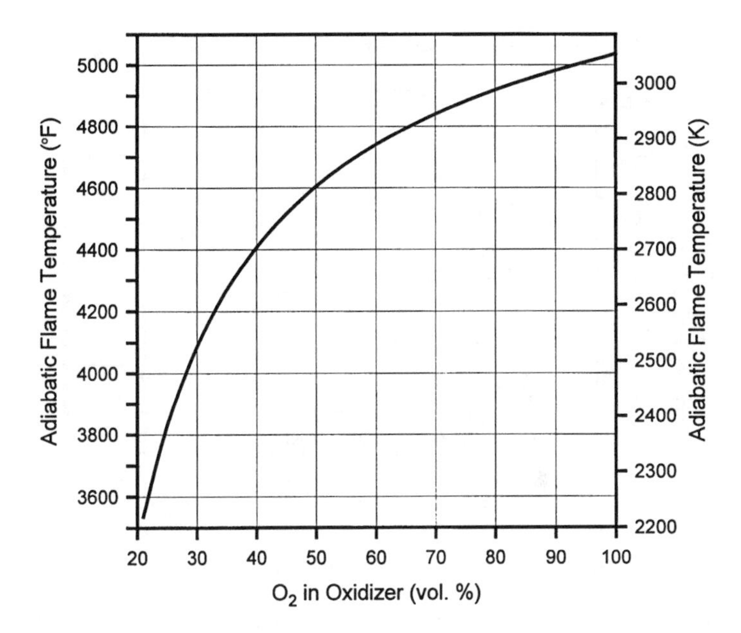

FIGURE 1.13 Adiabatic flame temperature vs. oxidizer composition, for an adiabatic equilibrium stoichiometric CH_4 flame.

TABLE 1.1
Adiabatic Flame Temperatures

	Air		O_2	
Fuel	**°F**	**K**	**°F**	**K**
H_2	3807	2370	5082	3079
CH_4	3542	2223	5036	3053
C_2H_2	4104	2535	5556	3342
C_2H_4	3790	2361	5256	3175
C_2H_6	3607	2259	5095	3086
C_3H_6	4725	2334	5203	3138
C_3H_8	3610	2261	5112	3095
C_4H_{10}	3583	2246	5121	3100
CO	3826	2381	4901	2978

heat of 30% and the available heat using the new technology is 45%, then the fuel savings = (1 – 45/30) × 100 = –50%, which means that 50% less fuel is needed for process 2 compared with process 1.

1.3.2.6 Ignition Characteristics

Flammability Limits. As the oxygen concentration in the oxidizer increases, the flammability limits for the fuel increase. Figure 1.19 shows the increase for CH_4

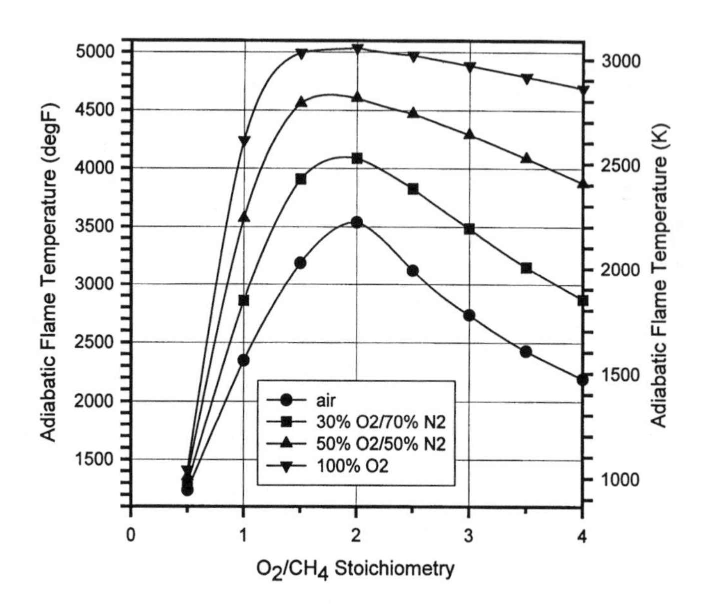

FIGURE 1.14 Adiabatic flame temperature vs. stoichiometry for a CH_4 flame and various oxidizers.

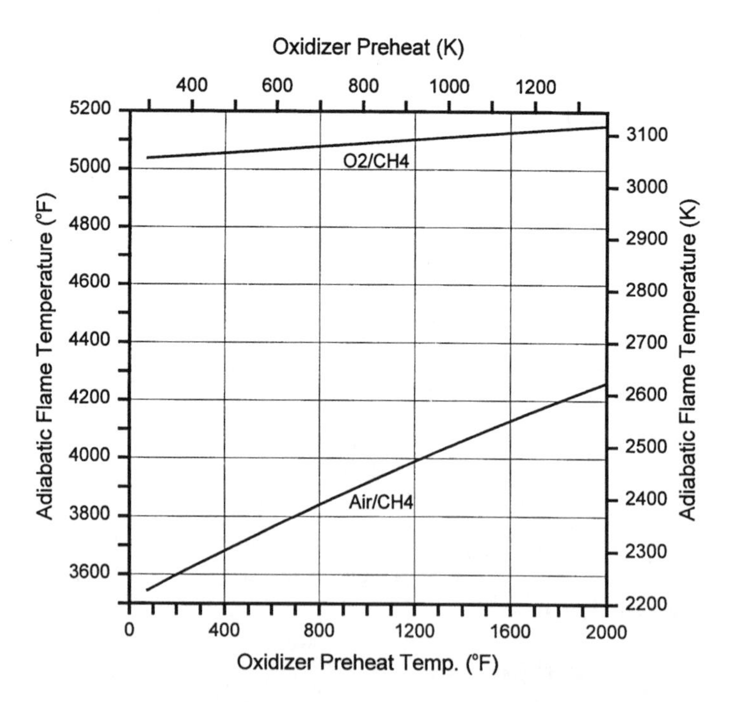

FIGURE 1.15 Adiabatic flame temperature vs. oxidizer preheat temperature for stoichiometric air/CH_4 and O_2/CH_4 flames.

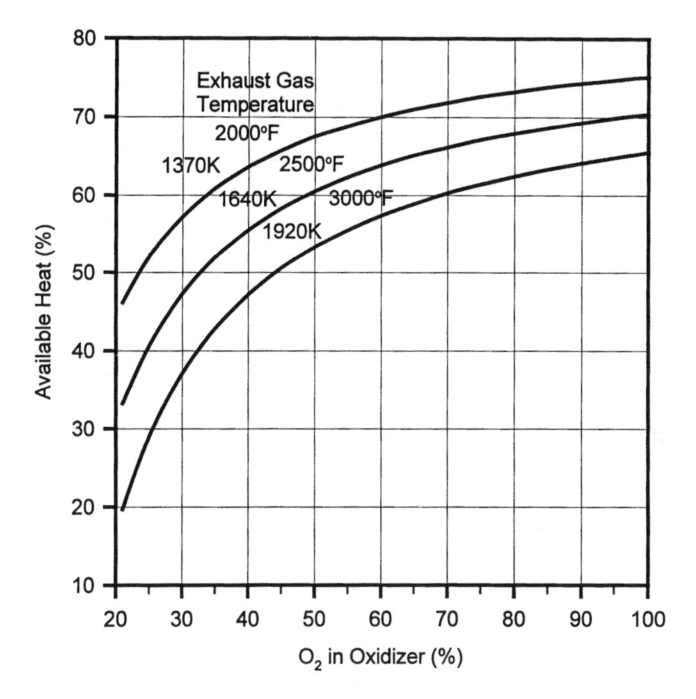

FIGURE 1.16 Available heat vs. oxidizer composition, for a stoichiometric CH_4 flame, at exhaust temperatures of 2000, 2500, and 3000°F.

combustion with oxidizers having a range of O_2 concentrations.[23] The upper flammability limit increases linearly with the O_2 concentration in the oxidizer, while the lower flammability limit is nearly constant for oxidizers with more than about 35% O_2 in the oxidizer. Table 1.2 shows examples of the way using O_2 instead of air widens the flammability range for a given fuel.

Flame Speeds. OEC increases the flame speed as shown in Figure 1.20. In a flame, the flame front is located where the gas velocity going away from the burner equals the flame velocity going toward the burner. The gas velocity exiting the burner must be at least equal to the flame speed. If not, the flame will flashback inside the burner leading to either flame extinguishment or to an explosion. Since flame speeds are higher using OEC, the burner exit velocities in an OEC system are usually higher than in air/fuel systems.

Ignition Energy. Less energy is required for ignition using OEC as shown in Figure 1.21.[24] This means that it is easier to ignite flames with OEC compared with air/fuel flames where much of the ignition energy is absorbed by the diluent nitrogen. The disadvantage is that it is also easier to inadvertently ignite an OEC system, which is the reason the proper safety precautions must be followed (see Chapters 9 and 10).

Ignition Temperature. Using OEC also reduces the ignition temperature. Figure 1.22 shows the ignition temperature for CH_4 as a function of the O_2 concentration in the oxidizer ranging from 15 to 35% O_2.[24] Table 1.3 shows ignition temperatures for gaseous fuels combusted with air and with pure O_2.[26] The graph and the table both show that it is easier to ignite flames that are enhanced with oxygen.

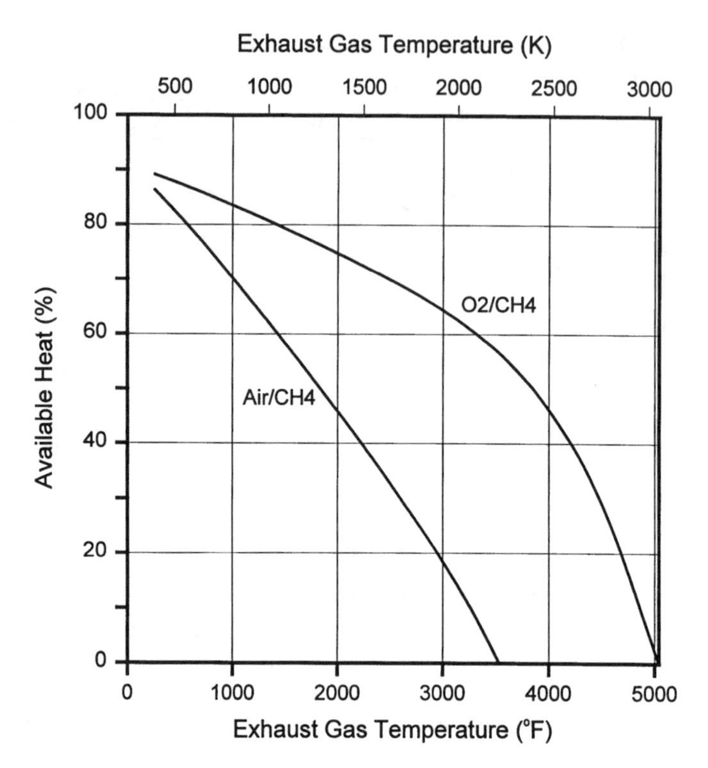

FIGURE 1.17 Available heat vs. exhaust gas temperature, for stoichiometric air/CH_4 and O_2/CH_4 flames.

1.3.2.7 Flue Gas

Reduced Volume. OEC basically involves removing N_2 from the oxidizer. One obvious change compared with air/fuel combustion is the reduction in the flue gas volume. Figure 1.23 shows the exhaust gas flow rate, normalized to the fuel flow rate, for the stoichiometric combustion of CH_4 where it has been assumed that all the combustion products are CO_2, H_2O, and N_2 (except when the oxidizer is pure O_2). This means that for each unit volume of fuel, three normalized volumes of gas are produced for oxy/fuel compared with 10.5 volumes for air/fuel. The reduction in flue gas volume is even larger when considering the increased fuel efficiency using OEC. Less fuel is required to process a given amount of material; therefore, fewer exhaust products are generated.

Different Composition. Another potential benefit of OEC related to the flue gas is that the composition is significantly different than that produced by air/fuel combustion, as was shown in Section 1.3.2.3. The products of combustion of a stoichiometric oxy/methane flame are approximately one third CO_2 and two thirds water by volume compared with air/methane, which has about 71% N_2, 10% CO_2, and 19% H_2O. Any pollutants, such as NOx or SOx, are easier to remove in an oxy/fuel exhaust because they are in much higher concentrations compared with the

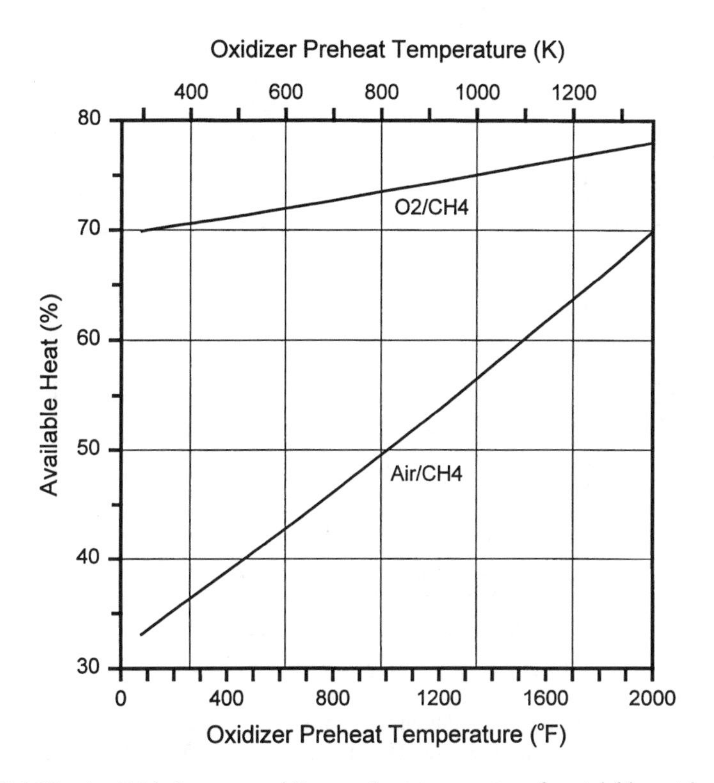

FIGURE 1.18 Available heat vs. oxidizer preheat temperature for stoichiometric, equilibrium air/CH_4 and O_2/CH_4 flames at an exhaust gas temperature of 2500°F (1644 K).

air/fuel exhaust stream which is highly diluted by N_2. If the water is condensed from the exhaust products, the remaining gas is nearly all CO_2 for an oxy/fuel exhaust. Since CO_2 has been identified as a global-warming gas (see Chapter 2), it can be more easily recovered from an oxy/fuel combustion system compared with an air/fuel system. The CO_2 can then either be used in another process, such as deep well injection for oil recovery, or it can be "disposed" of, such as in deep ocean injection.

1.3.3 General Benefits

Air consists of approximately 79% N_2 and 21% O_2 by volume. Only the oxygen is needed in the combustion reaction. By eliminating N_2, many benefits can be realized. These benefits include increased productivity, energy efficiency, turndown ratio, and flame stability, with reduced exhaust gas volume and pollutant emissions. These benefits are discussed next.

1.3.3.1 Increased Productivity

In most high-temperature heating processes, flame radiation is the dominant mode of heat transfer (see Chapter 4). Radiation is dependent on the fourth power of the absolute temperature. The higher temperatures associated with OEC increase the

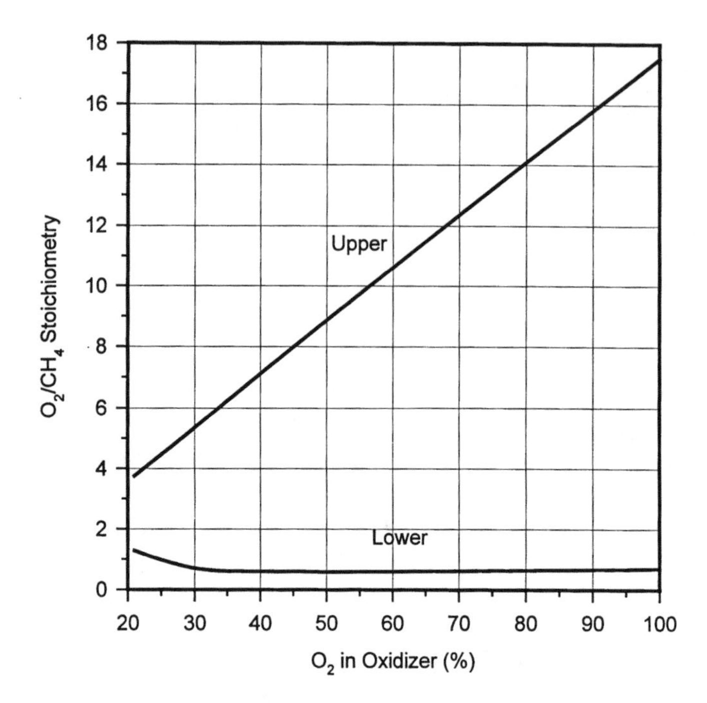

FIGURE 1.19 Upper and lower flammability limits vs. oxidizer composition for a CH_4 flame. (Adapted from Turin, J. J. and Huebler, J., Report no. I.G.R.-61, American Gas Association, Arlington, VA, 1951.)

TABLE 1.2
Flammability Limits[a] of Common Fuels in Air and in O_2

	Limit in Air		Limit in O_2	
Fuel	**Lower**	**Upper**	**Lower**	**Upper**
H_2	4.1	74	4.0	94
CH_4	5.3	14	5.4	59
C_2H_6	3.2	12.5	4.1	50
C_3H_8	2.4	9.5	—	—
C_2H_4	3.0	29	3.1	80
C_3H_6	2.0	11	2.1	53
CO	12.5	74	16	94

[a] Vol% fuel in fuel/oxidizer mixture.

Source: Data from Turin, J. J. and Huebler, J., Report no. J.G.R.-61, American Gas Association, Arlington, VA, 1951.

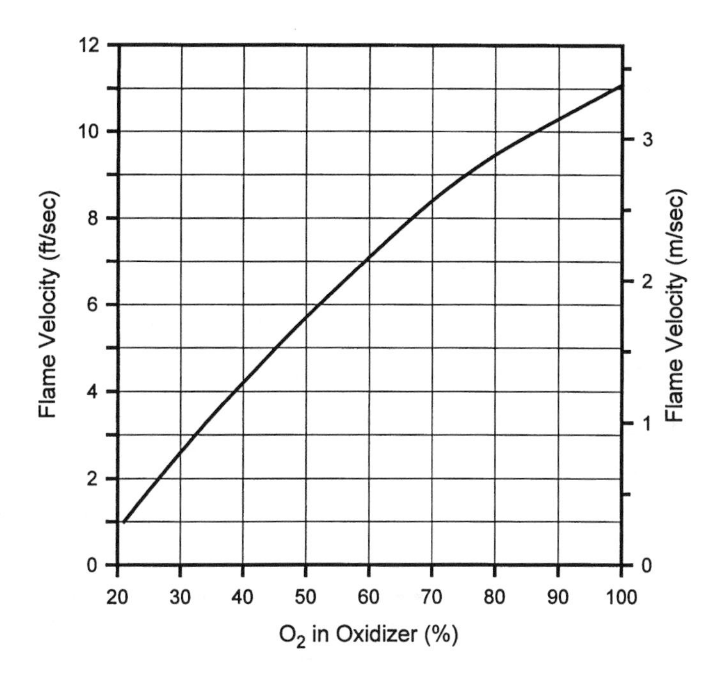

FIGURE 1.20 Normal flame propagation velocity vs. oxidizer composition for a stoichiometric CH_4 flame. (Adapted from Turin, J. J. and Huebler, J., Report no. I.G.R.-61, American Gas Association, Arlington, VA, 1951.)

radiation from the flame to the load. This then increases the heat transfer to the load, which leads to increased material processing rates through the combustion chamber. Therefore, more material can be processed in an existing system, or new systems can be made smaller for a given processing rate. This is particularly important when space in a plant is limited. The environment benefits because of the reduction in the size of the equipment since less material and energy are needed in fabricating the combustion system. In order to take advantage of increased processing rates, the rest of the production system must be capable of handling the increased throughput. For example, the material-handling system may need to be modified to handle the increased material flow rates.

The incremental costs of adding OEC to an existing combustion system are usually small compared with the capital costs of either expanding an existing system or adding new equipment in order to increase production. This has historically been one of the most popular reasons for using OEC. The advantage of OEC is that it can also be used intermittently to meet periodic demands for increased production. For example, if the demand for aluminum cans increases during the summer because of increased beverage consumption, OEC can be used at that time to meet the increased demand for aluminum. If the demand decreases in the winter, then the OEC system could be throttled back or turned off until it is needed again. In most cases, using OEC would be much more economical than adding new equipment for increased capacity if the increased production demands are only temporary.

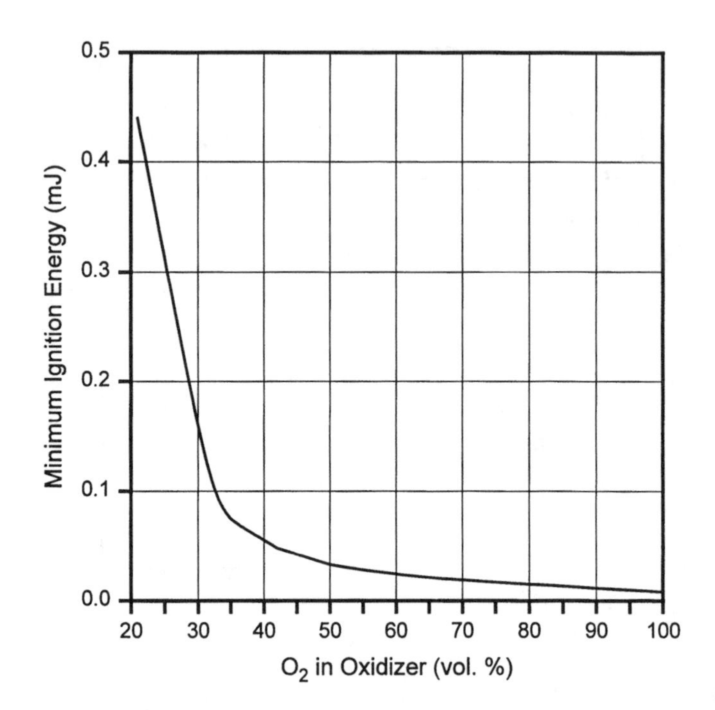

FIGURE 1.21 Minimum ignition energy vs. oxidizer composition for an atmospheric pressure, stoichiometric CH_4 flame. (Adapted from Lewis, B. and von Elbe, G., *Combustion, Flames and Explosions of Gases,* 3rd ed., Academic Press, New York, 1987.)

1.3.3.2 Higher Thermal Efficiencies

By using oxygen, instead of air, more energy goes into the load, instead of being wasted in heating up N_2. The energy needed to make O_2 from air is only a small fraction of the energy used in the combustion process. This is discussed further later. Therefore, the overall process uses less energy for a given amount of production due to the higher available heat. In some instances, the cost of the oxygen can be offset by the reduction in fuel costs because of the increase in energy efficiency. This is often the case when OEC is used to supplement a process that uses electricity, which is generally more expensive than fossil fuel combustion. In some cases, it is possible to substitute a less-expensive source of energy for an existing source of energy. For example, in the glass industry the furnaces are fueled primarily by oil or natural gas and commonly "boosted" with electrical energy. Some or all of the more-expensive electrical energy can be eliminated by the proper use of OEC.

One common reason for using OEC is a reduction in the specific fuel consumption when less fuel is required for a given unit of production because of the improvement in available heat. A recent study projects that, for example, the cost of natural gas is expected to rise by about 10% from 1993 to the year 2000.[15] At the same time, the cost of oxygen is expected to decrease by 10% due to lower electricity

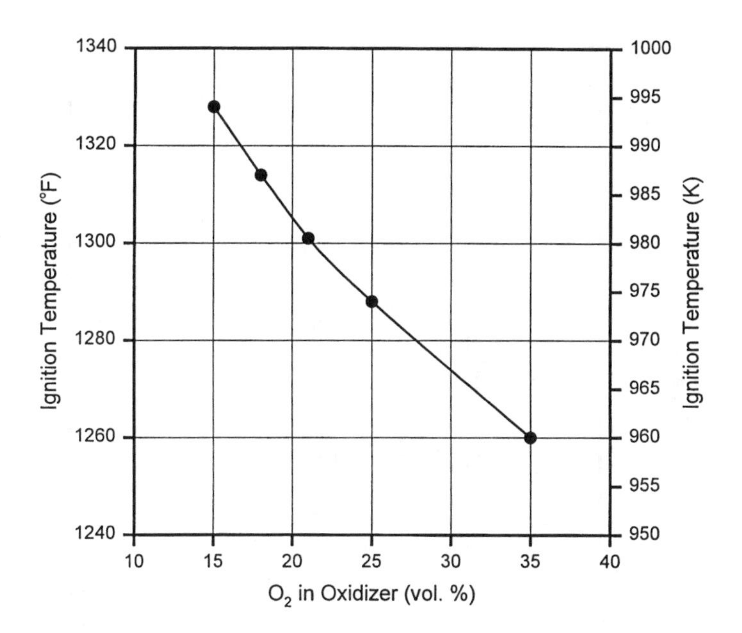

FIGURE 1.22 Ignition temperature vs. oxidizer composition for a stoichiometric CH_4 flame. (Adapted from American Gas Association, *Gas Engineers Handbook,* Industrial Press, New York, 1965.)

TABLE 1.3
Ignition Temperatures in Air and in Oxygen

	Air		O_2	
Fuel	°F	K	°F	K
H_2	1062	845	1040	833
CH_4	1170	905	1033	829
C_3H_8	919	766	874	741
C_4H_{10}	766	681	541	556
CO	1128	882	1090	861

Source: Data from Gibbs, B. M. and Williams, A., *J. Inst. Energ.*, June, 74, 1983.

costs and improvements in oxygen production technology. The rising cost of fuel and the declining cost of oxygen both make OEC a more attractive technology based solely on fuel savings.

Oxy/fuel combustion has been used in many applications in the steel industry including both continuous and batch reheat furnaces, soaking pits, and ladle pre-heaters. Fuel savings of up to 60% have been reported.[27] Typical fuel savings achieved by converting from air/fuel to oxy/fuel combustion are given in Table 1.4.[28]

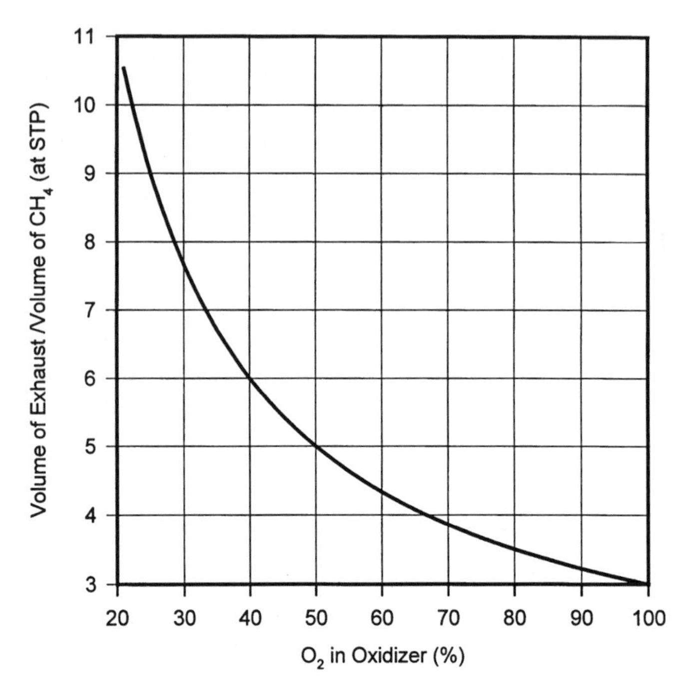

FIGURE 1.23 Normalized flue gas volume vs. oxidizer composition for a stoichiometric CH_4 flame.

Another example, from the glass industry, is Spectrum Glass, which reported fuel savings of 50% when the company converted from air/fuel to oxy/fuel in its glass-melting furnaces.[29]

1.3.3.3 Improved Flame Characteristics

Higher Turndown Ratio. As previously shown in Figure 1.19 and Table 1.2, as the oxygen concentration in the oxidizer increases, the flammability limits for the fuel increase. This leads to increased turndown ratio for the combustion system. A flame may exist under a wider range of conditions. For example, an air/CH_4 flame could exist at stoichiometries between about 1.3 and 3.8. An O_2/CH_4 flame could exist at stoichiometries between about 0.7 and almost 18. This is a consequence of removing the diluent N_2.

Increased Flame Stability. Oxygen-enhanced flames have higher flame speeds than air/fuel flames. This means that in order to prevent either flashback or flame extinguishment the minimum gas exit velocities for an OEC flame must be higher than those for an air/fuel flame. The potential for flashback is discussed further in Section 1.3.4.5. Harris et al.[30] have defined the critical boundary velocity gradient for a cylindrical tube as follows:

$$g = \frac{4V}{\pi R^3} \tag{1.9}$$

TABLE 1.4
Industrial Applications of Oxygen-Enhanced Combustion

Industry	Furnaces/Kilns	Primary Benefits[a]
Aluminum	Remelting	1, 2
	Coke calcining	1
Cement	Calcining	1
Chemical	Incineration	1, 2, 3, 4
Clay	Brick firing	1, 2, 3
Copper	Smelting	1, 2, 3
	Anode	2
Glass	Regenerative melters	1, 2, 4
	Unit melters	1, 2, 4
	Day tanks	1, 2, 4
Iron and steel	Soaking pits	2, 1
	Reheat furnaces	2, 1
	Ladle preheat	1
	Electric arc melters	1, 2
	Forging furnaces	1, 2
Petroleum	FCC regenerator	1
	Claus sulfur	1
Pulp and paper	Lime kilns	1, 2, 3
	Black liquor	1, 2

[a] Benefits of oxygen: 1 = productivity improvement, 2 = energy savings, 3 = quality improvement, 4 = emissions reduction.

Source: Reed, R. J., *North American Combustion Handbook,* 3rd ed., Vol. II, Part 13, North American Manufacturing Company, Cleveland, OH, 1997. With permission.

where V is the volumetric flow of a premixed mixture of fuel and oxidizer and R is the radius of the tube. For flashback, V is the minimum flow rate before the flame is extinguished as a result of flashback. The critical boundary velocity gradients for flashback were experimentally determined for premixed methane flames using a round tube. At stoichiometric conditions, the gradient was approximately 2100 and 510,000 s^{-1} for oxidizers of air and pure oxygen, respectively. This shows that much higher exit velocities can be used with OEC without blowing off the flame. These higher velocities generally produce more stable flames that are not as easily disturbed by flow patterns within a combustion chamber.

Better Ignition Characteristics. As shown in Figures 1.21 and 1.22, another benefit of using OEC compared with air is that less energy is required for ignition, which occurs at a lower temperature. This means that it is easier to ignite flames with OEC compared with air/fuel flames. The disadvantage is that it is also easier to inadvertently ignite an OEC system, which is the reason proper safety precautions must be followed (see Chapters 9 and 10). This improvement in ignition characteristics can be especially important when using liquid or solid fuels, which are generally more difficult to ignite compared with gaseous fuels.

Flame-Shape Control. OEC can be used to control the shape of a flame for an existing air/fuel system. For example, premix enrichment of oxygen into a combustion airstream has been used to shorten the flame length. Undershot enrichment can be used to lengthen a flame. Retrofitting an air/fuel system with oxygen in an air–oxy/fuel configuration has also been used to control the flame to produce a desired shape. Controlling the flame shape may be done to avoid overheating the refractory in a given location or to change the heat flux and temperature profile within the combustion chamber.

1.3.3.4 Lower Exhaust Gas Volumes

Eliminating ballast N_2 from air can reduce the exhaust gas flow rate by as much as 90% in aluminum melting.[31] In glass-manufacturing processes, flue gas reductions from 93 to 98% by mass have been measured.[32] This reduction in the flue gas volume has several advantages. The size of the exhaust gas ductwork can be reduced. Alternatively, OEC has been used effectively to increase production in combustion systems that are at the limit of their exhaust gas capacity. Normally, increasing capacity would also require a larger flue gas treatment system. However, the flue gas volume actually decreases with OEC so no new ductwork and treatment equipment are required. Another benefit of the reduction in the flue gas volume is that it can increase the efficiency of the existing flue gas treatment equipment.[33] This is because it is easier to treat the exhaust gases since the pollutant emissions are in higher concentrations and therefore easier to remove. The size of the posttreatment equipment can be proportionately reduced, because of the lower flow rates. This saves space, energy, materials, and money.

For a given combustion chamber, the flue gas volume reduction has the added benefit of reducing the average gas velocity by almost an order of magnitude. Lower gas velocities entrain fewer fine particles from the waste. This reduces particulate emissions. Other examples of particulate emission reductions are discussed in Chapter 2. Another potential advantage of the reduced velocity in the combustor chamber is the increased residence time. In an incinerator, increased residence time usually increases the level of destruction of undesired organic species in the off gases (see Chapter 8).

1.3.3.5 Higher Heat Transfer Efficiency

It has been argued that the efficiency of transferring heat from the flame to the load may be increased using OEC.[34] In a nonluminous flame, the flame emissivity is higher for an oxygen-enhanced flame, compared with an air/fuel flame. This is due to the higher concentrations of CO_2 and H_2O, which are the gases that radiate in a flame.[35] There is no radiation from the N_2 in the flame. These effects are discussed in Chapter 4.

1.3.3.6 Reduced Equipment Costs

Because OEC intensifies the combustion process and reduces the volume of flue gases, many combustion chambers and associated equipment can be significantly reduced in size. OEC can be used to reduce the cost of the off-gas treatment system.[32] Reducing the equipment size can reduce construction costs and the space required for the heating system.

1.3.3.7 Reduced Raw Material Costs

In some cases OEC can be used to reduce the costs of raw materials used in a process. Processes that have raw materials containing fine particles generally require some type of baghouse or scrubbing system to remove any particulates that might be entrained in the flue gases so they are not emitted into the atmosphere. For a given size combustion system, replacing air/fuel combustion with OEC can dramatically reduce the amount of particles entrained in the exhaust gases because the gas velocities are much lower as some or all of the nitrogen is removed from the system. Not only does this reduce the cost of cleaning up the exhaust gases, but it also reduces raw material costs because less is carried out of the system. Specific examples are given in the application section of this book (Chapters 5 to 8).

Another type of raw materials saving results from an improvement in the process where less material is needed for a given product. One example is in the glass industry where colorants are used to color the glass. Spectrum Glass reported savings of $20,000/year in colorants when the company converted from air/fuel to oxy/fuel because of the improvement in color stabilization using OEC.[29]

1.3.3.8 Increased Flexibility

There are many other benefits that may be achieved that are specific to a given process. OEC can increase the flexibility of a heating system.[36] In some cases, a wider range of materials can be processed with OEC compared with air/fuel combustion. In other cases, OEC may be required if very high melting temperatures are required. For example, some ceramic and refractory products require firing temperatures of 2900°F (1900 K) and higher.[37] Those temperatures are difficult if not impossible to achieve with standard air/fuel combustion with no air preheating. A heating system may also be brought up to operating conditions more quickly with OEC compared with air/fuel systems because of the higher heating intensity. For example, it has been shown that using OEC in metal reheat furnaces can substantially improve the ability to start up and shut down quickly.[46]

A combustion process can react more quickly to changes because of the higher heating intensity. This reduces the time, for example, that it takes to change processing rates or to change the product mix. OEC can give tighter control of the heating profile because of the higher intensity.

1.3.3.9 Improved Product Quality

OEC may improve product quality. For example, the proper use of OEC in a glass furnace can reduce the number of rejects. Quality improvements have been documented in nonferrous smelting, lamp making, chemical incineration, enamel fritting, and in both ceramic and lime kiln operations.[38]

1.3.3.10 Reduced Refractory Wear

By adding oxygen to an existing air/fuel combustion system, the flame length usually shortens. Some oxy/fuel burners are specifically designed to have a short flame length either to intensify the heat release or to fit better into a given furnace geometry to prevent flame impingement on the refractory wall opposite the burner. Shortening existing air/fuel flames or using short-flame-length oxy/fuel burners can reduce the wear on the refractory by reducing the effects of direct flame impingement on the walls.[39] As will be discussed in the next section, however, the improper use of oxygen in a combustion system can lead to refractory damage due to the higher flame temperatures associated with OEC.

1.3.4 Potential Problems

There are potential problems associated with the use of OEC if the system is not properly designed. Many of the potential problems can be generally attributed to the increased combustion intensity.

1.3.4.1 Refractory Damage

Overheating. As previously shown, oxygen-enhanced flames generally have significantly higher flame temperatures compared with conventional air/fuel flames. If the heat is not properly distributed, the intensified radiant output from the flame can cause refractory damage. Today's OEC burners are designed for uniform heat distribution to avoid overheating the refractory surrounding the burner. The burners normally are mounted in a refractory burner block, which is then mounted into the combustion chamber. The burner blocks are made of advanced refractory materials, such as zirconia or alumina, and are designed for long, maintenance-free operation. If the burner position and firing rate in the furnace are improperly chosen, refractory damage can result. For example, if the flame from an OEC burner is allowed to impinge directly on the wall of a furnace, most typical refractory materials would be damaged. This can be prevented by the proper choice of burner design and positioning. The flame length should not be so long that it impinges on the opposite wall. The burner mounting position in the furnace should be chosen to avoid aiming the flame directly at furnace refractories.

Corrosion. Another potential refractory problem can result from the increased volatile concentration in the combustion chamber by using OEC. This is a particular problem in the glass industry, for example, where corrosive volatile species are emitted during the glassmaking process. By removing the large quantity of diluent N_2 normally present in air/fuel combustion, these volatile species are now at much higher concentrations in the gas space. This can cause damage to the refractories by corrosion. This is discussed in more detail in Chapter 7.

1.3.4.2 Nonuniform Heating

Nonuniform heating is an important concern when retrofitting existing systems that were originally designed for air/fuel combustion. By intensifying the combustion process with OEC, there is the possibility of adversely affecting the heat and mass transfer characteristics within the combustion chamber. These issues are considered in more detail in Chapter 4 and are only briefly discussed here.

Hot Spots. OEC normally increases the flame temperature, which also increases the radiant heat flux from the flame to the load. If the increased radiant output is very localized, then there is the possibility of producing hot spots on the load. This could lead to overheating that might damage or degrade the product quality. Today, burners for OEC have been specifically designed to avoid this problem.

Reduction in Convection. As shown in Figure 1.23, the total volume flow rate of exhaust products is significantly reduced using OEC. However, the average gas temperature is usually higher, but not by enough to offset the reduced gas flow rate. The convective heat transfer from the exhaust gases to the load may be reduced as a result. Another important aspect of convection is mass transfer. In some heating processes, especially those related to drying or removing volatiles, the reduced gas flow rate in the combustion chamber could adversely affect the mass transfer process. This can be offset by using a burner that incorporates furnace gas recirculation, which increases the bulk volume flow inside the combustion chamber to help in removing volatiles that evolve from the load during the heating process.

1.3.4.3 Flame Disturbance

In recent years, the trend in burner design for OEC has been toward lower-momentum flames. One example is the Cleanfire® burner, which has been used extensively in the glass industry.[40] These lower-momentum flames tend to be longer and more luminous than the high-intensity burners that have traditionally been used in the past. One issue that needs to be considered is that these low-momentum flames can be more easily disturbed than high-momentum flames. One example is when low-momentum OEC flames are added or partially retrofitted into a furnace containing high-momentum air/fuel burners. The high-momentum flames can adversely affect the combustion characteristics of the low-momentum flames if the geometry is not properly designed. This may especially be a problem if a high-momentum burner is directly opposed to and firing at a low-momentum burner. Another example of possible flame disturbance is when low-momentum flames are located near the exhaust of the furnace. If there are many burners in the furnace, the large exhaust gas flow past a low-momentum flame can again disturb those flames. The problem of flame disturbance can be eliminated by the proper choice of burners, burner mounting positions, and burner operating conditions.

1.3.4.4 Increased Pollutant Emissions

NOx. When O_2 enrichment is used in an existing air/fuel combustion system, there may be an increase in NOx emissions,[41] which is due to the increased flame temperature that increases thermal NOx formation. This is discussed in more detail in Chapter 2.

Noise. As shown in Figure 1.20, the flame velocity increases with OEC compared with air/fuel combustion. This means that the gas velocities exiting the burner must be increased to compensate for the higher flame speed. These higher gas velocities

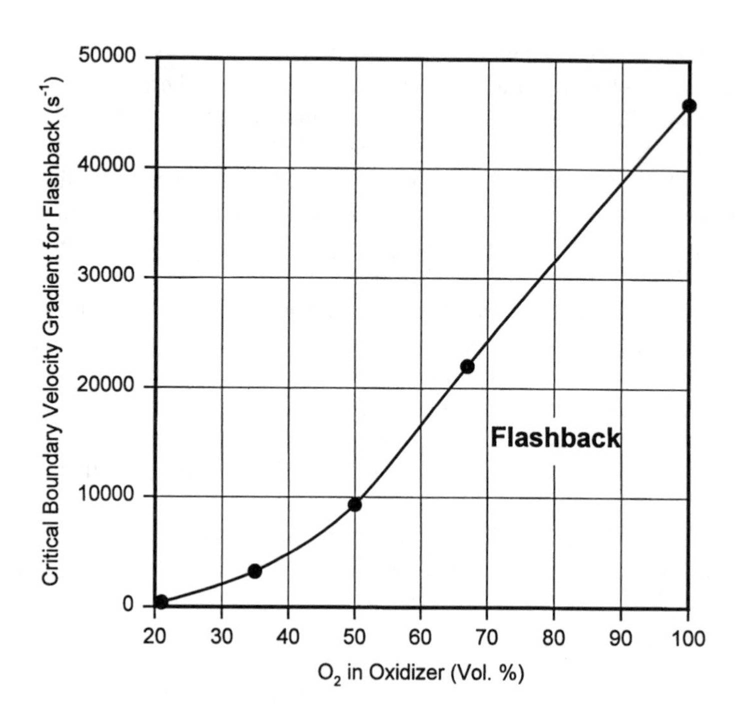

FIGURE 1.24 Critical boundary velocity gradient for flashback vs. oxidizer composition for a stoichiometric premixed methane flame through a cylindrical tube. (Adapted from Harris, M. E. et al., in *Third Symposium on Combustion, Flame and Explosion Phenomena,* Williams & Wilkens, Baltimore, 1949, 80.)

can increase the jet noise of the burner. The older-style high-intensity oxy/fuel burners used in metal-melting applications were generally very noisy in the open air. However, the noise was generally muffled by the refractory-lined combustion chamber. Also, although the gas velocities were high, the total volume flow rate was much lower by removing N_2 from the oxidizer, which also was a mitigating factor for the noise. Today, the low-momentum oxy/fuel burners are actually quieter than many air/fuel burners. Noise is discussed in more detail in Chapter 2.

1.3.4.5 Flashback

The use of OEC intensifies the combustion reactions. One consequence of this higher reactivity is the increased risk of flashback. Flashback occurs when the gas velocity exiting the burner is less than the flame velocity. This causes the flame front to move toward the burner. If the fuel and the oxidizer are premixed, the flame can burn inside the burner housing, which creates the potential for an explosion. The critical boundary velocity gradient was previously defined in Section 1.3.3.3. As shown in Figure 1.24, the gradient for flashback increases rapidly as the oxygen concentration in the oxidizer increases. This means that the minimum volumetric flow rate increases with OEC to prevent flashback. For blowoff, V is the maximum flow rate before the flame is extinguished due to blowoff.

In premixed systems, there is often a provision for arresting the flame if it should flash back. It may be a separate device known as a flame arrestor, or it may simply be incorporated into the burner design. The general idea in most flame arrestors is to remove enough heat from a flame that is flashing back to cool the flame down below its ignition point. In essence, the third leg of the combustion triangle (see Figure 9.1) is removed to extinguish the flame since neither the fuel nor the oxidizer can be removed quickly enough to prevent the flame from flashing back.

In virtually all combustion systems using OEC, the fuel and the oxidizer are not mixed until they exit the burner. This, commonly known as a nozzle-mix burner, essentially eliminates the potential for an explosion caused by flashback. If the flame were to flash back toward the burner, it would be extinguished at the burner nozzle. The flame would not continue to travel into the burner as there would no longer be a stoichiometric mixture since the fuel and the oxidizer are separated inside the burner. Therefore, the potential risk of flashback is eliminated by not premixing the fuel and the oxidizer inside the burner.

1.3.5 Industrial Heating Applications

OEC is used in a wide range of industrial heating applications. In general, OEC has been used in high-temperature heating and melting processes that are either very inefficient or not possible with air/fuel combustion. Table 1.4 shows some of the common reasons OEC has been used in a variety of industrial applications.[42] This section is only intended to give a brief introduction to those applications that are broadly categorized here as metals, minerals, incineration, and other. These applications are discussed in detail elsewhere in the book.

1.3.5.1 Metals

Heating and melting metal was one of the first industrial uses of OEC and continues to be an important application today. OEC has been widely used in both large integrated steel mills as well as in smaller minimills. It has also been used in the production of nonferrous metals such as aluminum, brass, copper, and lead. Chapter 5 specifically concerns ferrous metals, while Chapter 6 discusses applications in nonferrous metals.

1.3.5.2 Minerals

Here, mineral refers to glass, cement, lime, bricks, ceramics, and other related materials that require high-temperature heating and melting during their manufacture. OEC has been used in all of those applications. Chapter 7 discusses the use of OEC in glass in detail.

1.3.5.3 Incineration

This is a relatively new area for OEC. Initially, OEC was used to enhance the performance of portable incinerators used to clean up contaminated soil. It has also been used in municipal solid waste incinerators and in boilers burning waste fuels. The use of OEC in incineration is discussed in Chapter 8.

1.3.5.4 Other

OEC has been used in a wide variety of specialty applications which are not discussed in detail in this book. Some of these include gasifying organic materials and vitrifying residual ash,[43] removing unburned carbon from fly ash,[44] and oxygen enrichment of fluid catalytic crackers.[45]

1.4 ECONOMICS

As with any purchasing decision, benefits are expected to outweigh costs. There are many factors to be considered when calculating the costs and benefits associated with OEC. This section will use a more generalized approach. Specific examples pertaining to a particular industry are given elsewhere in the chapters on the applications of OEC. Because of the rapidly changing cost of oxygen, no attempt will made to use specific pricing. The interested reader should contact industrial gas suppliers for detailed cost information. The purpose of this section is to provide the factors that are commonly considered before implementing OEC. The actual value placed on these factors varies widely from company to company, from place to place, and between different locations even for the same company. Therefore, no attempt will be made to quantify each factor. Another important consideration is whether the OEC system will be retrofitted to an existing air/fuel system or whether a new combustion chamber will be built specifically for OEC.

1.4.1 Costs

The purpose of this section is to identify the common elements that should be considered in evaluating the economic viability of using OEC. It is difficult to give generalized conclusions regarding costs because they vary considerably as a function of both time and geography. Therefore, examples will be given as illustrations, but actual costs should be used to evaluate a specific application. More-detailed examples for specific industries are given elsewhere in the book.

The two components of cost that are considered here are ongoing operating expenses and the capital equipment costs that are normally incurred during the initial implementation of a technology and are amortized over some length of time. It has been assumed here that the base case for comparison is conventional air/fuel operation. However, in some applications the base case may be another technology, like electrical induction heating in a steel reheat furnace. The economic analysis should be modified accordingly.

1.4.1.1 Operating Costs

The two most important operating costs to be considered for OEC are for the fuel and the oxygen. It has been shown that OEC can dramatically reduce fuel consumption for a given unit of production. Typical reductions are shown in Table 1.5.

TABLE 1.5
Fuel Savings by Retrofitting Air/Fuel Systems with Oxy/Fuel

Industry	Fuel Savings (%)
Steel	40–60
Aluminum, lead	40
Copper	40–60
Glass	30–40
Waste incineration	50
Sulfuric acid recovery	50

Source: Data from Ding, M. G. and Du, Z., in *Prog. 1995 Int. Conf. on Energy and Environment,* Shanghai, China, Begell House, New York, May 1995, 674.

Because of the increased fuel efficiency using OEC, the fuel consumption for a given unit of production will be reduced. For example, if 1500 Btu/ton of aluminum are required to melt scrap aluminum using conventional air/fuel combustion, only 1000 Btu/ton of aluminum may be required using OEC. However, the objective of many secondary aluminum producers is to increase production so that instead of reducing the firing rate in the furnace to get the same production rate, they would prefer to maintain the same firing rate and increase the production rate using OEC. Therefore, it is important to normalize the fuel consumption per unit of production since there may be no actual change in the total fuel consumption, while there may be a large increase in the production rate.

The cost of oxygen is probably the most variable operating expense. It is very difficult to specify a "typical" cost, but some general principles may be given. For very large oxygen requirements, an on-site oxygen generator is often justified. The least expensive method for very large requirements is usually a cryogenic air separation plant. For medium to large requirements, an adsorption oxygen generator is usually the most cost-effective method of O_2 supply. For smaller requirements, the cheapest supply method is usually on-site storage of LOX trucked to the site in tanker trucks. For that method, the cost is dependent on how far the site is from the merchant oxygen production facility since the shipping costs are a significant fraction of the overall oxygen cost. One of the largest components in the cost of oxygen is electrical power. Therefore, the oxygen cost will be affected by the power costs in a given geographic location. As with any product, the oxygen cost is dependent on the supply and demand in the overall industry, as well as in a specific geographic location. The oxygen cost may also be dependent on the overall requirement and usage pattern. The oxygen price is generally less for larger requirements as opposed to smaller ones. The oxygen price is generally higher for cyclical requirements, in contrast to steady demands. More details on oxygen supply methods are given in Chapter 3. Interested parties should contact the appropriate industrial gas suppliers to get specific costs for a given location.

With OEC, there is a trade-off between the savings in fuel and the cost of the oxygen. A simplified analysis can be used to determine what the change in operating costs, if any, would be using OEC if only the fuel and oxygen costs are considered. Considering only the oxidizer and fuel, the normalized cost can be given as follows:

$$\mathrm{NC} = E(C \cdot F + 1) \tag{1.10}$$

where NC is the normalized cost, E is an efficiency factor, C is the ratio of the cost of the oxygen to the cost of the fuel (in the same units), and F is the ratio of the flow of oxygen to the flow of fuel (in the same units). For the base case of air/fuel operation, NC = 1. If NC < 1, then the fuel and operating costs using OEC would be less than the base case of air/fuel operation. If NC > 1, then the fuel and operating costs would be greater using OEC than those for air/fuel. The efficiency factor is defined as follows:

$$f = \frac{\eta(O_2) - \eta(O_2 = 0)}{\eta(O_2 = 0)} \tag{1.11}$$

where f is the fractional increase in fuel efficiency, $\eta(0)$ is the fuel efficiency of the base case, and $\eta(O_2)$ is the fuel efficiency at a given flow of oxygen per unit of fuel. An example will serve to illustrate the use of this equation. Assume that the thermal efficiency of the air/fuel base case operation is 40% and that the efficiency with oxy/fuel would be 60%. Then f would be (60 – 40)/40 = 0.50, which means that oxy/fuel is 50% more efficient than the air/fuel base case.

The dimensionless efficiency factor is then defined as

$$E = 1 - f = \frac{2 \cdot \eta(O_2 = 0) - \eta(O_2)}{\eta(O_2 = 0)} \tag{1.12}$$

Using the example efficiencies just given, the efficiency factor $E = 1 - 0.50 = 0.50$, which could be alternatively calculated by [(2 · 40) — 60]/40 = 0.50, which means that only 50% of the base case fuel flow is required for the same unit of production using oxy/fuel.

An example will be given to show how to use the normalized cost equation to compute NC. Assume that the fuel is natural gas with a cost of \$2.68/1000 ft^3, which was the average price for natural gas in the U.S. in 1993. Assume that the oxygen cost is \$35/ton which is equivalent to \$1.45/1000 ft^3. Then the normalized cost of oxygen to fuel, C, would be \$1.45/\$2.68 = 0.541. Depending upon the specific composition of natural gas, approximately two volumes of O_2 are required for each volume of natural gas for complete combustion; therefore, the ratio of the oxygen to fuel flow, F, would be 2. Then the normalized cost would be computed as follows: NC = 0.50[(0.541)(2.0) + 1] = 1.04, which would mean that the cost of the fuel and oxygen for the oxy/fuel case would be slightly more expensive than the cost of the fuel in the air/fuel case. For this case, the oxy/fuel would have to be justified based on other factors besides fuel savings.

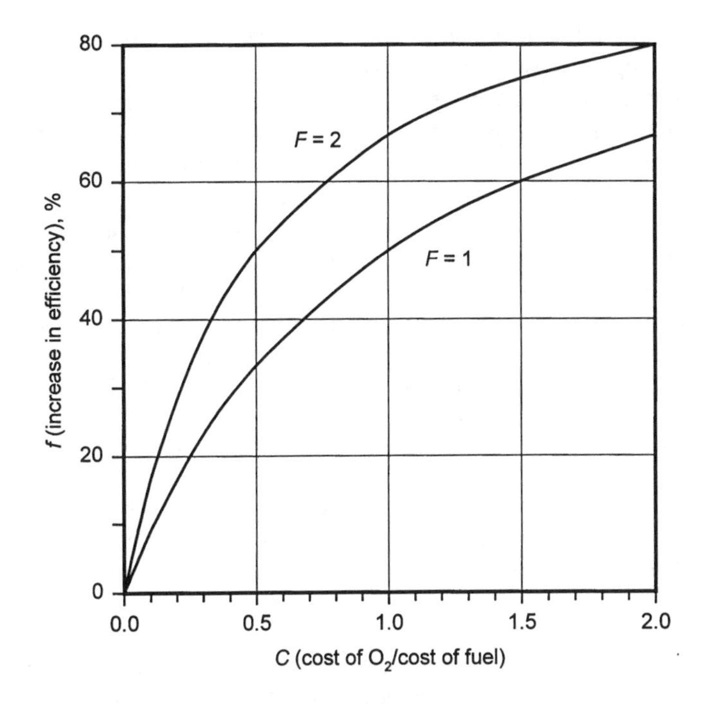

FIGURE 1.25 Breakeven cost curve for the ratio of the cost of O_2 to the cost of fuel (*C*) vs. the required increase in fuel efficiency (*f*) for two different ratios of the oxygen to fuel flow rate (*F*).

In one study, the cost of natural gas is expected to rise to \$3.29/1000 ft^3 by the year 2000, while the cost of O_2 is expected to decline by 10% as a result of a decrease in the cost of electricity and an increased production efficiency of generating oxygen.[15] In the previous example, the normalized cost of oxygen to fuel would then become (\$1.45)(0.9)/(\$3.29) = 0.397, which is a reduction of 27% compared with 0.541. Then the normalized cost would be NC = (0.50)[(0.397)(2.0) + 1] = 0.90, which is now lower than the base case.

The efficiency factor will be a function of the specific application. For example, an old leaky furnace with poorly adjusted burners will have a low efficiency compared with a newly lined furnace with new, properly adjusted burners. The fuel and oxygen costs are highly geographically dependent. In areas where fuel costs are high and oxygen costs are low, OEC will be a competitive technology with air/fuel. In areas where fuel costs are low and oxygen prices are high, other benefits will be needed to justify the use of OEC. The graph in Figure 1.25 shows how to determine whether or not the savings in fuel costs can offset the oxygen costs, for two different levels of oxygen use (F = 1 and 2). If a particular set of factors falls below the appropriate curve, then the cost of oxygen will be more than offset by the savings in fuel. If a set of factors falls above the curve, then the cost of oxygen will not be offset by the fuel savings. In the example used above, where $F = 2$ and $C = 0.541$, the minimum improvement in fuel efficiency would be 52% in order to break even with the air/fuel base case condition. For a higher fuel efficiency, the fuel and oxygen costs of OEC would be less than the cost of the fuel in the base case. For a lower efficiency, the opposite would be the case.

Other minor operating expenses that may be considered are potential cooling water requirements and changes in electrical power consumption. In the past, most oxy/fuel burners were water cooled. Some of those burners are still used today, primarily in the heating and melting of ferrous metals. There may be costs associated with the water consumption and possible water treatment to ensure clean, particle-free cooling of the burners, if required. New advances in burner design have eliminated the requirement for water cooling in most OEC technologies today. There is generally a small electrical requirement for the blower that supplies air to the air/fuel combustion system which may no longer be needed using OEC. In most cases, this cost is small and often neglected.

1.4.1.2 Capital Costs

Capital costs need to be considered for two types of installations: new and retrofit. There are many costs that are common to both, but there are also some costs that are unique to each. The exact capital costs are also highly dependent on how oxygen will be used in a particular application. If only low-level enrichment in the form of lancing will be used, then the costs will be lower than those associated with complete replacement of the combustion air with oxygen. In this section, the items normally considered as capital expenditures are described. However, many of these items may actually be part of the operating expenses, depending on how the equipment is being financed. If, for example, the industrial gas supplier will own and maintain the burner and flow controls, those costs may be part of a monthly service charge and would become operating expenses rather than capital costs.

When starting with the oxygen supply system, there are a number of factors to consider in estimating capital costs. If LOX will be trucked in to the plant, the LOX is stored in a cryogenic vessel that is typically owned and maintained by the industrial gas supplier. If the requirement is large enough, an on-site oxygen generation system, such as a vacuum swing adsorption or cryogenic liquid distillation plant, may be used. In some cases, the end user may purchase the oxygen generation plant. However, it is much more common for the plant to be owned and operated by the industrial gas supplier. In either of these types of installations, the equipment required to regulate the oxygen pressure and temperature is typically provided by the industrial gas supplier. The end user usually provides the land, the concrete pad where either the LOX storage system or the on-site generation system will be located, and all the necessary utilities. The main utility is electricity. In some applications, steam or a fuel-fired heat exchanger may be used to vaporize the LOX to gaseous oxygen. In the case of a LOX storage system, it has also become more common to have a phone line connected to the storage system which sends data back to the industrial gas supplier concerning the conditions of the system and the level of LOX in the tank. This telemetry information helps minimize runouts of oxygen and eliminates the need for the end user to order LOX when the level becomes low. In the case of an on-site oxygen generation plant, phone lines and water are generally required for personnel and equipment at the plant. For the LOX storage system, the end user also needs to maintain a clear, unobstructed path for the LOX tanker truck to get to the LOX storage system. A new access road may be needed if it does not already exist. The specific capital costs are highly dependent on the actual installation. The industrial gas supplier can provide typical costs based on previous experience.

Piping must be provided from the oxygen supply system to the use points inside the end user's plant. Again, the costs are dependent on the flow rate of oxygen that will be used, the distance between the supply system and the end-use points, the oxygen pressure in the pipe that helps to determine what type of pipe should be used, and how the piping will be routed. Large flows will require a larger pipe size. Long distances may also require a larger pipe size to minimize the pressure drop through the system. Smaller-diameter piping can be used if the oxygen pressure is higher, but the pipe thickness may need to be larger and the pipe material may need to be more expensive compared with that required for low-pressure oxygen supply. The piping may be buried underground or may need to be run overhead, depending on the needs of the customer.

The next capital cost to be considered is for the flow control equipment at the use points. This is highly dependent on the type of OEC and whether the installation is new or a retrofit to an existing flow control system. If only low-level oxygen enrichment is being used, then only a simple oxygen flow run is required. If a combustion system is being installed, then flow runs for oxygen, fuel, water (for water-cooled burners), and air or nitrogen (for purging) may be required. These flow runs are discussed in more detail in Chapter 10. For retrofit installations, there may be some duplication. For example, an existing air/fuel system should already have flow controls for fuel. Additional flow controls for the fuel may be required when the system is retrofitted for OEC.

The final capital cost to be considered is the equipment to deliver the oxygen into the combustion chamber. One component is the piping between the flow controls and the use point. If oxygen enrichment of the combustion air is being used, then the oxygen is typically injected into the airstream through a diffuser. The costs of modifying the air piping and the diffuser itself should be considered. If oxygen will be lanced into the furnace, then the cost of the lances and the installation of the lances are capital costs to be considered. If higher levels of OEC will be used, then there will be some cost for the combustion equipment. This may be in the form of a modification of existing burners designed for air/fuel, or it may be for burners specially designed for OEC.

1.4.2 Other Benefits

This section will discuss some other benefits of using OEC that are usually specific to a given application. The value of these benefits varies with each installation. Typical benefits for certain types of applications in the different industries are discussed in more detail in Chapters 5 through 8.

1.4.2.1 Increased Yields

Considerable cost savings may be achieved by increasing the amount of product that can be sold using a given manufacturing process. One way is to reduce the number of rejects. An example in the glass industry is the reduction in inclusions or bubbles, referred to as seeds or stones, contained in the glass (see Chapter 7). This defective glass cannot be sold and must be recycled back into the glass furnace for reprocessing, which adds costs to the product that is sold. Profits can be increased by reducing the amount of rejects.

Another way that OEC has been used to increase the yield in a process is by reducing the waste. One example is in the reheating of metal prior to forming. Oxy/fuel burners have been used to heat metal rapidly by direct flame impingement. One of the problems with traditional metal reheating is that an oxide scale layer forms on the surface of the metal, due to the extended time at high temperatures. It has been shown that oxy/fuel direct flame impingement can reduce the amount of scale that forms on the metal so that less metal is wasted and more can be sold.[46]

1.4.2.2 Reduced Maintenance

OEC has been used in the waste incineration industry as a method to reduce the amount of downtime for equipment repairs (see Chapter 8). This has produced dramatic savings in the overall cost to process waste materials.

1.4.2.3 Incremental Production

There are two basic choices to consider when more production is desired. One is to build additional capacity. This is generally very costly and usually time-consuming. Space limitations may prohibit this type of expansion. Building a new production line may be the appropriate choice if a very large increase in capacity is required which cannot be reasonably obtained by expanding existing lines. If the existing production facilities are fairly old and inefficient, it may be wise to replace these with more-modern and more-efficient equipment. Also, the existing equipment may already be at or near its limit of production and cannot be further improved. The incremental costs of manufacturing, using a completely new production line, are generally very high. Also, the time delay involved before the new line is operational could result in lost business to competitors who may be able to supply the increased demand.

Another way to increase capacity is to improve the existing production equipment. This is generally much less expensive and can usually be done more quickly with only a minimal impact on space requirements. If only an incremental increase in production is required, then the limiting equipment needs to be determined. For example, if the thermal output of the combustion chamber is the limiting factor, then adding more burners or increasing the firing rates of the existing burners may be an option. However, this could cause other problems with the associated equipment. The existing burners, the combustion air blower, the air flow controls, the flue gas–handling system, and the pollution control system may not be able to handle the increased thermal input. It has been previously shown that OEC may be a good option to consider because the incremental costs are small compared with building new production facilities.

1.4.2.4 New or Improved Product

OEC has been used to improve the product quality in some manufacturing processes. This may mean that the improved product can be sold at a higher price, which means a higher profit, or it may mean that a larger market share can be obtained, leading to increased sales.

In some cases it may be possible to make a new product using OEC that could not be made previously with an air/fuel system. This is particularly true when very high temperatures are required.

1.5 FUTURE OF OXYGEN-ENHANCED COMBUSTION

There are a number of factors that make the future look bright for OEC. Separate reports on the industrial uses of OEC sponsored by the U.S. Department of Energy in 1987[34] and the Gas Research Institute in 1989[1] predict the growing importance of OEC. New advancements are continually being made in existing and new (see, for example, Reference 47) oxygen generation technologies (see Chapter 3). This should reduce the cost to the end user, which should expand the number of amenable applications. Studies are being conducted to improve the energy efficiency of OEC processes further by using novel methods of heat recovery[48] and by integrating the air separation process with the combustion system.[28]

There are many areas related to the use of OEC that need further research. In a recent report done for the U.S. Department of Energy, the following were recommended for further research:

- Use of oxygen enrichment to increase furnace efficiency in the petroleum industry;
- Investigation of the benefits and disadvantages of varying the oxidizer composition ranging from air to pure O_2;
- Measurement of the NOx from oxygen-enriched burners in steel reheat furnaces;
- Optimization of the amount of O_2 injection into cupolas used in the foundry industry to recuperate energy from the CO produced in the process;
- Optimization of glass-melting processes that use oxygen/fuel combustion, including the development of computer models;
- Quantification of furnace efficiency and productivity improvements using OEC in aluminum melting; and
- Development of NOx reduction technologies for OEC in aluminum processes.[49]

Another major area for further research in the use of OEC is the design of the combustion chambers. In nearly all cases, OEC has been adapted and retrofitted to existing furnace designs. In the future, the design of new furnaces that will use OEC should be investigated to optimize the increased radiant heat transfer, reduced convective heat transfer, and reduced gas volume flow rates.

1.6 CONCLUSIONS

There are numerous examples of using OEC in a wide variety of industrial heating applications. In some cases, low-level O_2 enrichment of an existing air/fuel combustion system can have dramatic results. In other cases, oxy/fuel burners may be used to provide needed solutions.

Lower-cost O_2, coupled with the significant environmental and operating benefits of OEC, make OEC an attractive technology. Research continues to assess its effectiveness in a wide range of applications.

REFERENCES

1. Williams, S. J., Cuervo, L. A., and Chapman, M. A., High-Temperature Industrial Process Heating: Oxygen-Gas Combustion and Plasma Heating Systems, Gas Research Institute report GRI-89/0256, Chicago, July 1989.
2. Chace, A. S., Hazard, H. R., Levy, A., Thekdi, A. C., and Ungar, E. W., Combustion Research Opportunities for Industrial Applications — Phase II, U.S. Department of Energy report DOE/ID-10204-2, Washington, D.C., 1989.
3. Marshall, N., Oxygen, in *Gas Encyclopaedia*, Elsevier, New York, 1976, 1079.
4. Braker, W. and Mossman, A. L., Oxygen, in *Matheson Gas Data Book*, 6th ed., Matheson, Lyndhurst, NJ, 1980, 562.
5. Ahlberg, K., Oxygen, in *AGA Gas Handbook*, Almqvist & Wiksell, Stockholm, Sweden, 1985, 438.
6. Sychev, V. V., Vasserman, A. A., Kozlov, A. D., Spiridonov, G. A., and Tsymarny, V. A., *Thermodynamic Properties of Oxygen*, Hemisphere Publishing, Washington, D.C., 1987.
7. Compressed Gas Association, Oxygen, in *Handbook of Compressed Gases*, 3rd ed., Van Nostrand Reinhold, New York, 1990, 526.
8. Austin, G. T., Oxygen and nitrogen, in *Shreve's Chemical Process Industries*, 5th ed., McGraw-Hill, New York, 1984, 115.
9. Blakey, P. G., Liquified (cryogenic) gases, in *Encyclopedia of Chemical Processing and Design*, Vol. 28, McKetta, J. J. and Cunningham, W. A., Eds., Marcel Dekker, New York, 1988, 166.
10. Kirschner, M. J., Oxygen, in *Ullmann's Encylcopedia of Industrial Chemistry*, 5th ed., Vol. A18, Elvers, B., Hawkins, S., and Schulz, G., Eds., VCH Verlagsgesellschaft, Weinheim, Germany, 1991, 329.
11. Mattern, G. W., Industrial Gases, in *Riegel's Handbook of Industrial Chemistry*, 9th ed., Kent, J. A., Ed., Van Nostrand Reinhold, New York, 1992, 442.
12. Francis, A. W., Oxygen, in *McGraw-Hill Encyclopedia of Science & Technology*, 6th ed., Vol. 12, McGraw-Hill, New York, 1987, 604.
13. Kornblum, Z. C., Oxygen, in *The Encyclopedia Americana*, International Edition, Grolier, Danbury, CT, 1996, 164.
14. Baukal, C. E., Eleazer, P. B., and Farmer, L. K., Basis for enhancing combustion by oxygen enrichment, *Ind. Heating*, LIX (2), 22, 1992.
15. Benedek, K. R. and Wilson, R. P., The Competitive Position of Natural Gas in Oxy-Fuel Burner Applications, Gas Research Institute report no. GRI-96-0350, Chicago, September 1996.
16. Joshi, S. V., Becker, J. S., and Lytle, G. C., Effects of oxygen enrichment on the performance of air-fuel burners, in *Industrial Combustion Technologies*, M. A. Lukasiewicz, Ed., American Society of Metals, Materials Park, OH, 1986, 165.

17. U.S. Environmental Protection Agency, Alternative Control Techniques Document — NOx Emissions from Utility Boilers, EPA report EPA-453/R-94-023, Research Triangle Park, NC, 1994.
18. Kobayashi, H., Segregated Zoning Combustion, U.S. Patent 5,076,779, December 31, 1991.
19. Bazarian, E. R., Heffron, J. F., and Baukal, C. E., Method for Reducing NOx Production during Air-Fuel Combustion Processes, U.S. Patent 5,308,239, May 3, 1994.
20. Gitman, G. M., Method and Apparatus for Generating Highly Luminous Flame, U.S. Patent 4,797,087, January 10, 1989.
21. Dalton, A. I. and Tyndall, D. W., Oxygen Enriched Air/Natural Gas Burner System Development, NTIS report PB91-167510, Springfield, VA, 1989.
22. Gordon, S. and McBride, B. J., Computer Program for Calculation of Complex Chemical Equilibrium Compositions, Rocket Performance, Incident and Reflected Shocks, and Chapman–Jouguet Detonations, NASA report SP-273, 1971.
23. Turin, J. J. and Huebler, J., Gas-Air-Oxygen Combustion Studies, Report no. I.G.R.-61, American Gas Association, Arlington, VA, 1951.
24. Lewis, B. and von Elbe, G., *Combustion, Flames and Explosions of Gases*, 3rd ed., Academic Press, New York, 1987.
25. American Gas Association, *Gas Engineers Handbook*, Industrial Press, New York, 2/72, 1965.
26. Gibbs, B. M. and Williams, A., Fundamental aspects on the use of oxygen in combustion processes — a review, *J. Inst. Energ.*, 56(427), 74, 1983.
27. Farrell, L. M., Pavlack, T. T., and Rich, L., Operational and environmental benefits of oxy-fuel combustion in the steel industry, *Iron Steel Eng.*, 72(3), 35, 1995.
28. Ding, M. G. and Du, Z., Energy and environmental benefits of oxy-fuel combustion, in *Proceedings of the 1995 International Conference on Energy & Environment,* Shanghai, China, May 1995, Begell House, New York, 1995, 674.
29. Grisham, S., Oxy-conversion for a smaller furnace yields big results, *Am. Glass Rev.*, 117(6), 12, 1997.
30. Harris, M. E., Grumer, J., Von Elbe, G., and Lewis, B., Burning velocities, quenching, and stability data on nonturbulent flames of methane and propane with oxygen and nitrogen, in *Third Symposium on Combustion, Flame and Explosion Phenomena,* Williams & Wilkins, Baltimore, 1949, 80.
31. Abernathy, R., McElroy, J., and Yap, L. T., The performance of current oxy-fuel combustion technology for secondary aluminum melting, *Light Metals 1996,* Hale, W., Ed., The Minerals, Metals, and Materials Society, Warrendale, PA, 1996, 1233.
32. Brown, J. T., Glass melting: the elegance of direct-fired oxy-fuel, *Ceram. Ind.*, 138(3), 47, 1992.
33. Gill, J. H. and Quiel, J. M., *Incineration of Hazardous, Toxic, and Mixed Wastes*, North American Manufacturing Company, Cleveland, 1993.
34. Kobayashi, H., Oxygen Enriched Combustion System Performance Study, Phase I Interim/Final Report, Volume I — Technical and Economic Analysis, U.S. Department of Energy report DOE/ID/12597, Washington, D.C., March 1987.
35. Hottel, H. C. and Sarofim, A. F., *Radiative Transfer*, McGraw-Hill, New York, 1967.
36. DeLucia, M., Oxygen enrichment in combustion processes: comparative experimental results from several application fields, *J. Energ. Resour. Tech.*, 113, 122, 1991.
37. Industrial Heating Equipment Association, *Combustion Technology Manual*, 5th ed., Arlington, VA, 1994.
38. Dafoe, B. M., *Report of Committee F: Industrial and Commercial Utilization of Gases*, International Gas Union, Zurich, Switzerland, 1991.

39. Cornforth, J. R., *Combustion Engineering and Gas Utilisation*, 3rd ed., E&FN Spon, London, 1992.
40. Slavejkov, A. G., Baukal, C. E., Joshi, M. L., and Nabors, J. K., Oxy-fuel glass melting with a high-performance burner, *Ceram. Bull.*, 71(3), 340, 1992.
41. Guerrero, P. S., Rebello, W. J., Ally, M. R., and Jain, R., Combustion characteristics of fuels with preheated oxygen enriched air, in *Industrial Combustion Technologies*, M. A. Lukasiewicz, Ed., American Society of Metals, Materials Park, OH, 1986, 187.
42. Reed, R. J., *North American Combustion Handbook*, 3rd ed., Vol. II, Part 13, North American Manufacturing Company, Cleveland, OH, 1997.
43. Bishop, N. G. and Taylor, G., Method and Apparatus for Gasifying Organic Materials and Vitrifying Residual Ash, U.S. Patent 5,584,255, Dec. 17, 1996.
44. Martinez, M. P., Apparatus and Process for Removing Unburned Carbon in Fly Ash, U.S. Patent 5,555,821, Sept. 17, 1996.
45. Tamhankar, S., Menon, R., Chou, T., Ramachandran, R., Hull, R., and Watson, R., Enrichment can decrease NOx, SOx formation, *Oil Gas J.*, 94(10), 60, 1996.
46. Becker, J. S. and Farmer, L. K., Rapid fire heating system uses oxy-gas burners for efficient metal heating, *Ind. Heating*, 62(3), 74, 1995.
47. Assessment of Thermal Swing Absorption Alternatives for Producing Oxygen Enriched Combustion Air, Report no. DOE/CE-040762T-H1, U.S. Department of Energy, Washington, D.C., April 1990.
48. Air Products and Chemicals, Inc., Development of an Advanced Glass Melting System: The Thermally Efficient Alternative Melter "Team," Report no. DOE/CE/40917, U.S. Department of Energy, Washington, D.C., February 1992.
49. Keller, J. G., Soelberg, N. R., and Kessinger, G. F., Industry-Identified Combustion Research Needs, Report no. INEL-95/0578 (DE96003785), U.S. Department of Energy, Washington, D.C., 1995.

2 Pollutant Emissions

Charles E. Baukal, Jr.

CONTENTS

2.1 INTRODUCTION

The purpose of this chapter is to alert the interested reader to the potential effects on pollutant emissions by using oxygen-enhanced combustion (OEC). As will be shown, depending on the level of OEC used in a given combustion process, pollutant emissions may either increase or decrease. Specific examples for a given industry are discussed in the application chapters elsewhere in the book. This chapter will give some generalizations and discuss some of the theory that explains the tendencies.

0-8493-1695-2/98/$0.00+$.50

2.2 NOx

NOx refers to oxides of nitrogen. These generally include nitrogen monoxide, also known as nitric oxide (NO), and nitrogen dioxide (NO_2). They may also include nitrous oxide, also known as laughing gas (N_2O), as well as other, less common combinations of nitrogen and oxygen, such as nitrogen tetroxide (N_2O_4). In most high-temperature heating applications, the majority of the NOx exiting the exhaust stack is in the form of NO.[1] NO is a colorless gas that rapidly combines with O_2 in the atmosphere to form NO_2. NO is poisonous to humans and can cause irritation of the eyes and throat, tightness of the chest, nausea, headache, and gradual loss of strength. Prolonged exposure to NO can cause violent coughing, difficulty in breathing, and cyanosis, and could be fatal. It is interesting to note that *Science* magazine named NO as its 1993 Molecule of the Year. The reason is that NO is absolutely essential in human physiology. A growing body of research indicates its importance in everything from aiding digestion and regulating blood pressure to acting as a messenger in the nervous system. It is also a promising drug in the treatment of persistent pulmonary hypertension, which is a life-threatening lung condition that affects about 4000 babies each year.

NO_2 is a reddish brown gas which has a suffocating odor. It is highly toxic and hazardous because of its ability to cause delayed chemical pneumonitis and pulmonary edema. NO_2 vapors are a strong irritant to the pulmonary tract. Inhalation may also cause irritation of the eyes and throat, tightness of the chest, headache, nausea, and gradual loss of strength. Severe symptoms may be delayed and include cyanosis, increased difficulty in breathing, irregular respiration, lassitude, and possible death due to pulmonary edema. Chronic or repeated exposure to NO_2 could cause a permanent decrease in pulmonary function.

In addition to the poisoning effect that NOx has on humans, there are also other problems associated with these chemicals. In the lower atmosphere, NO reacts with oxygen to form ozone, in addition to NO_2. Ozone (O_3) is also a health hazard that can cause respiratory problems in humans. NO_2 is extremely reactive and is a strong oxidizing agent. It explodes on contact with alcohols, hydrocarbons, organic materials, and fuels. NO_2 decomposes on contact with water to produce nitrous acid (HNO_2) and nitric acid (HNO_3), which are highly corrosive. When NO_2 forms in the atmosphere and comes in contact with rain, acid rain is produced, which is destructive to whatever it comes in contact with, including plants, trees, and man-made structures like buildings, bridges, and the like. Besides acid rain, another problem with NO_2 is its contribution to smog. When sunlight contacts a mixture of NO_2 and unburned hydrocarbons in the atmosphere, photochemical smog is produced.

Many combustion processes are operated at elevated temperatures and high excess air levels. The combustion products may have long residence times in the combustion chamber. These conditions produce high thermal efficiencies and product throughput rates. Unfortunately, those conditions also favor the formation of NOx. NOx emissions are among the primary air pollutants because of their contribution to smog formation, acid rain, and ozone depletion in the upper atmosphere.

Oxy/fuel combustion is a recognized method for reducing NOx emissions under carefully controlled conditions.[2] The glass industry in particular has rapidly converted from air/fuel to oxy/fuel primarily to reduce NOx emissions from 70 to 95%.[3] The next section discusses the theory to explain how OEC reduces NOx.

2.2.1 Theory

There are three generally accepted mechanisms for NOx production: thermal, prompt, and fuel. Thermal NOx is formed by the high-temperature reaction of nitrogen with oxygen, by the well-known Zeldovich mechanism.[4] It is given by the simplified reaction:

$$N_2 + O_2 \rightarrow NO, NO_2 \tag{2.1}$$

Thermal NOx increases exponentially with temperature. Above about 2000°F (1400 K), it is generally the predominant mechanism in combustion processes. This makes it important in most high-temperature heating applications. Prompt NOx is formed by the relatively fast reaction between nitrogen, oxygen, and hydrocarbon radicals. It is given by the overall reaction:

$$CH_4 + O_2 + N_2 \rightarrow NO, NO_2, CO_2, H_2O, \text{trace species} \tag{2.2}$$

In reality, this very complicated process consists of hundreds of reactions. The hydrocarbon radicals are intermediate species formed during the combustion process. Prompt NOx is generally an important mechanism at lower-temperature combustion processes. Fuel NOx is formed by the direct oxidation of organo-nitrogen compounds contained in the fuel. It is given by the overall reaction:

$$R_xN + O_2 \rightarrow NO, NO_2, CO_2, H_2O, \text{trace species} \tag{2.3}$$

Fuel NOx is not a concern for high-quality gaseous fuels like natural gas or propane, which normally have no organically bound nitrogen. However, fuel NOx may be important when oil, coal, or waste fuels that may contain significant amounts of organically bound nitrogen are used.

NOx trends can be predicted by calculating the adiabatic equilibrium temperature and composition for methane combusted stoichiometrically with an oxidizer consisting of oxygen and a variable quantity of nitrogen. Figure 2.1 shows the results of those calculations for NO as a function of the oxygen concentration in the oxidizer. NO is given in both mass (lb NO/MMBtu) and volume units (ppmvd), for comparison. The mass unit (lb/MMBtu) has been normalized to a unit flow rate of fuel which also equates to a given unit of energy based on the higher heating value of the fuel. The volume unit (ppmvd) has not been corrected to any specific O_2 level. There are two competing effects that produce the parabolic shape of the NO curves. As the oxygen concentration increases, the flame temperature increases. This accelerates NOx formation, because of the exponential dependence of the thermal NOx

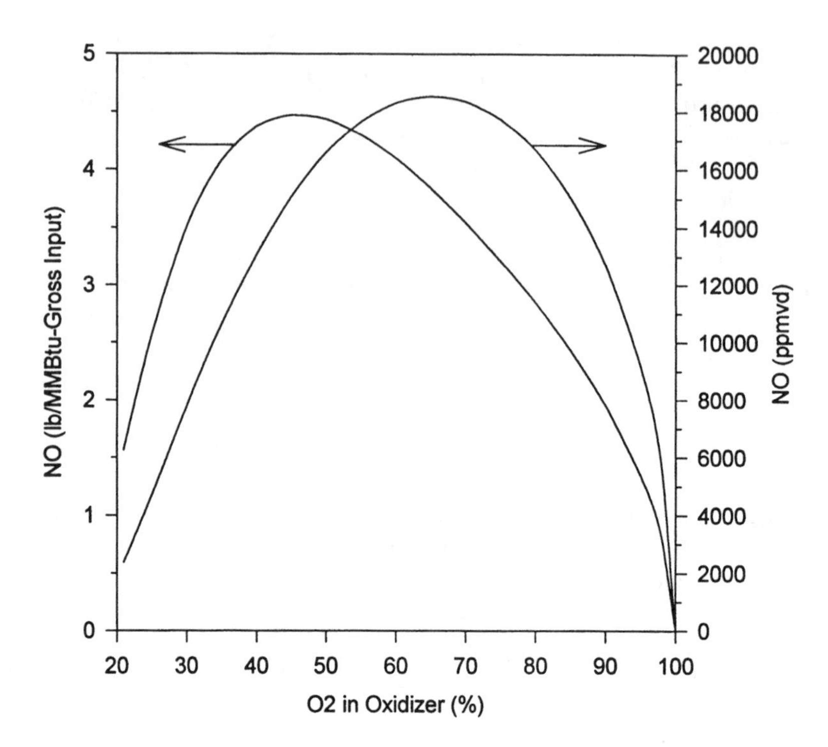

FIGURE 2.1 NO concentration (in units of lb/MMBtu and ppmvd) vs. oxidizer composition for an adiabatic equilibrium stoichiometric CH_4 flame.

reactions on temperature. However, as the O_2 concentration increases, the N_2 concentration simultaneously decreases. This lowers NOx because less N_2 is available to make NOx. The simultaneous effects of increasing flame temperature and reduction in the amount of N_2 produce peak NOx values in the middle range of oxygen enrichment.

Figure 2.2 shows NO and CO as functions of the O_2/CH_4 stoichiometry for CH_4 combusted with an oxidizer consisting of 95% O_2 and 5% N_2. This represents one type of oxygen-enhanced system. The graph shows the strong dependence of both pollutants on the stoichiometry. At fuel-rich conditions (stoichiometry < 2), NO decreases, while CO increases. Also, the fuel efficiency is reduced since the fuel is not fully combusted. At fuel-lean conditions (stoichiometry > 2), NO increases, while CO decreases. Again, the fuel efficiency is reduced because the excess oxidizer carries heat out of the process. In order to minimize both CO and NOx, the combustion system should be operated close to stoichiometrically, which also maximizes the fuel efficiency.

Figure 2.3 shows a plot of the adiabatic equilibrium flame temperature for an air/CH_4 flame and an O_2/CH_4 flame, as functions of the flame stoichiometry. There are several things to notice. The flame temperature for the air/CH_4 flame is very dependent on the stoichiometry. For the O_2/CH_4 flame, the temperature is very dependent on the stoichiometry only under fuel-rich conditions. The temperature is not very dependent on the stoichiometry when the O_2/CH_4 flame is fuel lean.

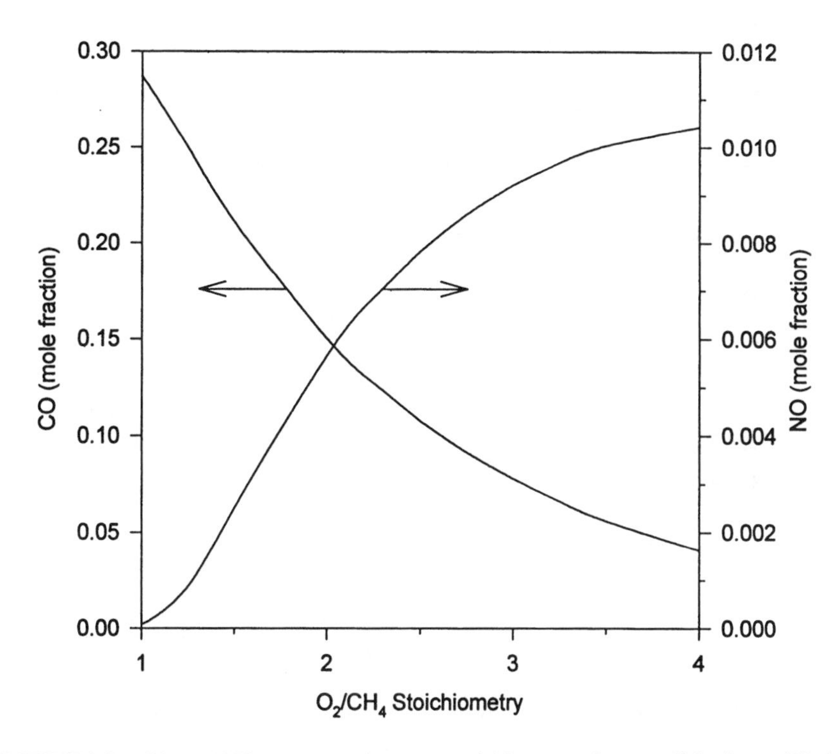

FIGURE 2.2 CO and NO concentrations vs. stoichiometry for an adiabatic equilibrium 95% O_2/5% N_2/CH_4 flame.

The figure helps to explain why, for example, NOx is reduced dramatically under fuel-rich conditions. One reason is the dramatic reduction in the flame temperature. Another reason concerns the chemistry. In a reducing atmosphere, CO is formed preferentially to NO. This is exploited in some of the NOx reduction techniques. An example is methane reburn.[5] The exhaust gases from the combustion process flow through a reduction zone which is at reducing conditions. NOx is reduced back to N_2. Any CO that may have formed in the reduction zone and other unburned fuels are then combusted downstream of the reduction zone. However, they are combusted at temperatures well below those found in the main combustion process. These lower temperatures are not as favorable to NOx formation. Figure 2.4 shows the importance of the gas temperature on thermal NOx formation. Many combustion modification strategies for reducing NOx involve reducing the flame temperature.

2.2.2 Abatement Strategies

Before air quality regulations, the flue gases from combustion processes were vented directly to the atmosphere. As air quality laws tightened and the public's awareness increased, industry began looking for new strategies to curb NOx emissions. The three strategies for reducing NOx are discussed next.

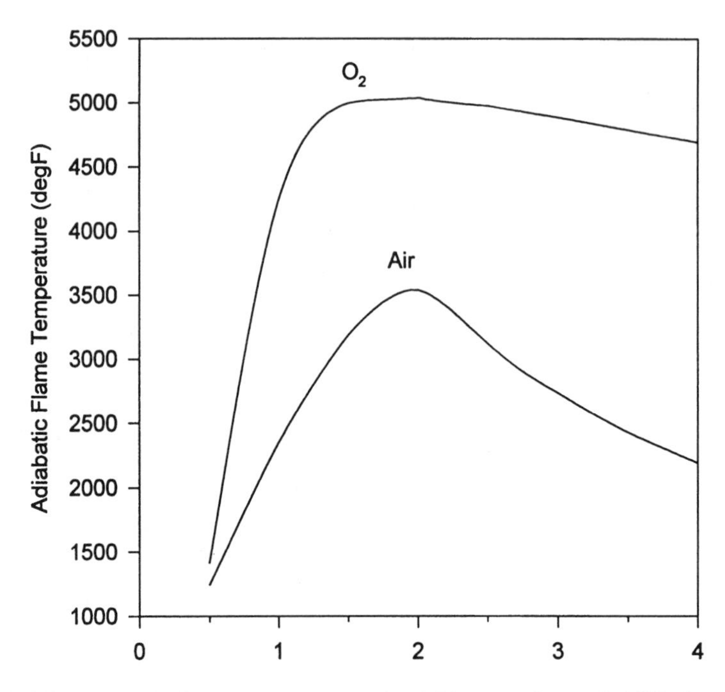

FIGURE 2.3 Adiabatic flame temperature vs. stoichiometry for an O_2/CH_4 flame and an air/CH_4 flame.

2.2.2.1 Pretreatment

The first NOx reduction strategy, which can be referred to as pretreatment, is a preventative technique to minimize NOx generation. In pretreatment, the incoming feed materials are treated in such a way as to reduce NOx. Some of these include fuel switching, using additives, fuel treatment, and oxidizer switching. Fuel switching is simply replacing a more-polluting fuel with a less-polluting fuel. For example, fuel oils generally contain some organically bound nitrogen which produces fuel NOx. Natural gas does not normally contain any organically bound nitrogen and usually has only low levels of molecular nitrogen (N_2). Partial or complete substitution of natural gas for fuel oil can significantly reduce NOx emissions by reducing the amount of nitrogen in the fuel. Another type of pretreatment involves adding a chemical to the incoming feed materials (raw materials, fuel, or oxidizer) to reduce emissions by changing the chemistry of the combustion process. One example would be injecting ammonia into the combustion airstream as a type of *in situ* de-NOx process. A third type of pretreatment involves treating the incoming fuel prior to its use in the combustion process. An example would be removing fuel-bound nitrogen from fuel oil or removing N_2 from natural gas. The fourth type of pretreatment is oxidizer switching, where a different oxidizer is used. Air is the most commonly used oxidizer. It will be shown later in this chapter that substantial NOx reduction can be achieved by using pure oxygen, instead of air, for combustion. By drastically reducing the N_2 content in the system, NOx is minimized. In addition, other process benefits

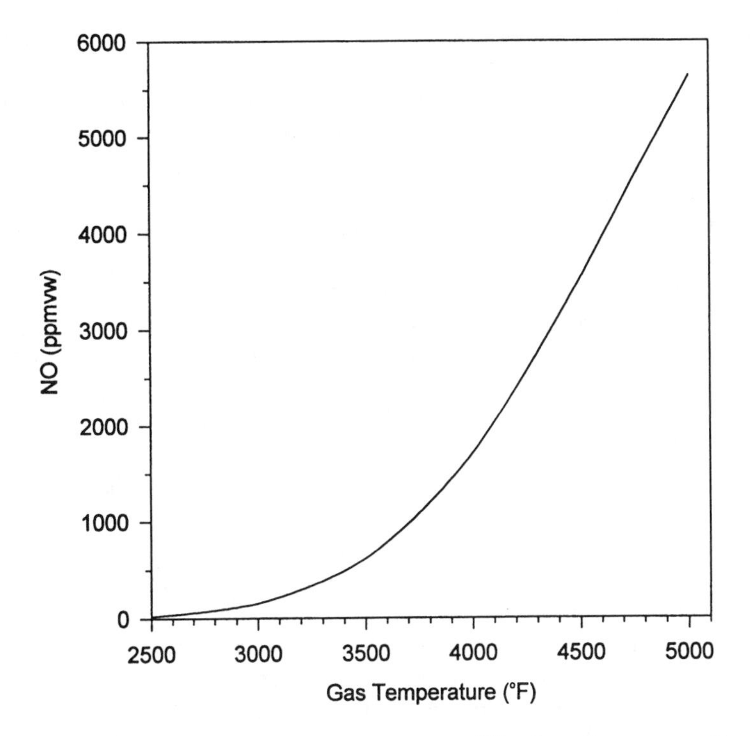

FIGURE 2.4 NO concentration (ppmvw = parts per million by volume on a wet basis) vs. exhaust gas temperature for an equilibrium stoichiometric 95% O_2/5% N_2/CH_4 flame.

may be gained, as previously discussed in Chapter 1. This method has the potential to lower NOx while simultaneously enhancing the overall system performance.

2.2.2.2 Combustion Modification

The second strategy, known as combustion modification, is to prevent NOx from ever forming, which is done by modifying the combustion process. There are numerous methods that have been used to modify the combustion process for low NOx. A popular method is known as low-NOx burners where specially designed burners generate less NOx than the previous burner technology. Low-NOx burners incorporate a number of techniques for minimizing NOx including, for example, flue gas recirculation, staging, pulse combustion, and advanced mixing. Another combustion modification technique is to reduce the air preheat, which greatly increases NOx, for processes that use heat recuperation. However, this also reduces the overall system efficiency. Steam or water may be injected into the flame to lower the flame temperature, which reduces NOx, but which also reduces the system efficiency. By reducing air infiltration (leakage) into the furnace, NOx can be reduced because excess O_2 generally increases NOx. Pulse combustion has been shown to reduce NOx because the alternately very fuel-rich and very fuel-lean combustion minimizes NOx formation. The overall stoichiometry of the oxidizer and fuel is maintained by controlling the pulsations. In many cases, NOx can be reduced by simply reducing the excess air through the burners. Again, excess O_2 tends to produce more NOx. Also, reducing excess O_2 tends to increase the system efficiency, which is an added benefit. The limit on reducing the excess air is CO emissions. If the excess O_2 is reduced too much, then CO emissions will increase. CO is not only a pollutant, but it is an indication that the fuel is not being fully combusted which results in lower system efficiencies. If the mass of NOx emitted from a combustor is too high, an alternative is to reduce the firing rate which means a corresponding reduction in production. This is generally not a preferred alternative for obvious reasons.

Staged combustion is an effective technique for lowering NOx. Staging means that some of the fuel or oxidizer is added downstream of the main combustion zone. The fuel, oxidizer, or both may be staged in the flame. For example, there may be primary and secondary fuel inlets where a portion of the fuel is injected into the main flame zone and the balance of the fuel is injected downstream of that main flame zone. Fuel staging is effective for two reasons: the peak flame temperatures are reduced, which reduces NOx, and the fuel-rich chemistry in the primary flame zone also reduces NOx. Similar benefits are achieved with oxidizer staging. In furnace gas recirculation the products of combustion inside the combustion chamber are recirculated back into the flame. Flue gas recirculation is similar, in that the exhaust gases in the flue are recirculated back through the burner into the flame via ductwork external to the furnace. Although the furnace or flue gases are hot, they are considerably cooler than the flame itself and, therefore, act as a diluent which reduces the overall flame temperature which in turn reduces NOx. Advanced mixing techniques use carefully designed burner aerodynamics to control the mixing of the fuel and the oxidizer. The goals of most of these techniques are to avoid hot spots and make the flame temperature uniform, to increase the heat release from the flame which lowers the flame temperature, and to control the chemistry in the flame zone to minimize NOx formation.

Many of the combustion modification methods attempt to reduce the temperature of the flame to lower NOx emissions. In many cases, this may result in a reduction of the combustion efficiency.[6] For example, if water is injected into the flame to lower NOx, the water absorbs heat from the flame and carries most of that energy out with the exhaust gases and does not transfer much of that energy to the load. Combustion modification methods are usually less capital intensive than most post-treatment methods. In many cases, there is a limit to how much NOx reduction can be achieved using these combustion modification methods.

2.2.2.3 Posttreatment

The third strategy for minimizing NOx, known as posttreatment, involves removing NOx from the exhaust gases after the NOx has already been formed in the combustion chamber. Two of the most common methods of posttreatment are selective catalytic reduction (SCR) and selective noncatalytic reduction (SNCR).[7] Wet techniques for posttreatment include oxidation/absorption, oxidation/absorption/reduction, absorption/oxidation, and absorption/reduction. Dry techniques for posttreatment, besides SCR and SNCR, include activated carbon beds, electron beam radiation, and reaction with hydrocarbons.

Many of these techniques are fairly sophisticated and are not trivial to operate and maintain in industrial furnace environments. For example, the catalytic reduction techniques require a catalyst which may become plugged or poisoned fairly quickly by dirty flue gases. Posttreatment methods are often capital intensive. They usually require halting production if there is a malfunction of the treatment equipment. Also, posttreatment does not normally benefit the combustion process in any way. For example, it does not increase production or energy efficiency. It is strictly an add-on cost.

2.2.3 Laboratory Results

To assess the validity of the theoretical NOx reduction using oxygen enrichment, a comprehensive R&D program was conducted, with partial funding from the Gas Research Institute.[8] The two regimes of low- and high-level oxygen enrichment were studied. Low-level enrichment typically involves adding pure oxygen to air to increase the total O_2 concentration from 21% to as high as 35%. In high-level enrichment, air is replaced with oxygen of varying purity, depending on the oxygen production method. These two regimes are important because they encompass most industrial applications.

These two regimes were studied by conducting an extensive set of experiments in a pilot-scale furnace. Several parameters were varied to study their effects on NOx. The furnace pressure was positive to prevent air infiltration. This was done for two reasons. Many industrial furnaces are run under positive pressure to prevent infiltrating air from affecting the process. In addition, excluding air infiltration effects isolates the effect of the burner on NOx production. The natural gas used for the tests consisted of 96.6% CH_4, 0.4% N_2, plus higher hydrocarbons. Low-level enrichment was studied using a standard North American Manufacturing Co. (Cleveland, OH) model 4425 air/fuel burner, as shown in Figure 2.5. High-level enrichment was studied using a standard Air Products and Chemicals, Inc. (Allentown, PA) model KT-3 oxy/fuel burner as shown in Figure 2.6. This burner consisted of three concentric tubes with the fuel going through the inner and outer passages. O_2 flowed through the middle passage.

Figure 2.7 shows a comparison of the normalized flue NO as a function of the burner firing rate. For both burners, the normalized NOx emissions were not dependent on the firing rate for the firing rate ranges that were tested. Note that with pure oxygen and no air infiltration there was still some flue NO for the oxy/fuel burner. This was caused by the small amount of N_2 in the natural gas.

Figure 2.8 compares the normalized flue NO as a function of the stoichiometry. For both burners, NOx emissions were much higher under fuel-lean conditions compared with fuel-rich conditions. This is as predicted by the theoretical calculations (see Figure 2.2).

Figure 2.9 is a plot of flue NO as a function of the oxidizer composition. For low-level O_2 enrichment, NOx increased rapidly as the oxygen concentration increased. For high-level O_2 enrichment, NOx increased rapidly as the oxidizer purity decreased. Both of these trends validated the theoretical calculations (see Figure 2.1). Those results are important because some oxygen generation methods produce O_2

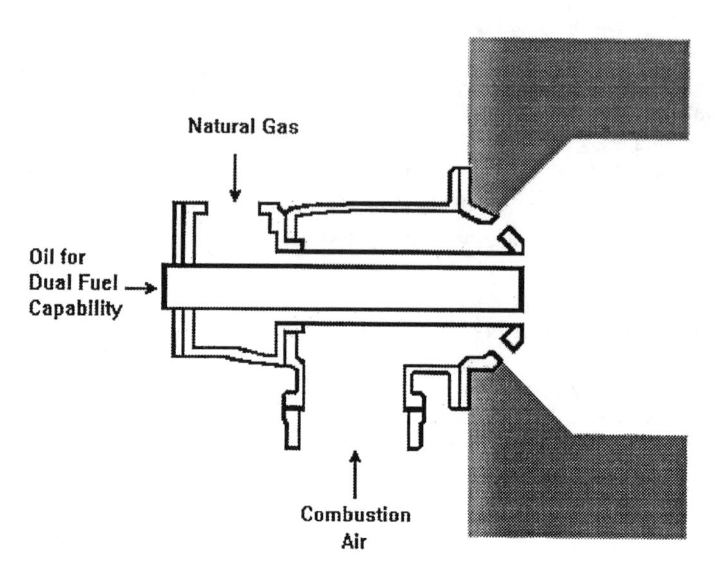

FIGURE 2.5 North American 4425-8A air/fuel burner.

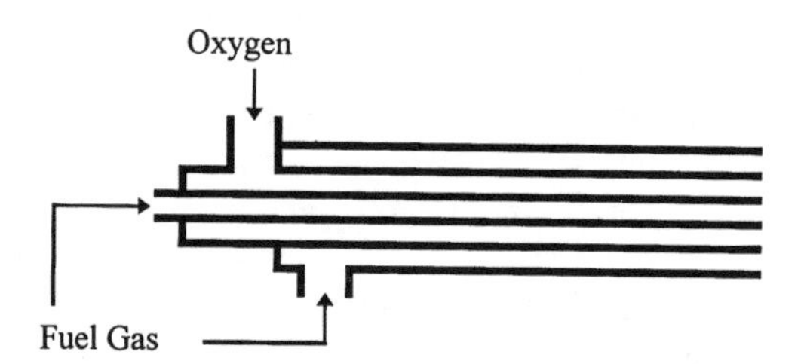

FIGURE 2.6 Schematic of Air Products KT-3 oxy/fuel burner.

that may contain several percent nitrogen and because some natural gas sources can have as high as 15% nitrogen by volume.

There were several important results from those experiments. The first is that the NOx emissions for low-level oxygen enrichment were nearly an order of magnitude higher than for high-level enrichment. The second is that the experimental NOx trends were the same as those predicted by theory. However, the experimental measurements were about an order of magnitude lower than the theoretical predictions. This is due to the fact that actual flames are not adiabatic processes, since a large amount of heat is radiated from the flames. The actual flame temperature is usually much lower than the adiabatic equilibrium flame temperature.

2.2.4 Regulations

Baukal and Eleazer[9] have discussed potential sources of confusion in the existing NOx regulations. These sources of confusion may be classified as either general or specific. General sources of confusion include, for example, the wide variety of units

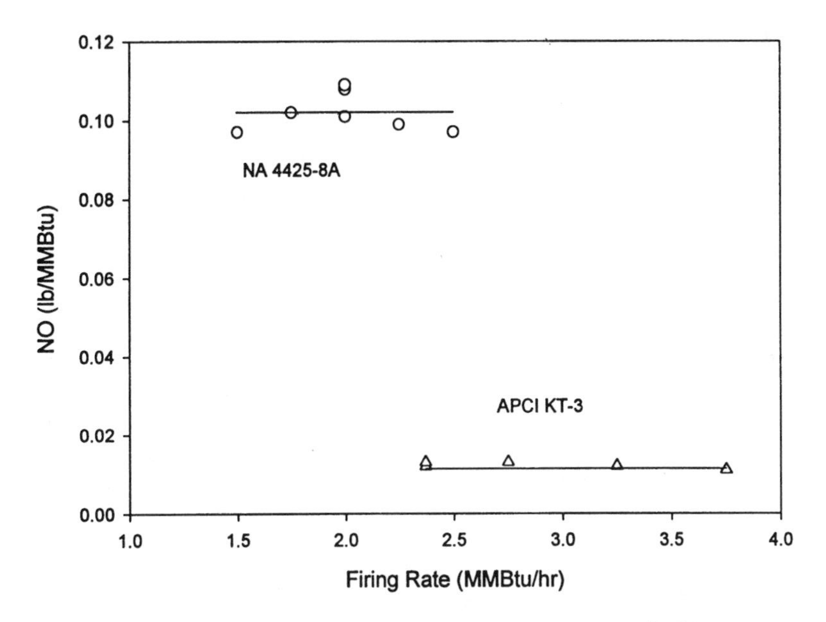

FIGURE 2.7 NO vs. firing rate for an air/fuel and an oxy/fuel burner.

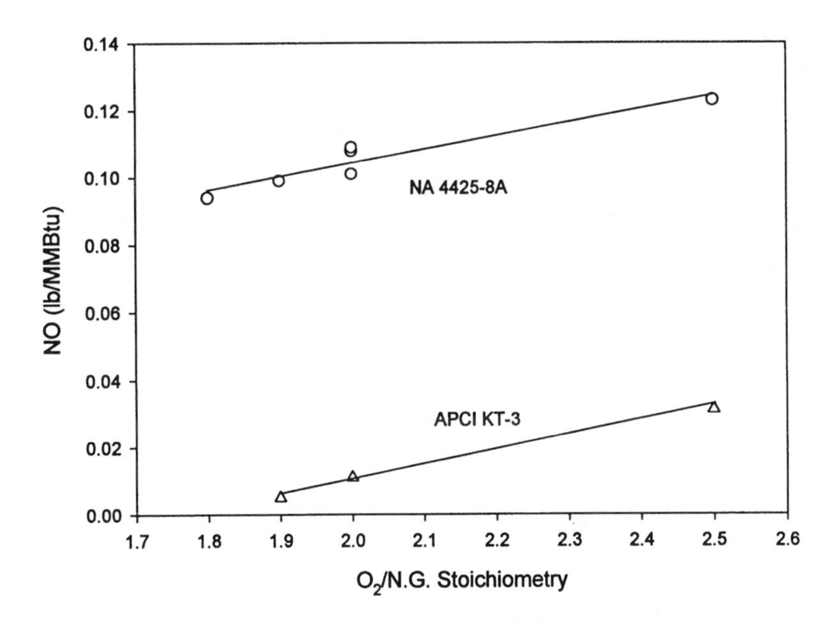

FIGURE 2.8 NO vs. stoichiometry for an air/fuel and an oxy/fuel burner.

that have been used, reporting on either a dry or wet sample basis, measuring NO but reporting NO_2, and reporting on a volume vs. a mass basis. These will not be discussed here. In addition to these, there are some specific sources of confusion when OEC systems are used, which are discussed here.

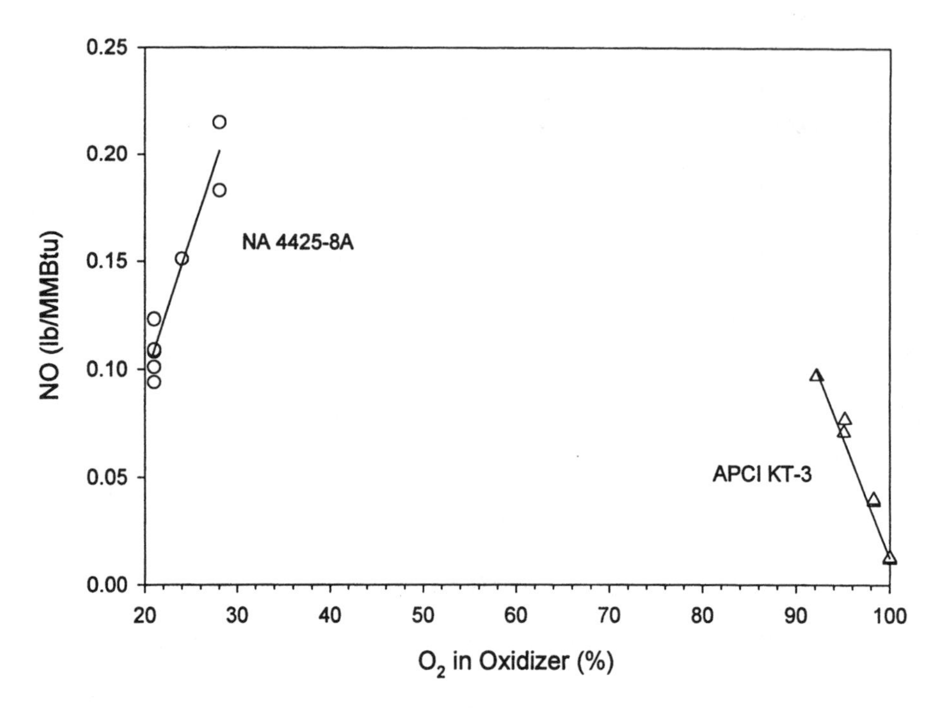

FIGURE 2.9 NO vs. oxidizer composition for an air/fuel and an oxy/fuel burner.

The first source of confusion in NOx reporting for an OEC process pertains to converting NOx measurements. One problem with current regulations, written on a volume basis, is how to correct the measurements for oxidizers other than air. The method for converting NOx measurements to a standard basis is given by[10]

$$ppm_{CORR} = ppm_{MEAS}\left[\frac{20.95 - O_{2_{BASIS}}}{20.95 - O_{2_{MEAS}}}\right] \tag{2.4}$$

where ppm_{MEAS} = measured pollutant concentration in flue gases (ppmvd)
ppm_{CORR} = pollutant concentration corrected to a standard O_2 basis (ppmvd)
$O_{2_{MEAS}}$ = measured O_2 concentration in flue gases (vol %, dry basis)
$O_{2_{BASIS}}$ = standard O_2 basis (vol%, dry basis)

The correction in Equation 2.4 assumes that the excess O_2 comes from air. This may or may not be the case with OEC. If the excess O_2 measured in the flue gases came from an oxygen-enriched oxidizer, then Equation 2.4 should be modified as follows:

$$ppm_{CORR} = ppm_{MEAS}\left[\frac{O_{2_{OXID}} - O_{2_{BASIS}}}{O_{2_{OXID}} - O_{2_{MEAS}}}\right] \tag{2.5}$$

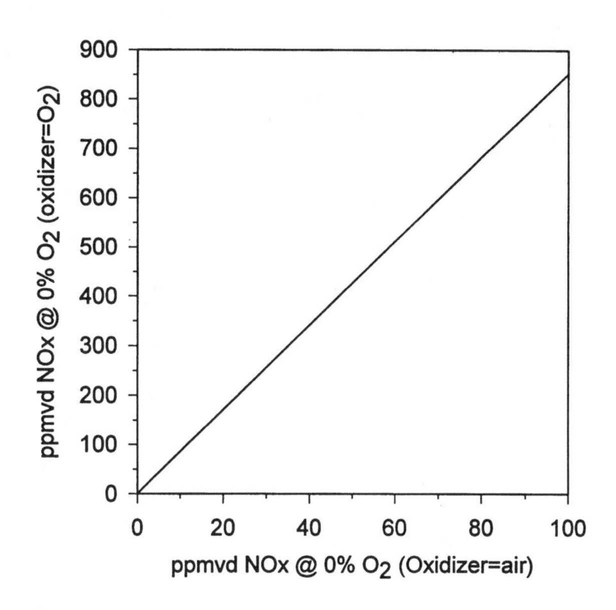

FIGURE 2.10 NOx (ppmvd @ 0% O_2) comparison when the oxidizer is air and pure O_2.

where ppm_{MEAS} = measured pollutant concentration in flue gases (ppmvd)
ppm_{CORR} = pollutant concentration corrected to a standard O_2 basis (ppmvd)
O_{2OXID} = O_2 concentration in oxidizer (vol%)
O_{2MEAS} = measured O_2 concentration in flue gases (vol%, dry basis)
O_{2BASIS} = standard O_2 basis (vol%, dry basis)

Many industrial heating systems are operated at negative pressures to prevent process gases from leaking into the work environment. Therefore, some amount of air generally leaks into the combustion chamber. In an OEC system, the "oxidizer" is then a combination of the air leaking into the process and the oxidizer supplied through the burner(s). Air leakage into an OEC process further complicates the problem of determining what O_2 concentration should be used for the oxidant.

From Figure 1.23, it may be seen that the exhaust volume per unit of fuel input is dramatically reduced as the O_2 in the oxidizer increases. It may also be seen that the available heat increases as well (see Figure 1.16). These both have a dramatic impact on the relevance of some of the NOx units. Because the flue gas volume may be reduced by >90% when replacing air with pure oxygen, comparing NOx from an air/fuel system with the NOx from an oxy/fuel system on a ppmv basis does not make sense. For example, 200 ppmvd NOx in an oxy/fuel system is actually less NOx by mass than 100 ppmvd NOx in an air/fuel system. This is because of the vast differences in flue gas volumes. Figure 2.10 shows a comparison of NOx in ppmvd @ 0% O_2 for systems using air and pure oxygen.

As can be seen, care must be exercised when comparing NOx on a volume basis for systems using different oxidizers. Hence, if it is assumed that air/fuel and oxy/fuel

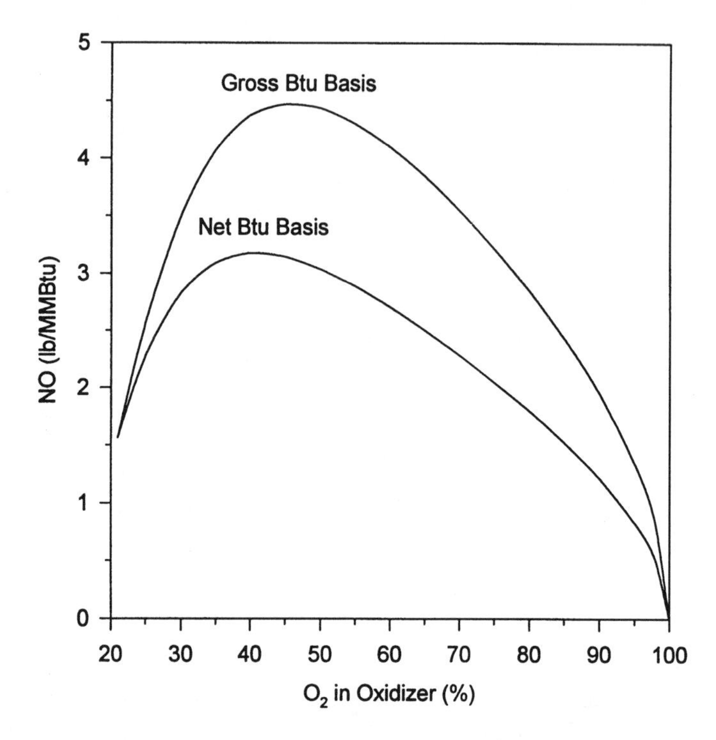

FIGURE 2.11 NO concentration (on net and gross Btu bases) vs. oxidizer composition for an adiabatic equilibrium stoichiometric O_2/CH_4 flame.

systems are equally efficient, it makes more sense to compare the NOx on a mass basis. Figure 2.11 shows NOx in both mass and volume units, as a function of the oxidizer composition. However, as shown in Figure 1.16, oxy/fuel systems are significantly more efficient than air/fuel systems. Less fuel is required for a given unit of production. Therefore, a better NOx unit would be the mass of NOx produced per unit mass of material processed.

This concept can be illustrated with a unit like lb/MMBtu on a net basis. This unit is derived by combining the NOx/MMBtu (on a gross basis) curve in Figure 2.1 and the available heat curve in Figure 1.16. It is assumed that the base case is air/fuel combustion. Then, as the O_2 in the oxidizer increases, the thermal efficiency increases. The gross and net Btu curves are shown in Figure 2.11. The net Btu curve is equivalent to a product throughput basis. It is assumed that the heat losses from the system and the heat transfer to the product would not vary significantly for different oxidizers. This figure shows that it may be very deceiving to compare NOx based only on the gross firing rate when the oxidizer compositions are different.

2.2.5 Measurement Techniques

Accurate measurements of pollutants, such as NO and CO, from industrial sources are of increasing importance in view of strict air quality regulations. Based on such measurements, companies may have to pay significant fines, stop production, install

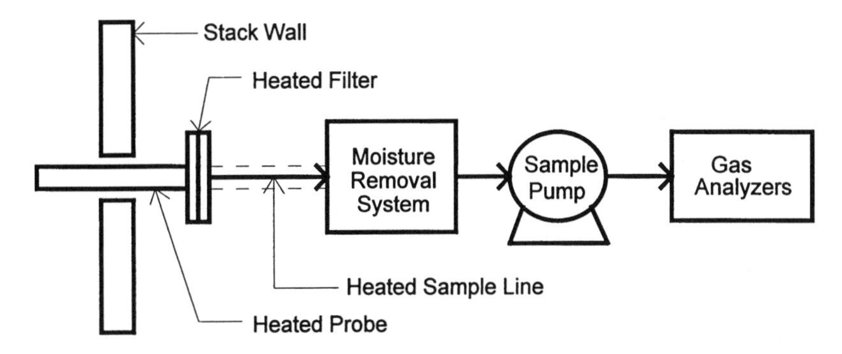

FIGURE 2.12 Sampling system schematic as recommended by the EPA.

expensive flue gas treatment systems, buy NOx credits in certain non-attainment areas, or change the production process to a less-polluting technology. If compliance is achieved, however, the company may continue its processes without interruption and, sometimes, sell its NOx credits.

Numerous studies have been done and recommendations made on the best ways to sample hot gases from high-temperature furnaces. For example, EPA Method 7E[11] applies to gas samples extracted from an exhaust stack that are analyzed with a chemiluminescent analyzer. A typical sampling system is shown in Figure 2.12. The major components are a heated sampling probe, heated filter, heated sample line, moisture removal system, pump, flow control valve (not shown), and then the analyzer. The EPA method states that the sample probe may be made of glass, stainless steel, or other equivalent materials. The probe should be heated to prevent water in the combustion products from condensing inside the probe.

The EPA method is appropriate for a lower-temperature, nonreactive gas sample obtained, for example, from a utility boiler. However, this method should not be used to obtain samples from higher-temperature industrial furnaces used in glass or metal production. Flue gas temperatures from such furnaces, as well as from some incinerators, can be greater than 2400°F (1600 K). This would cause the probe to overheat and affect the measurements because of high temperature surface reactions.

The effects of probe materials, such as metal and quartz, as well as the probe-cooling requirements, have been investigated for sampling gases in combustion systems.[12] Several studies have found that both metal and quartz probe materials can significantly affect NO measurements in air/fuel combustion systems, especially under fuel-rich conditions with high CO concentrations.[13,14] However, the NO readings were not affected under fuel-lean conditions.

The probe materials and cooling requirements are of even greater importance in oxy/fuel combustion systems, where pollutant concentrations are much higher, because of the lack of diluent N_2. It has been shown that the use of non-water-cooled probes can lead to false readings of both NO and CO, when sampling high-temperature flue gases.[15] A series of tests was done in a large-scale research furnace. Both quartz and inconel sampling probes were used. The results showed a discrepancy in the NOx measurements between these two probes. This discrepancy, as well as a temperature dependence, indicated that the probe material played a key role.

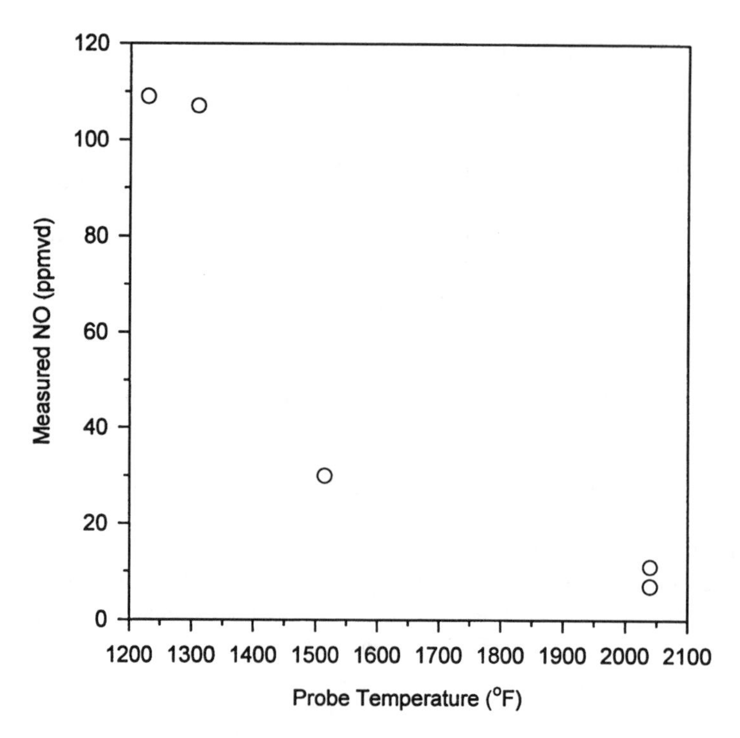

FIGURE 2.13 NO vs. probe temperature for an inconel probe with high CO concentrations in the exhaust gas.

Inconel is a good catalyst of the following reactions at temperatures above 1300°F (980 K):

$$CO + NO \rightarrow CO_2 + N \tag{2.6}$$

$$CO + O_2 \rightarrow CO_2 + O \tag{2.7}$$

The quartz tube also catalyzed these reactions, but with a much lower efficiency.

Excess quantities of oxidizers in Reactions 2.6 and 2.7 resulted in complete consumption of CO. In a fuel-rich case, a high NO concentration was almost completely eliminated by sampling through a hot inconel tube. However, the NO started increasing after the inconel tube temperature was lowered by pulling the tube out of the furnace. This temperature dependence is shown in Figure 2.13.

The higher NO readings at lower temperatures, especially below 1500°F (1100 K), indicate a decrease in the inconel catalytic efficiency for the reactions in 2.6 and 2.7. Other test data showed that the reactions in 2.6 and 2.7 were weakly catalyzed in the quartz tube, too, which lowered the NO readings.

These experiments demonstrated some of the problems that may be encountered in using a hot probe for sampling gases in high-temperature processes. For reducing conditions inside a combustion chamber, the measured NO may be much lower than the actual NO. This should not be the case for most combustors which are normally operated under at least slightly oxidizing conditions. Under oxidizing conditions, the exhaust gas measurements may show much lower CO readings than are actually present. The experiments have shown that both metal and quartz uncooled probes can affect the readings, provided the probe surface temperature exceeds 1300°F (980 K).

To avoid the surface catalytic effects, a water-cooled probe should be used when sampling high-temperature combustion products. This is of particular importance to oxy/fuel combustion, where measured pollutant species concentrations are much higher, due to the elimination of N_2. Recommended probe materials and cooling requirements, to avoid reaction of the various gases inside the probe, are given elsewhere.[10] It should be noted that the exhaust gas sample should not be cooled below the condensation point of water. This could condense the water in the gas sample inside the sampling probe or sample lines. Since NO_2 is water soluble, false low readings of NO_2 could result.

2.2.6 Field Results

Oxygen may be used in several ways to reduce NOx emissions from a combustion process. Some field NOx measurements are given in Chapters 6 to 8. As was previously shown, very low NOx levels can be achieved by replacing air with pure oxygen. The lowest emissions can be obtained when the nitrogen in the combustion chamber is minimized. This nitrogen can come from several sources. One common source is the fuel. However, for clean fuels such as natural gas, the nitrogen content is generally small, depending on the supply source. Another source of nitrogen is air infiltration. This can be caused by air leakage through open doors and view ports and through cracks in the walls of the combustion chamber, and by air being carried in with the feed materials. Air leakage can be minimized by operating the furnace at slightly positive pressure. However, this is not always an option, especially if there are any hazardous pollutants in the gases exiting the furnace. Another source of nitrogen can be the feed material. For example, one of the raw materials used in making glass is called "niter" which contains nitrogen.[16] The final potential source of nitrogen in the combustion process is from the oxidizer.

One method of reducing NOx using oxygen is known as oxygen-enriched air staging, or OEAS. In this technology, the air/fuel burners in the combustion system are run at slightly fuel-rich conditions to minimize NOx formation. Then, oxygen is injected into the gaseous combustion products prior to their exit from the furnace. This second stage of combustion accomplishes two things. Any unburned hydrocarbons and carbon monoxide in the exhaust stream, which are also regulated pollutants, are destroyed before entering the exhaust. Also, the heating value of those gases is captured in the combustion chamber, rather than burning them in the exhaust system. NOx reductions ranging from 55 to 75% have been achieved in a large end-port-fired regenerative glass container furnace.[17]

2.3 COMBUSTIBLES

This section has been divided into discussion of two types of combustibles. The first involves the incomplete combustion of the fuel, which usually produces carbon monoxide and, in some limited cases, not all of the hydrocarbon fuel is consumed and passes through the combustor unreacted. The second type of combustible is volatile organic compounds (VOCs), which are generally only important in a limited number of processes, typically involving contaminated or otherwise hazardous waste streams.

2.3.1 CO and Unburned Fuel

CO is generally produced in trace quantities in many combustion processes as a product of incomplete combustion (see Figure 2.1). CO is a flammable gas that is nonirritating, colorless, odorless, tasteless, and normally noncorrosive. CO is highly toxic and acts as a chemical asphyxiant by combining with hemoglobin in the blood, which transports oxygen inside the body. The affinity of carbon monoxide for hemoglobin is approximately 300 times more than the affinity of oxygen for hemoglobin.[18] CO preferentially combines with hemoglobin to the exclusion of oxygen so that the body becomes starved for oxygen which can eventually lead to asphyxiation. Therefore, CO is a regulated pollutant with specific emissions guidelines depending on the application and the geographic location.

CO is generally produced by the incomplete combustion of a carbon-containing fuel. Normally, a combustion system is operated slightly fuel lean (excess O_2) to ensure complete combustion and to minimize CO emissions. OEC generally reduces CO emissions compared with air/fuel systems, due to more-complete combustion. Some specific examples of using oxygen to reduce CO emissions are given in Chapter 8 on waste incineration where CO is particularly a problem because of the varying hydrocarbon content of the waste material. One potential problem of using high levels of oxygen enrichment is thermal dissociation, where CO is thermodynamically preferred to CO_2 at high temperatures (see Figure 1.10), even if there is some excess O_2 available. However, this is not usually a problem in most industrial heating processes because the temperatures of actual oxy/fuel flames are generally much less than the adiabatic flame temperature. If sufficient O_2 is available, CO_2 is thermodynamically preferred at those lower temperatures, instead of CO.

2.3.2 Volatile Organic Compounds

VOCs are generally low-molecular-weight aliphatic and aromatic hydrocarbons like alcohols, ketones, esters, and aldehydes.[19] Typical VOCs include benzene, acetone, acetaldehyde, chloroform, toluene, methanol, and formaldehyde. These compounds are typically considered to be regulated pollutants, as they can cause photochemical smog and depletion of the ozone layer if they are released into the atmosphere. They are not normally produced in the combustion process, but they may be contained in the material that is being heated, such as in the case of a contaminated hazardous waste in a waste incinerator. In that case, the objective of the heating process is usually to volatilize the VOCs out of the waste and to combust them before they can be emitted to the atmosphere.

There are two strategies for removing VOCs from the off-gases of a combustion process.[19] One is to separate and recover them using techniques like carbon adsorption or condensation. The other method involves oxidizing the VOCs to CO_2 and H_2O, and includes techniques like thermal oxidation, catalytic oxidation, and bio-oxidation. One common way to ensure complete destruction of VOCs is to add an afterburner or secondary combustion chamber, which may or may not have a catalyst, after the main or primary combustion chamber.[20]

VOC emissions would be expected to decrease as a result of using OEC because of the higher flame temperatures, lower dilution, and increased residence time within the combustor. The destruction and removal efficiency (DRE) in an incinerator has been shown to increase dramatically with OEC and is discussed in more detail in Chapter 8. In a municipal solid waste incinerator, VOC emissions decreased by using low levels of oxygen enrichment of the combustion air.[21] At a Superfund cleanup site in New Jersey, sandy soil containing VOCs including benzene derivatives, chlorinated hydrocarbons, and semivolatile organics such as naphthalene and phthalates, was effectively cleaned using OEC.[22]

2.4 PARTICULATES

There are two primary sources of particulates which may be carried out of a combustion process with the exhaust gases. One is entrainment and carryover of incoming raw materials, and the other is the production of particles as a result of the combustion process.

2.4.1 PARTICLE ENTRAINMENT

The gas flow through the combustor may entrain particles from the raw materials used in the process. This is often referred to as carryover. For example, in the production of glass, fine batch materials are fed into a high-temperature furnace. The gas flow within the furnace is generally high enough that some of the batch may be entrained into the gas flow and carried out the exhaust stack where it must then be removed before the gases exit to the atmosphere. This creates two problems. The batch material carried out of the combustor is wasted raw material which is an added cost in the production process. A second cost is incurred because the carryover must be removed from the exhaust gases. In the production of leaded glass, the lead is an expensive component in the raw material, and it is also a highly toxic heavy metal that is a highly regulated pollutant. OEC can simultaneously reduce the loss of an expensive raw material as well as minimize or eliminate the emissions of a very hazardous pollutant.

OEC can dramatically reduce the carryover in a process originally designed for air/fuel combustion because of the reduction in the average gas velocity through the combustor that results from removing some or all of the diluent nitrogen from the system (see Figure 2.14). For example, OEC was specifically used in a mobile waste incinerator to reduce the carryover of fine material from the primary to the secondary combustion chamber.[23] When the incinerator was fired on air/fuel, the material that was carried over into the secondary chamber adhered to the wall and eventually clogged the chamber necessitating frequent shutdowns for maintenance. This problem was eliminated when the air/fuel system was replaced by an oxy/fuel system.

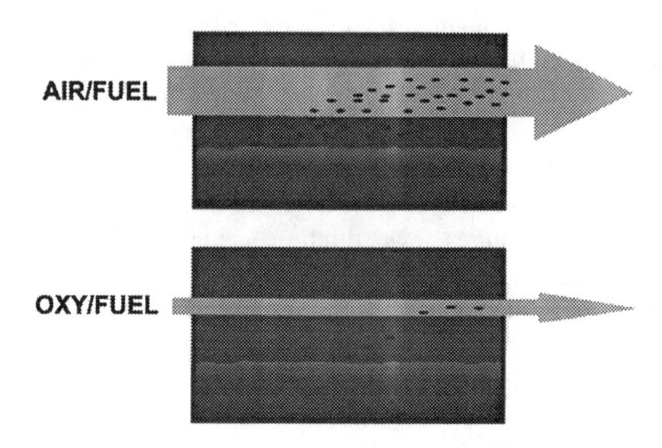

FIGURE 2.14 Comparison of particle entrainment in a furnace using air/fuel and oxy/fuel combustion.

In another mobile incinerator, oxy/fuel was successfully used, instead of the conventional air/fuel system, to meet very stringent particulate emissions regulations at a Superfund cleanup site where the soil contained a significant amount of sand.[22]

2.4.2 Combustion-Generated Particles

The second method that particles may be emitted from the combustion system is through the production of particles in the combustion process. For example, in the combustion of solid fuels, like coal, for example, ash is normally produced. The airborne portion of the ash, usually referred to as fly ash, may be carried out of the combustor by the exhaust gases. The use of OEC should reduce fly ash emissions because of more-complete combustion of the fuel compared with an air/fuel system.

Another source of combustion-generated particles is soot, which may be produced in a flame, even for gaseous fuels, under certain conditions. To a certain extent, soot is desirable in that it generally enhances the radiant heat transfer between the flame and the load. Fuels that have a higher ratio of carbon to hydrogen mass tend to produce more soot than fuels with a lower ratio. For example, propane (C_3H_8), which has a C:H mass ratio of about 4.5, is more likely to produce soot than methane (CH_4), which has a C:H mass ratio of about 3.0. For clean-burning fuels like natural gas, it is much more difficult to produce sooty flames compared with other fuels, like oil and coal, which have little or no hydrogen and a high concentration of carbon. Flames containing more soot are more luminous and tend to release their heat more efficiently than flames containing less soot, which tend to be nonluminous. Soot particles generally consist of high-molecular-weight polycyclic hydrocarbons and are sometimes referred to as "char."

Ideally, soot would be generated at the beginning of the flame so that it could radiate heat to the load and then it would be destroyed before exiting the flame so that no particles would be emitted. Soot can be produced by operating a combustion system in a very fuel-rich mode or by incomplete combustion of the fuel due to poor mixing. If the soot particles are quenched or "frozen," they are more difficult to incinerate and more likely to be emitted with the exhaust products. The quenching

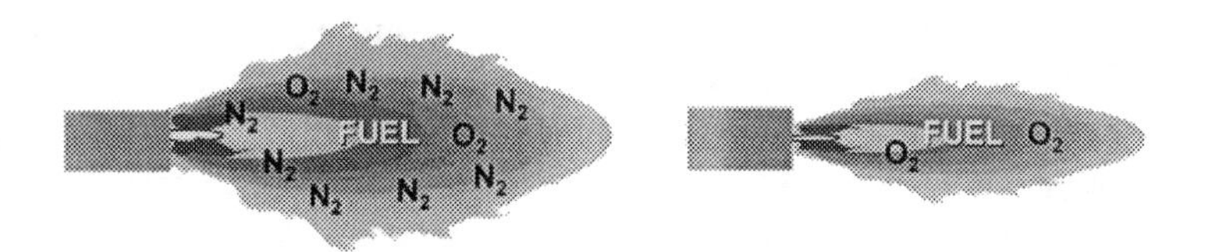

FIGURE 2.15 Enhanced chemical reactivity in oxy/fuel flames compared with air/fuel flames.

could be caused by contact with much colder gases or possibly by impingement on a cool surface, such as a boiler tube. Soot particles tend to be sticky and can cling to the exhaust ductwork, which can then clog up the ductwork and other pollution treatment equipment in the system. If the soot is emitted into the atmosphere, it can contribute to smog in addition to being dirty. The emitted soot particles become a pollutant because they produce a smoky exhaust which has a high opacity. Most industrial heating processes have a regulated limit for opacity.

The use of OEC generally reduces the likelihood of emitting soot into the exhaust products because of the intensified combustion (see Figure 2.15) and higher flame temperatures, compared with air/fuel systems. Older style, high-intensity oxy/fuel burners usually had a very hot, bluish flame which was very nonluminous. The newer style oxy/fuel burners are much lower intensity and have a much higher luminosity. However, soot carryover into the exhaust has not been identified as a problem for OEC.

2.5 CARBON DIOXIDE

CO_2 is a colorless, odorless, inert gas which does not support life since it can displace oxygen and act as an asphyxiant. CO_2 is found naturally in the atmosphere at concentrations averaging 0.03% or 300 ppmv. Concentrations of 3 to 6% can cause headaches, dyspnea, and perspiration. Concentrations of 6 to 10% can cause headaches, tremors, visual disturbance, and unconsciousness. Concentrations above 10% can cause unconsciousness eventually leading to death.

CO_2 emissions are produced when a fuel containing carbon is combusted near or above stoichiometric conditions. Studies indicate that CO_2 is a greenhouse gas that may contribute to global warming.[36] Many schemes have been suggested for "disposing" of CO_2, including injection deep into the ocean or deep-well injection for oil recovery. In Europe, CO_2 emissions are considered a pollutant and as such are regulated. OEC can significantly reduce CO_2 emissions because of the greatly increased fuel efficiency which means that less fuel needs to be burned for a given unit of available heat output (see Chapter 1). As an example, oxy/fuel burners were used in place of air/fuel burners to preheat the ladles used to transport molten metal in a steel mill, which reduced CO_2 emissions by 55%.[24]

2.6 SOx

Sulfur oxides, usually referred to as SOx, include SO, S_2O, S_nO, SO_2, SO_3, and SO_4 of which SO_2 and SO_3 are of particular importance in combustion processes.[25] SO_2

tends to be preferred at higher temperatures, while SO_3 is preferred at lower temperatures.[26] Since most combustion processes are at high temperatures, SO_2 is the predominant form of SOx emitted from systems containing sulfur. Sulfur dioxide (SO_2) is a colorless gas with a pungent odor which is used in a variety of chemical processes. Sulfur dioxide can be very corrosive in the presence of water. SO_2 is considered to be a pollutant because of the choking effect it can cause on the human repiratory system, as well as the damage that it can do to green plants which are more sensitive to SO_2 than people and animals. SO_2 can produce acid rain when it is released into the atmosphere by combining with water to produce sulfuric acid (H_2SO_4) which is very corrosive and can cause considerable damage to the environment.

It is often assumed that any sulfur in a combustor will be converted to SO_2 which is then carried out with the exhaust gases.[27] The sulfur may come from the fuel or from the raw materials used in the production process. Fuels like heavy oil and coal generally contain significant amounts of sulfur, while gaseous fuels like natural gas tend to contain little or no sulfur. The two strategies for minimizing or eliminating SOx are removing the sulfur from the incoming fuel, oxidizer, or raw materials or removing the SOx from the exhaust stream using a variety of dry and wet scrubbing techniques.[28] One dry scrubbing technique is to use limestone injection when the combined product can be used in gypsum board. Another reduction technique is the development of new membrane separation technologies. The use of OEC does not normally change the production of SOx in a combustion system. However, as previously shown in Chapter 1, since the flue gas volume is greatly reduced using OEC compared with conventional air/fuel combustion, any SOx in the exhaust gases should be easier to remove because it is in higher concentrations, which improves the efficiency of the scrubbing system.

In one application it has been shown that using OEC can increase the efficiency of SOx removal from exhaust gases from a combustion process. One way of removing SOx is to inject a transfer agent into the contaminated gases to collect SO_3 so that it can be released as H_2S in a stripper. Additives are used to convert the SO_2 in the gases to SO_3. In a fluid catalytic cracker (FCC), OEC was used to reduce SOx.[29]

2.7 DIOXINS AND FURANS

This class of pollutants includes the carbon–hydrogen–oxygen–halogen compounds and has received considerable attention from both the general public and from regulatory agencies because of the potential hazards associated with them. Dioxins generally refer to polychlorinated dibenzo-*p*-dioxin (PCDD) compounds, while furans generally refer to polychlorinated dibenzofuran (PCDF) compounds. Some of the potential health risks include toxicity because of the poisoning effect on cell tissues, carcinogenicity because cancerous growth may be stimulated, mutagenicity because of possible mutations in cell structure or function, and teratogenicity because of the potential changes to fetal tissue.[20] The over 200 dioxin/furan compounds are regulated in certain industries, particularly in waste incineration, and also in certain geographic locations for a wide range of applications, especially in Europe.

In the vast majority of cases, dioxin/furan emissions result from some contaminant in the load materials being heated in the combustor. A quick scan of most of the textbooks on combustion shows that these emissions are completely ignored because they are not generally produced in the flame, except in certain limited cases. This is primarily because there are not usually any halogens in either the fuel or the oxidizer to produce dioxins or furans. An exception to that is when waste materials are burned as a fuel by direct injection into a flame. One example is the destruction of waste solvents which may be injected into an incinerator through the burner.

In some applications, the waste material may directly contain dioxins and furans, or it may contain halogens that could lead to the formation of dioxins and furans. Many of the U.S. Superfund cleanup sites contain soil that has been contaminated with halogenated compounds. One common method to clean the soil is to process it through an incinerator which reduces the contaminants to trace levels (see Chapter 8).

At this time there is little information on the level of dioxin and furan emissions from OEC processes. Two applications are briefly considered here. Dioxins and furans are a concern in the nonferrous melting industry because of the halogens in the salt which is used to produce the slag covering on the molten bath in the melting process to prevent excess oxidation of the aluminum. Hydrocarbons from contaminated scrap or from unburned fuel can react with those halogens to form dioxins and furans. In a rotary furnace used to melt scrap aluminum, an air/oil system was replaced with an oxy/oil system, which resulted in over an order of magnitude reduction in dioxin and furan emissions (see Chapter 6 for more details). In a waste incineration application using oxy/fuel to incinerate PCBs in a mobile incinerator, dioxin and furan emissions were below the detectable limits of the analyzers (see Chapter 8 for more details). Therefore, OEC can be used to reduce the emissions of dioxins and furans. One potential explanation is that soot burnout is more efficient in OEC flames compared with air/fuel flames. One theory concerning dioxins and furans is that they form on the surface of carbon particles like soot.[30] More-complete soot burnout may reduce the number of sites for the dioxins and furans to form. However, further research is recommended to understand better the mechanisms for the improved destruction efficiency of dioxins and furans using OEC.

2.8 NOISE

2.8.1 MEASURING NOISE

Sound is a physical disturbance, measured in a frequency unit known as hertz (Hz), that can be detected by the human ear, which is normally capable of hearing from approximately 20 to 20,000 Hz. The human ear is most sensitive to sound between 2000 and 5000 Hz and is less sensitive at higher and lower frequencies. Noise is an unwanted or unpleasant sound which generally has a random nature. Noise is commonly measured in decibels which is ten times the logarithm of the ratio of the actual sound to some reference sound level, which may be in terms of sound pressure or sound power:

$$L_w = 10 \log_{10}\left(W/W_0\right) \quad \text{(dB)} \tag{2.8}$$

TABLE 2.1
U.S. Occupational Safety & Health Administration Standard 1910.95 for Noise[32]

Duration per Day (h)	Sound Level dBA Slow Response
8	90
6	92
4	95
3	97
2	100
1½	102
1	105
½	110
¼ or less	115

Source: OSHA Standard 1910.95, 1995.

where L_w is the sound power level, W is the power of the source in watts, and W_0 is the reference power in watts. The standard reference sound power is 1 pW (10^{-12} W). The sound power level depends only on the power from the source and not on the distance from the source. Similarly, the sound pressure level is defined as

$$L_p = 10 \log_{10} \left(p/p_0\right)^2 \quad \text{(dB)} \tag{2.9}$$

where L_p is the sound pressure level, p is the pressure of the source in pascals, and p_0 is the reference pressure level in pascals. The standard reference sound pressure is 20 μPa. A sound pressure of 20 μPa equals a sound pressure of 20 dB. A doubling of the sound pressure corresponds to an increase of approximately 6 dB, while multiplying the sound pressure by a factor of 10 equals an increase of 20 dB. The sound pressure level depends on both the sound source and the distance from the source, as well as on the acoustical characteristics of the space around the source.

In most industrialized nations, noise is a "pollutant" which is regulated in the work environment. Table 2.1 shows the allowable noise levels permitted by the U.S. Department of Labor as a function of the length of the exposure.[31] The table shows that as the length of exposure increases, the permitted sound level decreases. Noise is commonly measured with a sound level meter, which is an instrument designed to respond to sound in approximately the same way as the human ear. The meter is designed to give objective, repeatable measurements compared with the human ear. Human hearing varies from person to person and even varies for the same person as he or she ages. Different weighting systems are used over the frequency spectrum to closely match the frequency response of a typical human ear. The A-weighting network is the most widely used in noise work so that sound measured with the A-weighting is given as dB(A) or sometimes simply as dBA.

2.8.2 Combustion Noise

One major source of noise in a conventional air/fuel combustion system is the air-handling system.[32] The blower that moves the air is typically noisy and may need to be located outside the building or acoustically insulated in order to meet noise regulations. Alternatively, workers may need to wear hearing protection when in the vicinity of the blowers. This source of noise can be completely eliminated if the air is replaced by pure oxygen since a blower is not typically required for oxygen.[24] If a low-pressure oxygen source, like a VSA (see Chapter 3), is used, an oxygen blower, used to boost the supply pressure, may be a source of noise.

Another source of noise can be the high volumetric flow of gas through the piping system. The actual flow of gas through a pipe is commonly given as acfh, or actual cubic feet of gas per hour. This is the flow rate of gas at the pressure and temperature in the pipe. This actual flow is usually corrected to a standard temperature and pressure (STP) level and reported as scfh, or standard cubic feet of gas per hour. Although there are various definitions for STP, they are usually at or about 70°F (21°C) and atmospheric pressure (14.7 psia or 760 mm Hg). Since air is typically supplied at a low pressure, the actual flow of gas through the piping is high, which may produce a significant amount of noise. The noise produced by oxygen flowing through the supply piping is generally negligible for several reasons. In many cases, oxygen is supplied at a much higher pressure than air so that the actual volume of flow, commonly measured in actual cubic feet per hour, is very low compared with the flow of air. Oxygen safety guidelines (see Chapter 9) generally limit the maximum gas velocity through the piping. Lower gas velocities generate less noise. The piping used for oxygen service generally has a thicker wall, compared with air piping, which also helps to deaden the noise caused by the gas flowing through the piping.

Another source of noise in a combustion system comes from the burner and is sometimes referred to as "combustion roar."[33] This noise is a combination of the gas flow through the burner nozzles and also from the combustion process itself. There are many factors that affect the noise level produced by the combustion system. These include the firing rate, oxidizer-to-fuel ratio, turbulence intensity of the gas flows, combustion or mixing intensity, amount of swirl, preheat of the oxidizer or fuel, type of fuel and oxidizer, number of burners, geometry of the combustion chamber, insulation used in the combustor, and even the dampening effects of the material being heated.

It is difficult to generalize a comparison of the noise between air/fuel and oxy/fuel burners, because of the wide variety of designs that are available. The older style oxy/fuel burners that were used in the steel industry in the 1970s and 1980s typically had flames that were very high intensity and high momentum which made them very noisy. One study reported noise levels up to 125 dB for one such high-intensity oxy/fuel burner.[34] The same study also reported noise levels as low as 85 dB for a lower-intensity oxy/fuel burner at the lower end of the firing rate range for the burner. Many of the burners commonly used today for OEC are lower momentum and lower intensity and therefore significantly quieter than the older style high-momentum burners. In many cases, new design oxy/fuel burners can be significantly quieter than most air/fuel burners.

Noise generated by the burners in a combustion system may be greatly mitigated by the combustion chamber, which is usually a furnace of some type. The refractory linings in most furnaces generally significantly reduce any noise emitted from the burners. Noise is not commonly considered in many industrial heating applications for a variety of reasons. This is evidenced by the general lack of information available on the subject. It is difficult to predict the noise levels before installing the equipment because of the wide variety of factors that influence noise. Often, there are many other pieces of machinery which are much noisier than the combustion system so that the workers are already required to wear hearing protection. In the future, noise reduction may become more important and OEC may be one way to minimize the noise produced by the combustion system.

2.8.3 Noise Reduction Techniques

There are several strategies that can be used to reduce combustion noise. One strategy is either to move the source of the undesirable sound away from the people or to move the people away from the sound. However, this may not be practical for many industrial applications. Another strategy is to put some type of sound barrier between the noise and the people. The barrier can be either reflective or absorptive to minimize the noise. The noise source might be surrounded by an enclosure or the operators may be located inside a soundproofed enclosure. In some cases, it may be possible to use a silencer, which would act as a barrier, to reduce the noise. For example, the exhaust from a car is reduced by the muffler which acts like a silencer. The barrier could also be in the form of earplugs, ear phones, or some other sound-reducing safety device worn by people in the vicinity of the noise. Another technique is to reduce the exposure time to the noise since noise has a cumulative effect on human hearing. In some cases, it may be possible to replace noisy equipment with new equipment that has been specifically designed to produce less noise, or to retrofit existing equipment to produce less noise. For example, old combustion air blowers and fans could be replaced by new, quieter blowers and fans. Another way to reduce noise is to increase the pipe size and reduce the number of bends in the pipe to reduce the jet noise of the fluids flowing through the pipe. Resonance and instabilities can usually be designed out of a system if they are a problem. Noisy burners can be replaced by quieter burners. The burner noise is a function of the burner design, fuel, firing rate, stoichiometry, combustion intensity, and aerodynamics of the combustion chamber. A detailed discussion of reducing noise in combustion is given elsewhere.[35]

REFERENCES

1. U.S. Environmental Protection Agency, Nitrogen Oxide Control for Stationary Combustion Sources, U.S. EPA report EPA/625/5-86/020, 1986.
2. Reese, J. L., Batten, R., Moilanen, G. L., Baukal, C. E., Borkowicz, R., and Czerniak, D. O., State of the Art of NOx Emission Control Technology, American Society of Mechanical Engineers paper 94-JPGC-EC-15, presented at the ASME Joint International Power Generation Conference, Phoenix, October, 1994.

3. Sheppard, L. M., Reducing emissions from glass furnaces with combustion technology, *Ceram. Ind.,* 146(130), 37–40, 1996.
4. Zeldovich, Y. B., *Acta Physecochem.* (USSR), 21, 557, 1946.
5. U.S. Environmental Protection Agency, Alternative Control Techniques — NOx Emissions from Utility Boilers, U.S. EPA report EPA-453/R-94-023, 1994.
6. Shelton, H. L., Find the right low-NOx solution, *Environ. Eng. World,* 2(6), 24, 1996.
7. Bluestein, J., NOx Controls for Gas-Fired Industrial Boilers and Combustion Equipment: A Survey of Current Practices, Report GRI-92/0374, Gas Research Institute, Chicago, 1992.
8. Dalton, A. I. and Tyndall, D. W., Oxygen Enriched Air/Natural Gas Burner System Development, Final Report, July 1984 — September 1989, Report GRI-90/0140, Gas Research Institute, Chicago, 1989.
9. Baukal, C. E. and Eleazer, P. B., Quantifying NOx for industrial combustion processes, *J. Air Waste Manage. Assoc.,* 48, 52–58, 1998.
10. American National Standards Institute/American Society of Mechanical Engineers, Performance Test Code PTC 19.10, Part 10: Flue and Exhaust Gas Analyses, ASME, New York, 1981.
11. U.S. Government, *Code of Federal Regulations 40,* Part 60, Revised July 1, 1994.
12. Drake, M. C., Kinetics of Nitric Oxide Formation in Laminar and Turbulent Methane Combustion, Report no. GRI-85/0271, Gas Research Institute, Chicago, 1985.
13. Zabielski, M. F., Dodge, L. G., Colket, M. B., and Seery, D. J., The optical and probe measurement of NO: a comparative study, in *Eighteenth Symposium (Int.) on Combustion,* The Combustion Institute, Pittsburgh, 1981, 1591.
14. Berger, A. and Rotzoll, G., Kinetics of NO reduction by CO on quartz glass surfaces, *Fuel,* 74, 452, 1995.
15. Slavejkov, A. G. and Baukal, C. E., Flue gas sampling challenges in oxygen-fuel combustion processes, in *Tranport Phenomena in Combustion,* Vol. 2, S. H. Chan, Ed., Taylor & Francis, Washington, D.C., 1996, 1230.
16. Eriksson, R., Coe, D., and Eichler, R., Measurement and control of NOx in oxygen-fired glass furnaces, *Ceram. Eng. Sci. Proc.,* 13(3–4), 25, 1992.
17. Joshi, M. L., Mohr, P. J., Abbasi, H. A., and Grosman, R. E., Oxygen Enriched Air Staging on a Regenerative Endport Glass Container Furnace, Report GRI-96/0229, Gas Research Institute, Chicago, July 1996.
18. Ahlberg, K., Ed., *AGA Gas Handbook,* AGA AB, Lidingö, Sweden, 1985.
19. Setia, S., VOC emissions — hazards and techniques for their control, *Chem. Eng. World,* 31(9), 43–47, 1996.
20. Niessen, W. R., *Combustion and Incineration Processes,* 2nd ed., Marcel Dekker, New York, 1995.
21. Strauss, W. S., Lukens, J. A., Young, F. K., and Bingham, F. B., Oxygen enrichment of combustion air in a 360 TPD mass burn refuse-fired waterwall furnace, in *Proceedings of the 1988 National Waste Processing Conference, 13th Bi-Annual Conference,* Philadelphia, May 1–4, 1988, 315.
22. Romano, F. J. and McLeod, B. M., The use of oxygen to reduce particulate emissions without reducing throughput, in *Proceedings of 1990 Incineration Conference,* Paper 3.3, San Diego, May 14–18, 1990.
23. Griffith, C. R., PCB and PCP destruction using oxygen in mobile incinerators, in *Proceedings of the 1990 Incineration Conference,* San Diego, May 14–18, 1990.
24. Farrell, L. M., Pavlack, T. T., Selines, R. J., and Rich, L., Environmental benefits available through oxyfuel heating burners, *Steel Times,* 224(8), 259, 1996.

25. Weil, E. D., Sulfur compounds, in *Kirk-Othmer Encyclopedia of Chemical Technology*, 3rd ed., Vol. 22, John Wiley & Sons, New York, 1983.
26. Bowman, C. T., Chemistry of gaseous pollutant formation and destruction, in *Fossil Fuel Combustion*, W. Bartok and A. F. Sarofim, Eds., John Wiley & Sons, New York, 1991.
27. Bruner, C. R., *Handbook of Incineration Systems*, McGraw-Hill, New York, 1991.
28. Turns, S. R., *An Introduction to Combustion*, McGraw-Hill, New York, 1996.
29. Tamhankar, S., Menon, R., Chou, T., Ramachandran, R., Hull, R., and Watson, R., Enrichment can decrease NOx, SOx formation, *Oil & Gas J.*, 94(10), 60, 1996.
30. Huang, H. and Buekens, A., On the mechanisms of dioxin formation in combustion processes, *Chemosphere*, 31(9), 4099–4117, 1995.
31. U.S. Department of Labor, Occupational Safety and Health Administration, Occupation Noise Exposure, OSHA standard 1910.95, 1995.
32. Bruce, R. D., Noise pollution, in *Kirk-Othmer Encyclopedia of Chemical Technology*, 3rd ed., Vol. 16, John Wiley & Sons, New York, 1983.
33. Putnam, A. and Faulkner, L., An overview of combustion noise, *J. Energ.*, 7(6), 458–469, 1983.
34. Song, X.-R. C., Experimental study of combustion noise generated by oxygen-fuel burners, in *Proceedings of the 1993 National Conference on Noise Control Engineering: Noise Control in Aeroacoustics,* Williamsburg, VA, 1993, 97–102.
35. Reed, R. J., *North American Combustion Handbook,* 3rd ed., Vol. II, Part 12, North American Manufacturing Co., Cleveland, OH, 1997.
36. Mourelatos, A., Diakoulaki, D., and Papagiannakis, L., Impact of CO_2 reduction policies on the development of renewable energy sources, *Int. J. Hydrogen Energy,* 23(2), 139–149, 1998.

3 Oxygen Production

Roger M. McGuinness and
William T. Kleinberg

CONTENTS

0-8493-1695-2/98/$0.00+$.50

3.1 INTRODUCTION

The supply of oxygen is commercially limited to three alternatives:

- Delivery and storage of liquid oxygen,
- Vacuum swing adsorption (VSA) system for on-site supply, and
- Cryogenic air separation system on-site supply.

Figure 3.1 shows the approximate product application ranges of each of these technologies. In general, liquid supply is good when small quantities of product are required. The liquid oxygen produced at a central facility is distributed to a customer's storage tank. When needed, the liquid oxygen (LOX) is vaporized and warmed up to ambient conditions. The advantage of this type of supply is that there is no major capital investment. The disadvantage is that the unit price of the LOX is relatively high since it includes not only the production costs to make the LOX but also the distribution costs. Even at a high unit cost, liquid supply is generally economical for small supplies or even larger amounts for short duration. Commercial-grade LOX is available at 99.5% purity which is applicable through the range of use purities.

Adsorption has a limited range of application between 10 and 100 t/day of impure oxygen, above which limits on vessel size can occur. The limit of oxygen

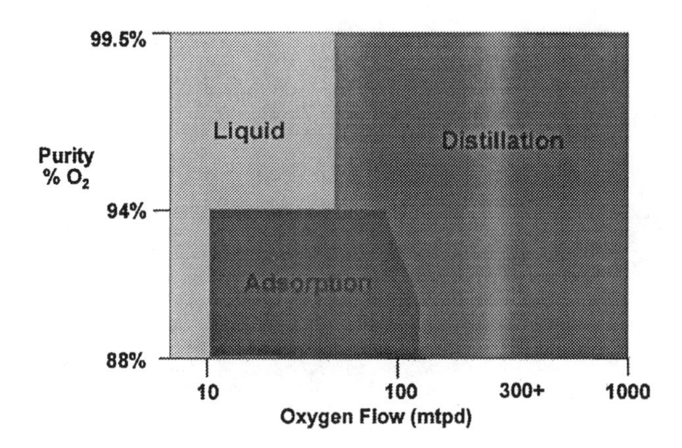

FIGURE 3.1 Technology selection chart. (Courtesy of Air Products.)

purity for an adsorption system is approximately 95% O_2. The advantage of an adsorption system is that the production costs are significantly lower than LOX supply, but again there is a capital investment required. Oxygen by adsorption is economical when the long-term quantity of oxygen needed justifies the capital and the purity is acceptable.

Cryogenic air separation has the widest range of economical production capability from low to 3000 t/day plus of oxygen with purity ranges from 90 to 99.9% O_2. Purities below this are possible but generally not required since they can be attained by air dilution. At the higher production requirements, this is the only choice for the production of oxygen. With the rapid evolution of technology in both adsorption and cryogenics the boundaries at the lower end are continually changing.

3.2 CRYOGENIC OXYGEN PRODUCTION

3.2.1 Fundamentals of Cryogenic Oxygen Production

The basic unit operations of an air separation unit (ASU), shown in Figure 3.2, are

- Compression of air,
- Pretreatment to remove CO_2 and water and some hydrocarbons,
- Cooling the air down to cryogenic temperatures to allow separation to occur,
- Separation of air into its components,
- Refrigeration to keep the ASU in energy balance, and
- Compression of gaseous products and storage of liquid products.

The cryogenic equipment is all contained in an insulated structure termed a *coldbox* to minimize the impact of heat leak into the process.

The simplified process flow diagram (PFD; Figure 3.3) shows the equipment configuration for a simple gaseous oxygen generator. Air is filtered, compressed, and passed through adsorbers to remove CO_2 and water before entering the cryogenic

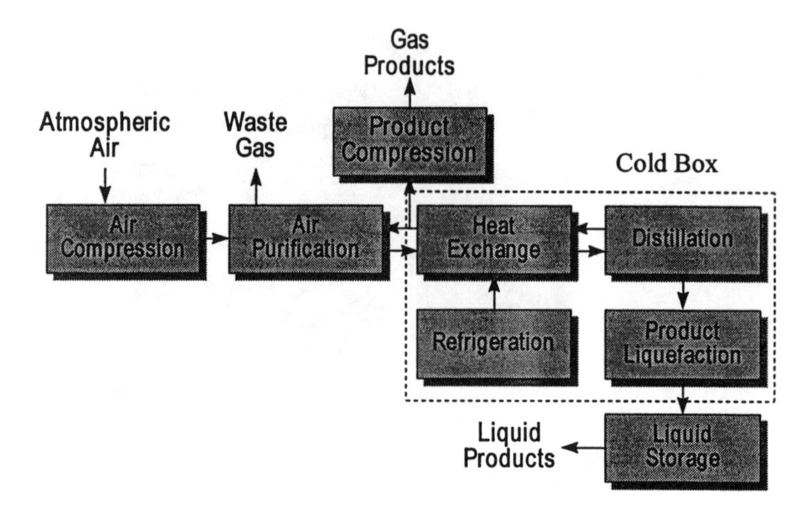

FIGURE 3.2 ASU block diagram.[2] (Courtesy of Academic Press, London, England.)

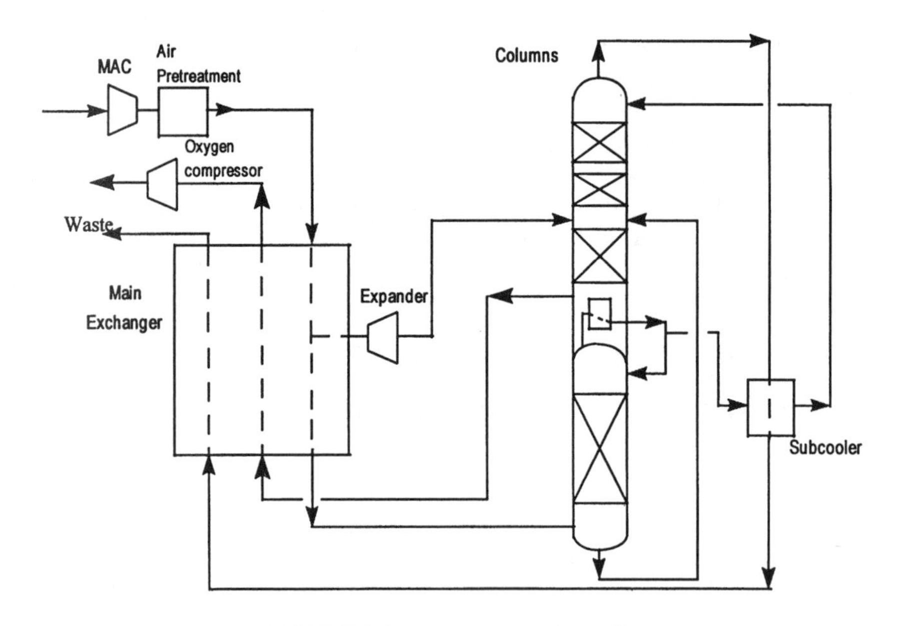

FIGURE 3.3 Simple PFD of an ASU.

portion of the plant. The air is then cooled to a temperature close to its dew point in the main exchanger by countercurrent heat exchange with oxygen product and a waste stream. The cooled air is then passed to the distillation system. In the first column (high-pressure column), operating at a pressure slightly lower than that of the air compressor, a rough separation of N_2 from the air occurs. The N_2 being the more volatile, concentrates, as the vapor passes up the column. A N_2-rich stream generated at the top of the column is condensed, providing reflux for both the high-pressure (HP) column and another column operating at low pressure. In the low-pressure (LP) column the oxygen-rich stream from the bottom of the HP column is further processed. A higher-purity oxygen stream is produced at the bottom of this column, and product oxygen is removed as a vapor and warmed up in the main exchanger. Vapor boil-up for the LP column is generated in a reboiler thermally linked to the HP column.

Liquid reflux for the LP column can be subcooled by cooling against waste vapor from the LP column. This minimizes flash when the reflux steam is dropped to the lower pressure and also warms up the waste stream.

A portion of the airstream is removed from the main exchanger and expanded in a turbine to generate refrigeration, and gaseous oxygen product from the main exchanger is compressed to its required delivery pressure.

Section 3.2.1.1 describes the process theory of each of these operations in detail, and Section 3.2.2 discusses the equipment used.

3.2.1.1 Explanation of Processes

3.2.1.1.1 Air compression and pretreatment

The main air compressor (MAC) on a cryogenic ASU provides air to the coldbox at pressures which range from 80 to 100 psia (552 to 690 kPa) depending upon the type of ASU and economic parameters associated with the plant design. For most ASUs, centrifugal compressors are used.

The compressed air is intercooled between stages with the final stage heat of compression removed in an aftercooler. As the air is compressed its saturation capacity for water vapor is reduced and water is condensed in the intercoolers and aftercooler. The water condensed in the aftercooler is normally removed in a phase separator.

Certain contaminants in the airstream have to be removed before entering the cryogenic portions of the plant. Carbon dioxide and water, for example, will freeze in the coldbox heat exchangers if not removed prior to entering the coldbox.

3.2.1.1.1.1 Reversing Heat Exchangers — The majority of air separation units built before 1980 were deliberately designed to freeze water, carbon dioxide, and other high-freezing-point, above –274°F (103 K), components in the main exchangers.[1,7] After a certain period of time, the heat exchanger needs to have the solid impurities removed. This is accomplished by physically switching the air and waste streams and passing the clean, warming waste gas through the passages that had previously contained the airstream. The solubility of water and CO_2 in the LP waste gas is significantly higher than that of the higher-pressure air so that the waste flow can be less than that of the airstream. Typically, waste flows of 50% of the airflow are required to effect the necessary clean-up of water and CO_2.

The flow switching or reversal is accomplished by a combination of switch and check valves as illustrated in Figure 3.4. Air initially passing through the "A" set of passages is switched to the "B" passages after a preset time period by opening and closing the appropriate air switch valves. At the same time, the waste is switched from the cleaned "B" passages to the "A" passages, again by opening and closing

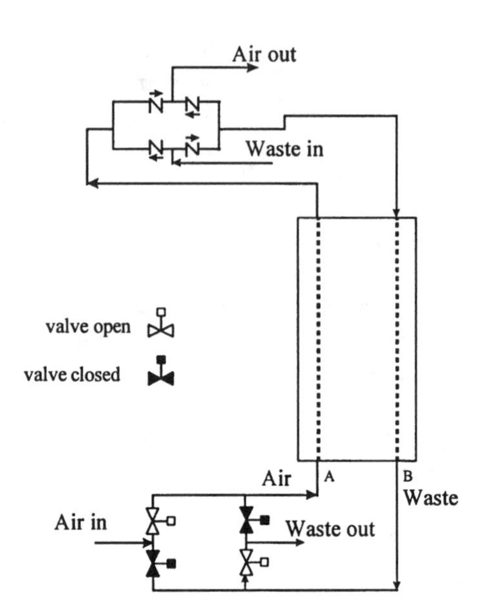

FIGURE 3.4 RHE switch and check valve assembly.

the appropriate switch valves. The check valves at the cold end of the heat exchanger direct the flows to the appropriate path.

3.2.1.1.1.2 Molecular Sieve Contaminant Removal — Most modern ASUs remove water and CO_2 in a warm-end adsorption system.[7,11] The adsorbent of choice is 13× molecular sieve (Na zeolite), which has a very high capacity for water and a good CO_2 capacity. Sometimes alumina is used for the water removal zone due to its low heat of desorption, which translates into lower energy requirements. Air is passed through one or more adsorbent beds for a predetermined time period, which is typically just prior to breakthrough of CO_2 from the end of the bed(s). At this time, a clean bed is brought on-line and the spent bed regenerated with hot, LP waste gas. The arrangement shown in Figure 3.5 is a two-bed system with one bed on-line and the other being regenerated. Bed A is on-line and Bed B is in regeneration mode. The molecular sieve capacity for CO_2 and water is significantly reduced at the lower pressure and higher temperature, and desorption energy is provided by the hot waste gas. The fact that the regeneration takes place at high temperatures, typically 300 to 400°F (420 – 480 K), means that the desorption energy can be provided by a flow significantly less than that of the airflow. A normal regeneration flow requirement would be in the 10 to 20% range depending upon regeneration temperature and CO_2 and water loading on the bed. This type of adsorption process is called "temperature swing adsorption" or TSA.

A sequence typically used for the configuration in Figure 3.5 consists of five steps:

Parallel Flow: Both beds are on-line for a short period of time.
Depressurization: The spent bed is depressurized to waste gas pressures.

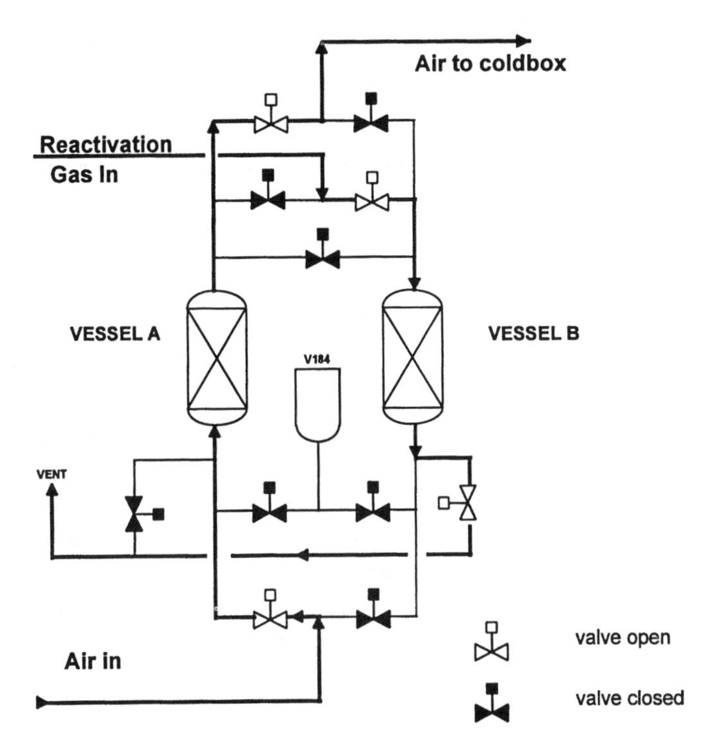

FIGURE 3.5 Two-bed adsorption system PFD.

Heating: Waste gas passed through the heater is used to heat the bed.

Cooling: When an appropriate amount of heat has been added, the heater is switched off and the bed cooled with waste gas.

Pressurization: The regenerated bed is pressurized from the on-line vessel and, when pressurized, is brought online.

After Bed B is regenerated and put on-line, Bed A is then depressurized and regenerated in the same sequence.

The amount of CO_2 loading is determined by the ambient CO_2 content of the air, which has been steadily climbing over the last 50 years, and the length of time the bed is onstream. A normal design concentration of CO_2 in air is 375 ppm, but can be a function of plant location. The water content in the air to the molesieve system is not a function of ambient air humidity since the air is compressed and water removed. The air to the adsorption system will be saturated with water at its saturation vapor pressure (SVP). The SVP of water is a function of temperature, and the water content is a function of its partial pressure which is the ratio of the SVP and the operating pressure. The water content of air at 90 psia (620 kPa) approximately doubles as the temperature changes from 40 to 60°F (278 to 289 K).

As the temperature of the airstream increases, the adsorbent quantities increase, as does the regeneration energy and flow. Historically, TSAs have been operated at 40 to 50°F (277 to 283 K). There is, however, a balance between the energy and capital needed to cool the air vs. the energy and capital needed in the adsorption system. When energy is cheap or can be recovered from the heat of compression, the optimum operating bed temperatures are warmer. This increases the cost of the adsorbent system, but may eliminate the need for cooling. The net result is a capital saving. When maximum product recovery is needed (minimum regeneration flow) or low-cost refrigeration is available, the optimum operating temperatures are cooler. The optimum bed onstream time follows the same logic.

3.2.1.1.1.3 Advantages of Adsorption Systems — The front-end adsorber systems have significant operating advantages over reversing heat exchangers. This is due to the fact that CO_2 and water do not enter the coldbox in an ASU with adsorbers, which makes plant cool down easier and simplifies the general operation of the plant. In a plant with reversing heat exchangers (RHE), CO_2 and water are deliberately allowed into the cryogenic areas of the plants. The cleanup capability of the cores is a strong function of waste flow and temperature profiles within the core. It, therefore, is very important to control closely temperatures and flows in the heat exchangers. Plant upsets can result in significant CO_2 breakthrough into the column system, plugging lines and heat exchangers requiring the plant to be defrosted.

Reversing exchangers are subject to rapid pressure and temperature reversals every 10 min, which results in a finite life of the heat exchangers due to stress induced leaks.

One of the major advantages of a molecular sieve system is the almost complete removal of heavy and unsaturated hydrocarbons, notably acetylene. Acetylene if precipitated out of LOX will detonate. It is, therefore, very important to monitor and keep acetylene levels significantly below the solubility limits, 8 ppm in LOX at 20 psia (138 kPa). In a reversing ASU it is necessary to add cryogenic silica gel adsorbers to scavenge any acetylene entering the column system.

When the RHE is switched, the air passages are rapidly depressurized and that air is lost. This can amount to 2% of the airflow and is an inefficiency of that design. The TSA system, however, needs regeneration energy and perhaps refrigeration energy.

The major advantage with a TSA system is that with a regeneration flow of 10 to 20% vs. 50% in an RHE design there is significantly more product available. A plant with a TSA system is capable of producing significant amounts of nitrogen in addition to oxygen.

3.2.1.1.1.4 Precooling — As discussed in the previous section, it may be necessary to chill the air before entering the adsorbers. This may be accomplished with an in-line mechanical refrigeration unit, a water chiller using evaporative cooling from waste gas from the plant, or a combination of both.

In a mechanical refrigeration system, the air is cooled directly in an exchanger against a boiling freon. Based on the Montreal Protocol, that choice of refrigerant has been limited but is still a practical system especially with the continual development of environmentally friendly refrigerants.

If only oxygen and a small amount of nitrogen are required as products, a significant portion of the air is available as a nitrogen-rich waste gas. Chilled water can be generated in a humidification tower by passing that waste gas countercurrently

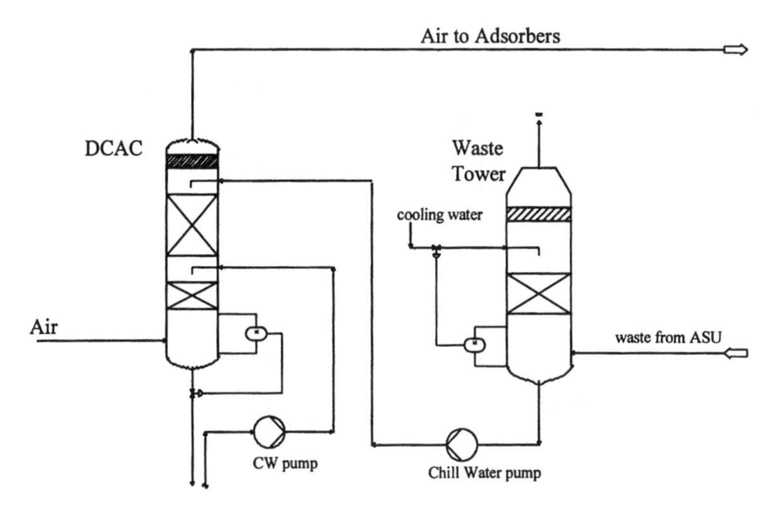

FIGURE 3.6 Direct contact tower and waste chill tower.

against cooling water. It is possible to chill water to a colder temperature than the waste gas; since the waste is dry, its wet bulb temperature will be several degrees colder. The chilled water can then be used to cool the airstream in a shell-and-tube heat exchanger or preferably by directly contacting the chilled water with air in a direct-contact aftercooler (DCAC). This has the added advantage of removing water-soluble acid gases before the molesieve that would otherwise degrade the adsorbent. Figure 3.6 shows a typical DCAC and waste-chilling tower arrangement. Water pumps are needed to feed the chilled water to the top of the DCAC.

3.2.1.1.2 Heat exchangers

In order to minimize the refrigeration losses (Section 3.2.1.1.3) from an ASU, it is very important to have efficient heat exchangers. It is important to maximize both the heat transfer coefficients and the surface area available for heat transfer. Heat transfer coefficients for sensible heat exchange between gases are poor and can only be improved at the expense of pressure drop. The ideal heat exchanger for cryogenic ASUs with gas/gas exchange has to have a high ratio of surface area to cross-sectional flow area with low resistance to flow. The standard heat exchanger type used is therefore the plate-and-fin type. Brazed aluminum plate-and-fin heat exchangers are used almost exclusively for cryogenic gas/gas heat exchange in air separation.

3.2.1.1.2.1 Main Heat Exchanger — The function of the main heat exchanger (MHE) is to maintain the plant refrigeration at the cold end of the plant and to ensure that the product and waste gas leave the heat exchanger at a temperature close to that of the air entering the exchanger. The typical average temperature difference between the air and the warming streams is of the order of 8°F (4.4 K). A lost work analysis of a typical MHE arrangement shows that it contains 10% of the total plant inefficiencies. It is therefore important to optimize the benefits of increasing heat exchanger size vs. cost.

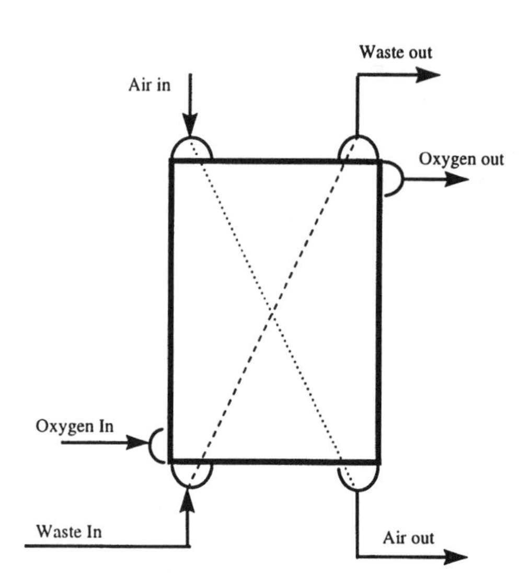

FIGURE 3.7 Simple MHE arrangement.

Figure 3.7 shows a simple MHE arrangement with air cooling against oxygen and a waste stream. Typically, there are more than one heat exchanger arranged in banks. The streams are introduced to the heat exchangers via manifolds and headering. The warming streams flow countercurrent to the airstream. For plants with adsorption systems, the heat exchangers are oriented vertically with the cold end of the heat exchanger at the bottom. In this way, any liquid air formed will drain from the heat exchanger.

3.2.1.1.2.2 Subcoolers — Subcoolers are also typically brazed aluminum heat exchangers with the function of warming the waste stream from the LP column and, in turn, subcooling the reflux streams. The subcooler accomplishes two things. Firstly, it minimizes the flash losses as the refluxes enter the LP column, and, secondly, it transfers heat to the waste stream which in turn allows a warmer airstream leaving the main exchanger. The degree of complication required in the subcooler is a function of the process flow sheet. For a simple plant with no liquid products and no pure gaseous nitrogen product, a simple two-stream subcooler can be used, as illustrated in Figure 2.1. For more-complicated and more-challenging designs the subcooler may consist of two warming vapor streams and three or four liquid streams.

If liquid products are generated, it may also be necessary to subcool them and minimize losses when transferred to storage tanks.

3.2.1.1.3 Refrigeration generation

Consider a hot process, for example, a reformer. To minimize fuel costs, it is important to minimize heat leaks, cool effluent gases close to ambient temperature, and eliminate losses of high-energy gases. The same is true for cryogenic processes. Losses in refrigeration normally translate into a process inefficiency. The generation

TABLE 3.1
Material and Energy Balance at Low Liquid Production

	Pressure psia	Temp. °F	Enthalpy Btu/mca	Btu/lb-mol	Flow (m/mca) N_2	Ar	O_2	Total lb-mol/h
LP Air to coldbox	**100**	**50**	**–211**	–211	0.7812	0.0093	0.2095	1.0000
GOX from coldbox	**20**	45	**–47**	–230	0.0000	0.0000	0.2050	0.2050
LOX from coldbox	**23**	**–300**	**–5.6**	–5598	0.0000	0.0000	0.0010	0.0010
Waste from coldbox (with GOX)	**19**	45	**–180**	–226	0.7812	0.0093	0.0035	0.7940
Heat leak			**20**					
Refrigeration loss			**41**					
Turboexpander			**41**	**600**				**0.069**

Note: mca denotes lb-mol per mol of air feed to the columns.

of refrigeration to offset losses requires additional power consumption or loss in product recovery.

Refrigeration losses in an ASU are typically

1. Heat leak into the coldbox,
2. Warm-end losses in the main exchanger, and
3. Liquid products.

This is illustrated in the energy and material balance in Table 3.1 for a simple ASU producing gaseous O_2 (GOX) and a small amount of LOX. The enthalpies of the GOX and waste streams exiting the main exchanger are slightly lower than incoming air and constitute a loss. The LOX, although a small flow, has a significant impact on the refrigeration balance. The enthalpy of the LOX per mole is significantly less than that of the incoming airstream. The higher the LOX requirement, the larger the refrigeration loss. Table 3.2 shows the enthalpy balance for a LOX requirement of close to 3% of the incoming air. The refrigeration requirement is close to six times that of the base case. In both cases the heat leak was assumed to be 20 Btu/lb-mol (46.5 kJ/kg-mol) of airflow and a warm-end temperature loss of 5°F (2.7 K) was assumed. It can be seen that liquid production has a significant impact on refrigeration requirements.

To understand the impact of a high refrigeration requirement upon the process, we need to examine how refrigeration is generated in an ASU. Early air separation plants (Linde or Hampson cycles[1,2]) used HP compressed air to feed the plant and generated liquid by cooling this air and dropping the pressure through a valve from the supercritical zone into the two-phase dome. This refrigeration system is illustrated using the simplified pressure enthalpy (PH) diagram shown in Figure 3.8. Air is compressed to point *A*, cooled to point *C*, and, when the pressure is dropped to point *D*, a two-phase mixture is produced. Refrigeration is generated only by the difference in enthalpy of the HP air (point *A*) and the lower pressure vapor streams

TABLE 3.2
Refrigeration Balance with High Liquid Production

	Pressure psia	Temp. °F	Enthalpy Btu/mca	Btu/lb-mol	Flow (m/mca) N_2	Ar	O_2	Total lb-mol/h
LP Air to coldbox	**100**	**50**	**–211**	–211	0.7812	0.0093	0.2095	1.0000
GOX from coldbox	**20**	45	**–41**	–230	0.0000	0.0000	0.1770	0.1770
LOX from coldbox	**23**	**–300**	**–156**	–5598	0.0000	0.0000	0.0280	0.0280
Waste from coldbox (with GOX)	**19**	45	**–180**	–226	0.7812	0.0093	0.0045	0.7950
Heat leak			**20**					
Refrigeration loss			**186**					
Turboexpander			**186**	**600**				**0.310**

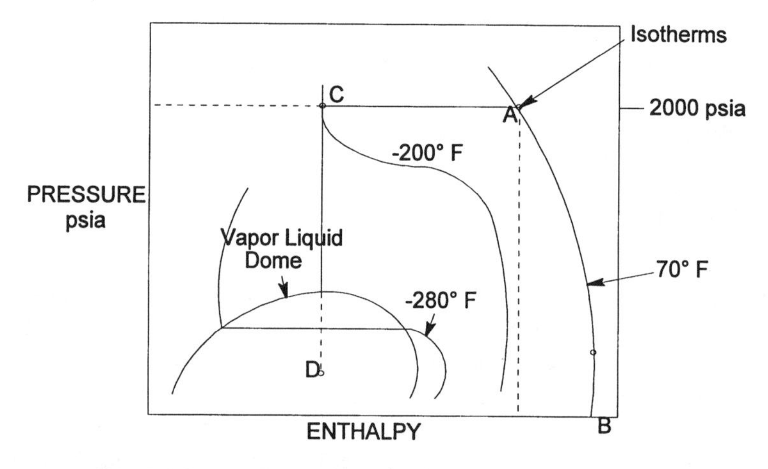

FIGURE 3.8 PH diagram showing JT effect of an HP airstream.

(point *B*). No net refrigeration is generated across the valve (between C and D), and the expansion is purely Joule Thompson expansion.

A significant improvement to the process outlined above was the use of an expansion engine. Typically, in ASUs turbo expanders are used. An ideal turbo expander is isentropic and reversible. Illustrated in Figure 3.9, air at –150°F (172 K) and 90 psia (620 kPa) is expanded to 20 psia (138 kPa). In an isentropic expansion *A–B*, the expansion follows the isentrope with a net change in enthalpy. In reality the expansion will not be reversible and will follow a curve similar to *A–C*. The actual enthalpy change divided by the isentropic enthalpy change is a measure of the expander efficiency.

$$\text{Efficiency} = (H_A - H_C)/(H_A - H_B)$$

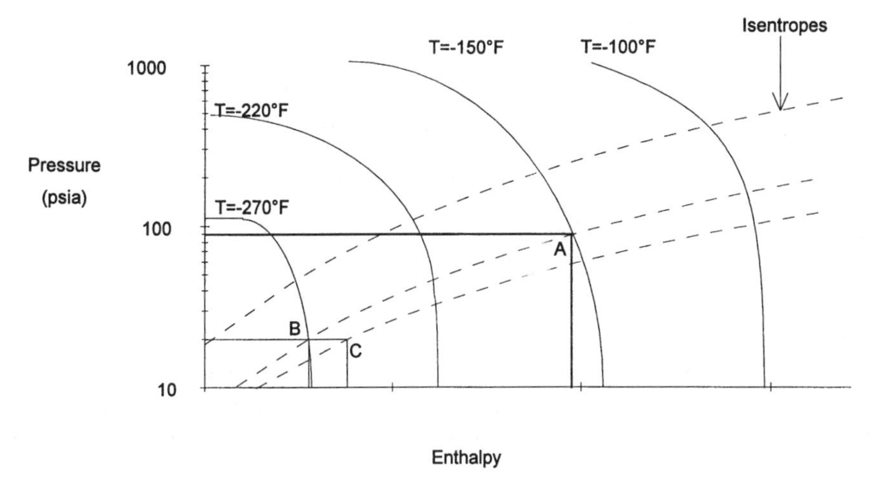

FIGURE 3.9 PH diagram showing enthalpy change across a turbine.

The work extracted from the expanding fluid (H_A–H_C), may be utilized to generate electricity, compress gas, or just be dissipated in an ambient blower or oil friction brake (see Section 3.2.2.4). The net result is that as refrigeration is generated across the turbine, less refrigeration is needed from the incoming air and the air compressor discharge pressure can be run at much lower pressures. Cycles termed *low-pressure cycles* ensued from this development.

The simple ASU PFD, Figure 3.3, shows the typical application of an expander in an ASU generating GOX and a small quantity of LOX. Air is compressed to a typical pressure of 90 psia (620 kPa), cooled in the main exchanger, and then a portion of the air is expanded to a lower pressure to generate the necessary refrigeration. The HP and LP airstreams are then passed to the distillation columns. As more flow passes through the turbine, less is available to the HP distillation. This impacts distillation efficiency and increases air compressor power. By incorporating the enthalpy balance from Table 3.1 and a typical refrigeration generation capability from the type of expansion turbine shown in Figure 3.3, which would be approximately 600 Btu/lb-mol expander flow (1400 kJ/kg-mol), an expander flow requirement can be determined. For the base refrigeration balance with low LOX make and a refrigeration requirement of 41 Btu/lb-mol (95 kJ/kg-mol) of airflow, the required expander flow is 7% of the airflow.

For the high-LOX case, where the refrigeration requirement is 186 Btu/lb-mol (434 kJ/kg-mol) of airflow, the required expander flow of 31% is needed to produce less than 3% of the airflow as liquid. This is a simplistic demonstration of the ability of an ASU to produce liquid. For small amounts of LOX, the ASU is efficient. Beyond 3% of the airflow, the distillation impacts of the high expander flow become prohibitive. Above these rates, the addition of compressors/expanders specifically needed for liquefaction are added. (See Section 3.2.1.2.5.)

TABLE 3.3
Boiling Points of Pure Air Components at 20 psia

Oxygen	–292°F (93 K)
Nitrogen	–316°F (80 K)
Argon	–297°F (90 K)

3.2.1.1.4 Distillation

The distillation columns are the heart of the ASU . The boiling points of nitrogen, argon, and oxygen are different enough to allow separation by distillation. The boiling points at 20 psia (138 kPa) are listed in Table 3.3. The separation of oxygen and nitrogen is relatively easy because of the fact that the boiling points are significantly different. The separation of argon and oxygen is relatively arduous for the opposite reason.

3.2.1.1.4.1 Distillation Theory — To illustrate how the separation of oxygen and nitrogen occurs in a cryogenic distillation column, refer to the phase equilibrium diagram for an O_2/N_2 mixture at 20 psia (138 kPa), Figure 3.10. These simplified curves treat air as a binary mixture of nitrogen and oxygen. At the 0% N_2 side of the curves, the temperature is –292°F (93 K) which is the boiling point of pure O_2. At the 100% N_2 side, the temperature is –316°F (80 K) which is the boiling point of pure N_2. The dew point curve (upper curve) represents the temperatures at which liquid droplets are formed in a cooling vapor for various O_2/N_2 compositions. The bubble point curve (lower curve) represents the temperature when gas bubbles start to form in a warming liquid. Note that a mixture of vapor and liquid at the same temperature is at equilibrium but will have different vapor and liquid compositions.

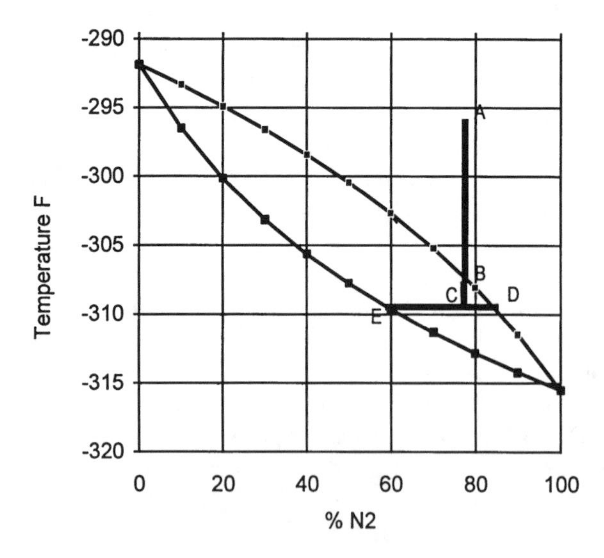

FIGURE 3.10 Simple equilibrium diagram with phase separation.

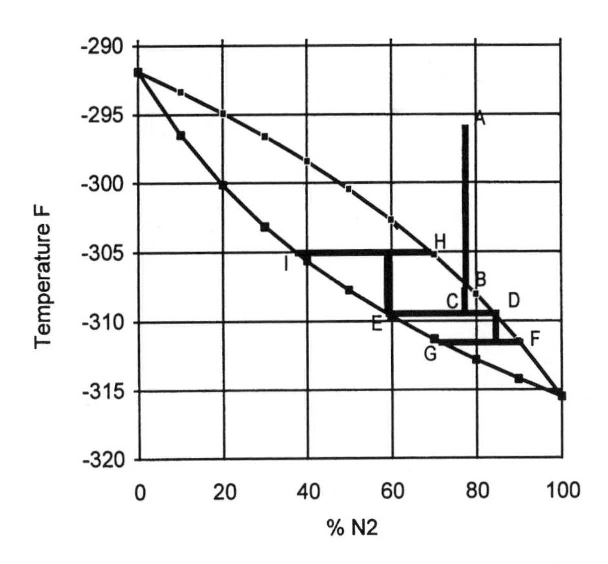

FIGURE 3.11 Equilibrium diagram with stage separation.

Figures 3.10 and 3.11 illustrate what happens when air is first cooled to its dew point (B), then partially liquefied (C) and separated into gas (D) and liquid (E) components as demonstrated on the curves. The liquid composition is significantly richer in oxygen, and the vapor stream richer in nitrogen. If the liquid stream (E) were to be heated and then separated, further separation will take place, (I), (H). Likewise, if the vapor stream (D) were to be cooled and then separated, additional separation will occur, (F), (G).

In an ideal distillation column, the concepts demonstrated in Figures 3.10 and 3.11 can be shown as different equilibrium stages with heat and mass transfer at each stage (see Figure 3.12). As the vapor rises up the column, it encounters liquid

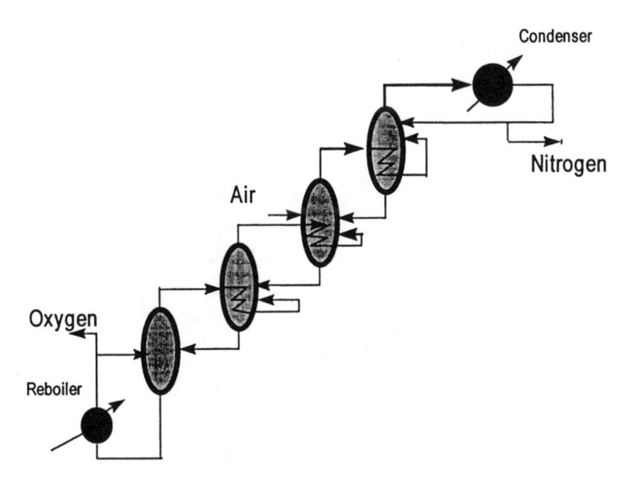

FIGURE 3.12 Ideal distillation system.

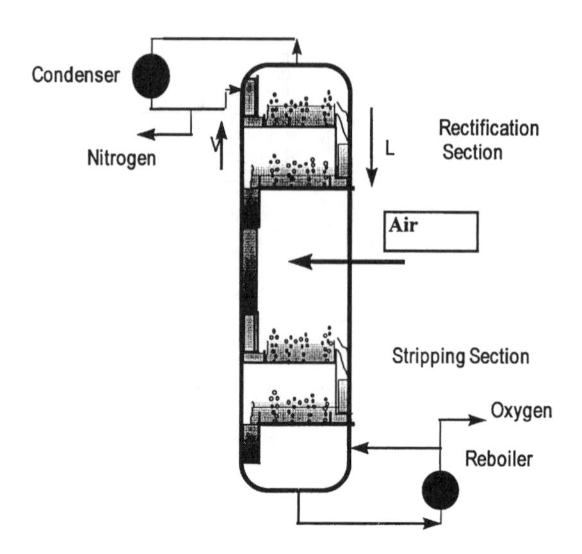

FIGURE 3.13 Actual distillation system.

flowing down. At each stage, heat is exchanged and a new equilibrium is reached. The vapor becomes richer in nitrogen as it moves up the column. The liquid, on the other hand, becomes richer in oxygen as it flows down the column.

For distillation to occur, vapor has to be contacting liquid, which is termed *reflux*. For a column with vapor feed to produce a pure N_2 stream at the top of the column, a top condenser is needed. This section in a column is called the rectification section. To produce pure O_2 at the bottom of the column, a reboiler is also needed to generate vapor or boil-up. This section in a column is referred to as the stripping section. A more realistic depiction of the ideal distillation column is shown in Figure 3.13. Gas is contacted with the liquid by bubbling through holes in distillation trays as shown or in contact with a thin liquid film on the surface of a structured packing. In a distillation sieve tray (Figure 3.14), liquid is introduced to the tray from the tray above through a downcomer. The liquid flows over the tray encountering gas flowing up through the holes. Mass and heat transfer occur in the resulting froth. After

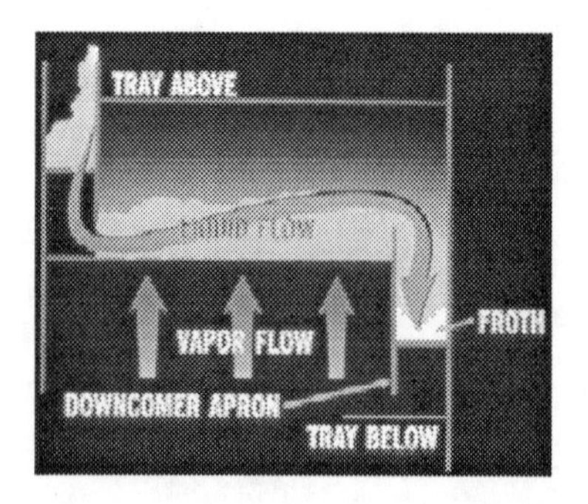

FIGURE 3.14 Simple distillation tray diagram.

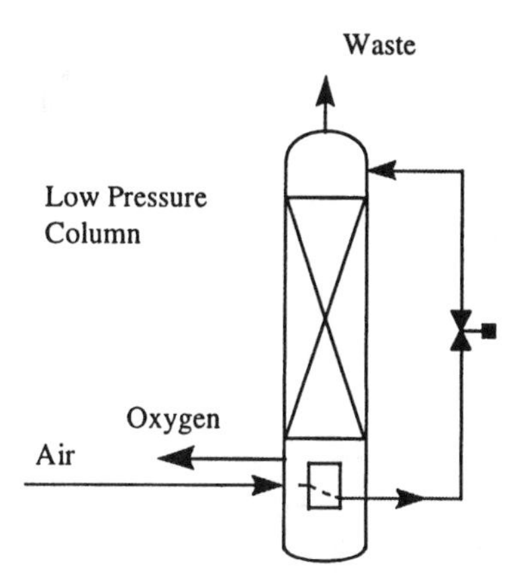

FIGURE 3.15 Single-column design.

passing over the tray, the liquid flows over an outlet weir and into another downcomer to the tray below. The different types of column are discussed in Section 3.2.2.5.

3.2.1.1.4.2 Single-Column Cycle — One of the first and simplest of air separation column systems was the Linde single-column design (Figure 3.15). Air is compressed and cooled in an MHE. The HP air is then used to boil the liquid in an LP column. The air is condensed and then transferred into the top of the LP column. As the liquid flows down the column, it contacts the vapor generated in the reboiler and becomes richer in oxygen. Product oxygen in a gaseous form is generated at the bottom of the column at a purity defined by the number of distillation stages in the column. Due to the fact that there is not a rectification section in this column, a pure nitrogen stream cannot be generated and the purity is limited to a vapor oxygen concentration close to the equilibrium value with liquid air. This can be approximated by referring to the equilibrium diagram. At the liquid air composition (21% O_2), the equilibrium dew point concentration is approximately 7%. The waste stream therefore contains a significant amount of oxygen that is not recovered. This limits the efficiency of this cycle.

3.2.1.1.4.3 Double-Column System — The double column (Figure 3.16) is by far the most common column configuration used in air separation plants. Air at a temperature close to its dew point is fed to the bottom of the HP column. This column consists of only a rectification section; thus, a nitrogen-rich stream can be generated at the top of the column. The reflux for this column is generated in the reboiler condenser. The fact that the HP column does not have a stripping section means that the liquid leaving the bottom of the column is no richer in oxygen than the liquid in equilibrium with the air vapor. This oxygen-rich reflux is then subcooled against warming waste gas and flashed into the LP column. This column has both stripping and rectification sections. Reflux for the rectification stream is provided from the top of the HP column. This reflux is also subcooled against the waste nitrogen stream.

The rectification section produces a nitrogen-rich stream allowing oxygen recovery to be significantly higher than that of a single-column design. The stripping

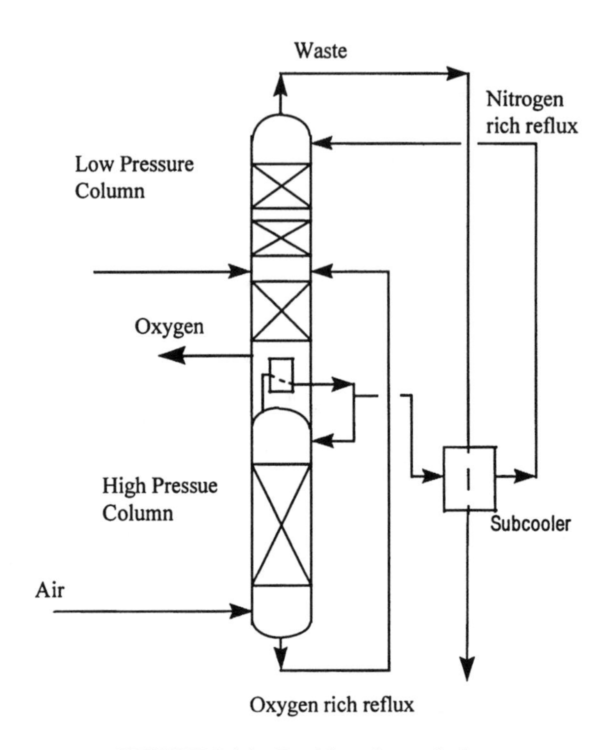

FIGURE 3.16 Double-column design.

section produces a bottoms product richer in oxygen. The purity of this product vapor stream is a function of the number of distillation stages available and the boil-up generated in the reboiler. The recovery of oxygen is a function of total distillation stages, boil-up, and reflux availability.

In Section 3.2.1.1.3, the production of small quantities of liquid product was discussed. Increasing need for liquid production requires increased expansion turbine flow. Referring to the PFD of the LP cycle (Figure 3.3), as expander flow is increased, less air is available for the HP column. This air generates boil-up and reflux from the HP column. Therefore, as less reflux and boil-up is available in the LP column, oxygen recovery is reduced. The purer the oxygen product, the more sensitive it is to boil-up and reflux effects. This is illustrated in Figure 3.17. For a high-purity product of 99.5% O_2 (with no argon production), the recovery is reduced even at low expander flows. As the product purity is reduced, the oxygen recovery is only affected at the higher expander flows. The purity of the product, therefore, has a significant affect upon the optimum liquid production of the plant.

3.2.1.1.5 Reboilers and reboiler safety

The reboiler is the thermal link between the HP column and the LP column. It provides vapor boil-up for the LP column and reflux for the HP column. The reboiler functions by condensing a nitrogen stream and boiling an oxygen stream. Based upon the boiling points presented in Table 3.3, this would not seem feasible. However, as demonstrated in Figure 3.18, as the pressure increases, the boiling point

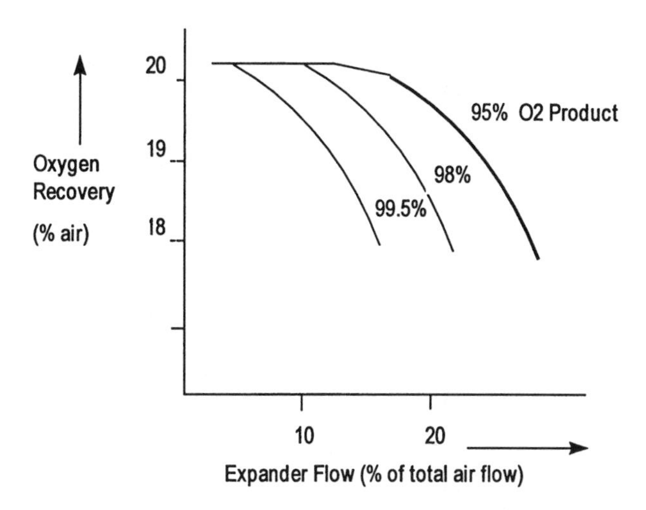

FIGURE 3.17 Oxygen recovery vs. expander flow for various purity levels.

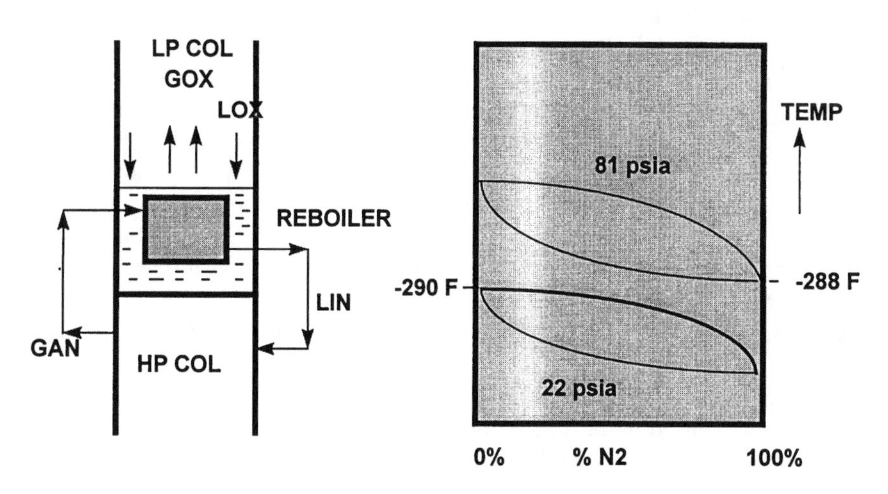

FIGURE 3.18 Reboiler diagram.

temperatures also increase. At a certain pressure the nitrogen will condense and boil the oxygen. A sustainable nitrogen pressure is defined by the LP column operating pressure, the oxygen composition and, hence, its boiling point, and the size and type of reboiler.

There are two types of reboilers used commercially. The first is termed the *thermo-syphon reboiler* and is submerged in a pool of liquid oxygen at the bottom of the LP column. The nitrogen passages are contained in a high pressure circuit. The oxygen passages, however, are open at the top and the bottom so LOX is free to flow into and out of the reboiler. As oxygen is boiled within the open side of the heat exchanger, it flows upward. The two-phase stream is buoyant, and as it rises it induces fresh liquid flow into the reboiler. By pressure balance, it is easily seen that

TABLE 3.4
Atmospheric Air Contaminants

Contaminant	Typical Levels (ppm)	Consequences of High Site-Specific Levels of Concentration
Hydrogen	1	Contamination of UHP N_2 product if required
Carbon monoxide	2	Contamination of UHP N_2 product if required
Carbon dioxide	350	Larger adsorbers
Methane	2	Increased LOX purge need
Acetylene	0.05	Slight increase in adsorber size
Ethane	0.05	Increased LOX purge need
Ethylene	0.1	Increased LOX purge need; use selective adsorbent
Propane	0.05	Increased LOX purge need
Higher hydrocarbons	1.0	Slight increase in adsorber size
Acid gases	0.1	Corrosion resistant materials in front-end equipment

the extent of recirculation is determined by the difference in density and height of the clear liquid outside the core and that of the two-phase stream inside. The higher the level of liquid on the outside, the more the recirculation.

Any hydrocarbons entering the coldbox will migrate to the bottom of the LP column and, hence, the LOX in the reboiler. It is important that hydrocarbons should not be allowed to concentrate. A constant liquid purge is necessary, as discussed in Section 3.2.1.1.6, but it is also necessary to design the reboiler area to minimize potential for dry boiling areas. It is important to maintain high liquid levels and recirculation rates through the reboiler to eliminate dry boiling zones and to sweep solids trapped in the core. It is recommended that the reboiler level be maintained at the top of a thermosyphon reboiler.

The other type of reboiler commonly used is termed the *downflow reboiler* due to the fact that LOX enters at the top of the core and flows down. Vapor and liquid exits at the base of the core. The exit stream should have sufficient liquid flow to wash contaminants from the core and prevent dry boiling. A constant purge is still needed, however, from the column sump.

3.2.1.1.6 Impurity removal and hydrocarbon safety

Typical impurities contained in air are shown in Table 3.4. Coarse particulate matter will be removed in the inlet air filter before the main air compressor. As the air is compressed and intercooled water is condensed, a significant portion of the water-soluble acid gas contaminants are removed in the condensate. If ambient concentrations of SO_2 and NOx are high, special alloys may be necessary in the intercoolers. A DCAC is also an efficient scrubber of acid gases. Any acid gases not removed in the coolers will be either adsorbed reversibly in the adsorbers, if in its vapor state, or chemically reacted with the adsorbent, if dissolved in water condensate. Normally, DCACs and separators are designed such that entrained condensate drops are unlikely to enter the adsorbers.

The molecular sieve adsorbers will also remove all the heavy hydrocarbons from butane through the heavier hydrocarbons. Acetylene and propylene are also very

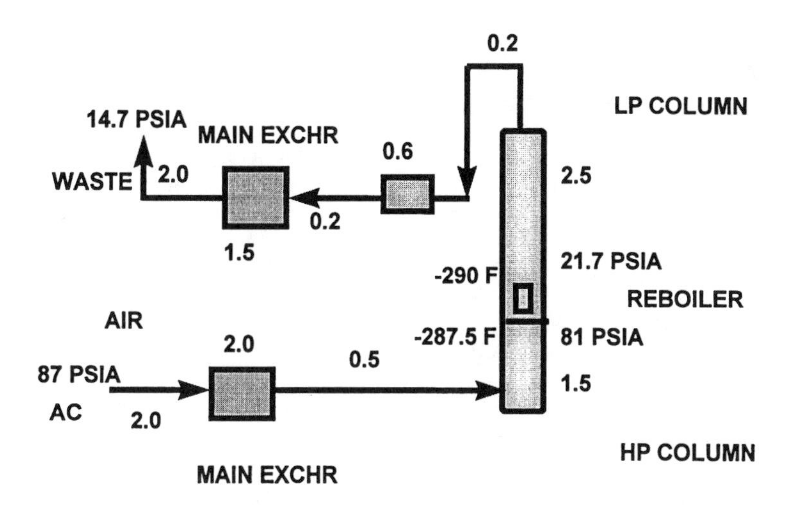

FIGURE 3.19 Pressure balance figure.

strongly adsorbed on 13×. Ethylene is only partially removed. Ethylene is, however, strongly adsorbed by cryogenic adsorption on silica gel beds. This type of adsorber was used extensively in reversing ASUs. The lighter saturated hydrocarbons (methane, ethane, and propane) will be eluted by 13× before CO_2 and will enter the coldbox. Since all three contaminants have boiling points warmer than O_2, they will concentrate in the LP column bottom. It is extremely important that a constant liquid purge be maintained from the LP column to prevent the accumulation of hydrocarbons. The concentrations of hydrocarbons in the LP column bottom, measured as a methane equivalent, are generally monitored to check if the purge rate is adequate or if local atmospheric levels are high. If concentrations reach higher levels than are acceptable, the purge rate is increased or the plant shut down depending on the values. A pumped LOX plant by nature has a very high liquid purge from the LP column.

Hydrogen and CO will not be adsorbed in the adsorbers and will pass into the column system and exit in the waste nitrogen stream. The hydrogen should not be allowed to concentrate in the HP column, and noncondensible vents in the condenser may be necessary. If a pure nitrogen product is generated, it will be contaminated by H_2 and CO. Nitrogen used by the electronics industry is required to have low CO and H_2 contaminant levels. ASUs supplying nitrogen to that industry will need to remove H_2 and CO catalytically.

Inert gases will exit the coldbox in the product or waste streams. The lighter gases, helium and neon, will be removed in the nitrogen or waste streams and the krypton and xenon in the oxygen streams. Argon, if not removed as a product, will exit in both the oxygen and waste streams.

3.2.1.1.7 Air compressor discharge pressure determination

The discharge pressure required by the main air compressor is determined from individual equipment and line resistances of the plant. This is best illustrated by referring to Figure 3.19. The LP column top pressure is set by the waste exit pressure and the resistances in the waste circuit. These consist of piping and heat exchanger

TABLE 3.5
Pressure Balance

Waste Circuit	Pressure Drop	Absolute Pressure
Atmospheric Pressure		14.7 psia (101 kPa)
Waste pressure needed to regenerate the adsorbers	1–2 psi	
Typical pressure of waste from coldbox		16.7 psia (115 kPa)
Frictional resistance in the MHE	1–2 psi	
Typical pressure at cold end of MHE		18.2 psia (125 kPa)
Subcooler resistance	0.5–1 psi	
Typical pressure to subcooler		18.7 psia (129 kPa)
Miscellaneous piping resistance	0.5 psi	
Typical pressure at LP column top		19.2 psia (132 kPa)
Differential pressure across the column	~2.5 psi	
Pressure at LP column bottom		21.7 psia (150 kPa)
Dew Point of oxygen at 98% purity and 21.7 psia	–290°F (94 K)	
Typical reboiler warm end approach temperature	2.5°F (1.4 K)	
Air Circuit		
Dew point of nitrogen	–288.5°F (95.4 K)	
Pressure of HP column top		81.0 psia (560 kPa)
Differential pressure across HP column	1.5 psi	
Pressure at HP column bottom	1.5 psi	82.5 psia (568 kPa)
Miscellaneous piping resistances	0.5 psi	
Typical pressure at MHE cold end		83.0 psia (572 kPa)
friction resistance in the MHE, typical	1–2 psi	
Typical pressure of air to coldbox		85.0 psia (586 kPa)
Frictional resistance of purification system	~2 psi	
Typical pressure at air compressor		87.0 psia (600 kPa)

frictional losses. The pressure at the bottom of the column includes the resistance of the column itself. The reboiler top approach temperature determines the HP column pressure. The main air compressor discharge is the result of this pressure and the resistances in the air circuit. The values in the example are tabulated in Table 3.5.

It can be easily seen that the equipment size, pressure drop, and hence cost have a direct bearing upon the air compressor power. It is important to obtain the correct optimization between equipment costs and power consumption. The values used in the example are for illustration purposes. An actual plant design would vary based upon the cost vs. power analysis. This is covered in more detail in Section 3.2.3.

3.2.1.2 Production of Gaseous Oxygen

There are two major commercial choices in the production of gaseous oxygen. The first is the standard LP design producing gaseous oxygen from the LP column and compressing this oxygen externally to the coldbox. For very small plants, a single-column cycle may be feasible, but normally these designs are limited by the very low recoveries of oxygen. The alternative design, which is quickly becoming the preferred option, is internally pumped LOX. This option eliminates the need for the oxygen compressor.

3.2.1.2.1 Oxygen compression cycles

The LP oxygen flow diagram is shown in Figure 3.3. Air compressed in the main air compressor is cleaned in the air pretreatment system and cooled in the main exchangers to a temperature close to its dew point, –277°F (101 K). A side stream is removed from the exchanger at a temperature of approximately –160°F (166 K) and expanded through the turboexpander to the LP column to generate the plant refrigeration. The remaining air is passed to the HP column. The distillation system discussed in Section 3.2.1.1.4.3 is a double-column system. Oxygen produced as a vapor from the LP column is warmed in the MHE and compressed in an oxygen compressor to the required usage pressure.

The turboexpander used in this example is a simple oil brake machine expanding air. This is limited in its refrigeration generation capability and can be improved by driving a process compressor instead of dissipating the recovered work in a friction system. The compressor can then precompress the air feeding the turbine and increase the refrigeration generated per unit expander flow. The compressor/expander (compander) configuration is discussed in Section 3.2.2.4.

This design has been a standard in the industry for many years due to its simplicity and ease of operation. In the last 10 years pumped LOX designs which eliminate the oxygen compressor have become favored for safety and cost reasons.

3.2.1.2.2 Pumped LOX cycles

Figure 3.20 shows a pumped LOX design. The flow sheet is very similar to an LP cycle with an air expander and a double-column system. The major difference is that oxygen is produced from the LP column as a liquid, pumped to the required pressure, and then vaporized in the main exchangers. To allow the oxygen to be vaporized at pressure, an HP condensing stream is needed to match the boiling curve of the oxygen stream. In Figure 3.21, a boiling curve of pure oxygen at 100 psia (690 kPa) is shown. The bubble point of oxygen at this pressure is –255°F (114 K). To vaporize this stream in an efficient heat exchanger, condensing air would need an air pressure that corresponds to a bubble point warmer than –255°F (114 K). With a 2°F (1 K) approach temperature, the bubble point would be –253°F (115 K) and the air would need to be at 235 psia (1620 kPa). As the oxygen is vaporized, HP air is condensed. The airstreams leaving the MHE will contain approximately 25% liquid. In this design the oxygen compressor has now been replaced by a pumped system with an air booster compressor. In certain instances the air booster duty can be combined with the main air compressor and significantly reduce costs.

The overall energy balance is given in Table 3.6. Although liquid oxygen is produced from the LP column, the column system is in energy balance since it is fed with liquid and vapor air. The main inefficiency of the cycle is the pump work which is a direct input into the process and any losses in the heat exchangers. The

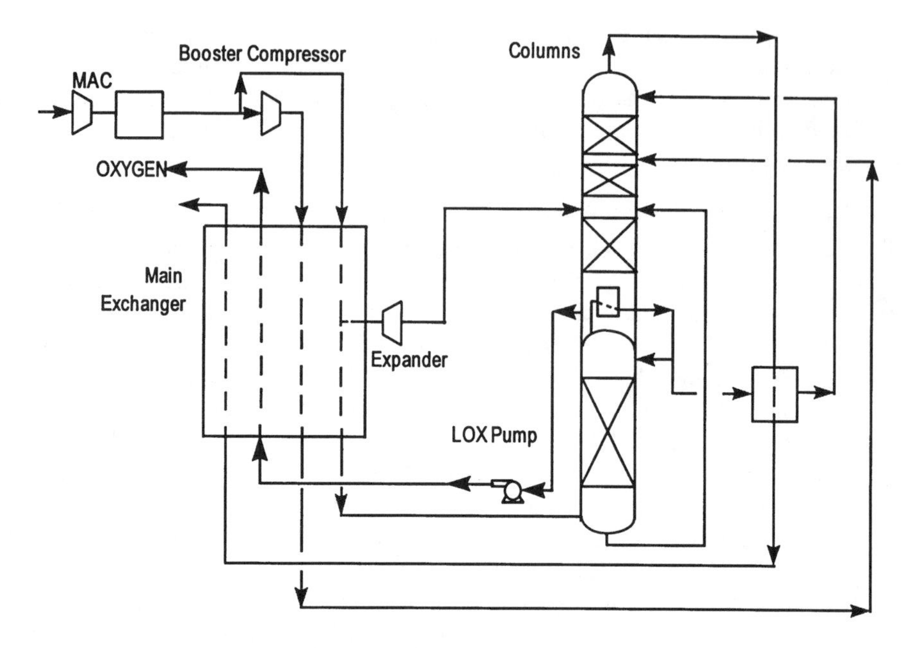

FIGURE 3.20 Pumped LOX cycle.

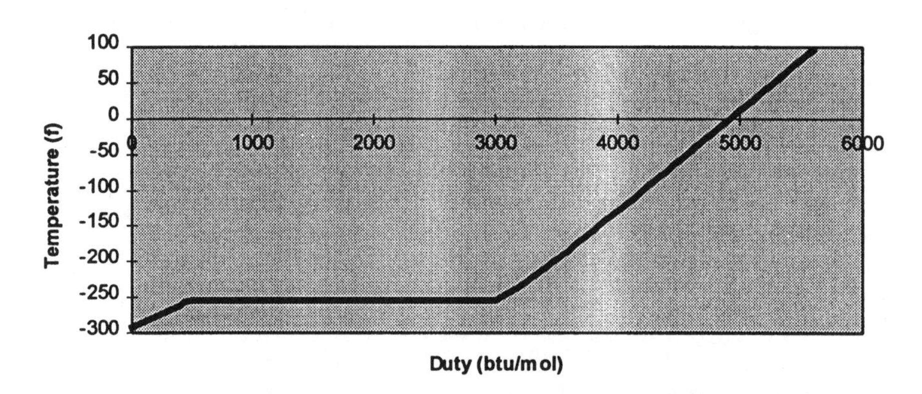

FIGURE 3.21 Boiling point curve of oxygen at 100 psia.

overall balance assumes a slightly higher warm-end temperature difference to the LP cycle and HP airflow of 26% of the total air. The net result is only a small difference in refrigeration requirement between the cycles.

The relative overall efficiency of the pumped LOX cycles vs. the LP cycle is a function of product pressure and the efficiency differences between the oxygen compressor and the air booster. Typically, the power consumption of both cycles is comparable. At higher pressures the pump work becomes more significant. The pumped LOX cycle is typically lower cost, due to the elimination of the oxygen compressor.

TABLE 3.6
Refrigeration Balance for Pumped LOX

	Pressure psia	Temp. °F	Enthalpy Btu/mca	Btu/lb-mol	Flow (m/mca) N_2	Ar	O_2	Total lb-mol/h
LP Air to coldbox	**100**	**50**	**–156**	–211	0.5781	0.0069	0.1550	0.7400
HP Air to coldbox	**235**	50	**–63**	–242	0.2031	0.0024	0.0545	0.2600
GOX from coldbox	**100**	45	**–52**	–252	0.0000	0.0000	0.2050	0.2050
LOX from coldbox	**23**	**–300**	**–5.6**	–5598	0.0000	0.0000	0.0010	0.0010
Waste from coldbox (with GOX)	**19**	45	**–180**	–226	0.7812	0.0093	0.0035	0.7940
Heat leak			**20**					
Pump work			**3**					
Refrigeration loss			**40**					
Turboexpander			**40**	**600**				**0.068**

3.2.1.2.3 Coproduction of nitrogen

From the material balances in Table 3.1, it is clear that an air separation plant designed for oxygen production has a waste nitrogen stream with a flow of approximately 80% of the airflow to the plant. Waste nitrogen purity varies based upon oxygen recovery, but would typically be in the 1 to 5% O_2 range. Commercial nitrogen purity levels for most usages is in the 1 to 10 ppm O_2 range with some electronic customers requiring ppb contaminant levels. The production of a pure gaseous nitrogen (GAN) product is only a function of adding distillation capacity, and, since, the molecules are already available, no significant additional air compressor power is required. Coproduction of nitrogen is therefore economically attractive.

The simplest method to produce nitrogen from the ASU is to add distillation trays in the HP column and produce nitrogen directly at a delivery pressure slightly lower than the HP column pressure, as illustrated in Figure 3.22. The capital additions needed to coproduce small quantities of nitrogen are modest, consisting of a section of distillation stages and piping to and from the MHE. In fact, there may be a slight reduction in MHE size since an LP waste stream has been replaced by an HP nitrogen stream with lower volumetric flow.

The disadvantage of producing nitrogen from the HP column is that by removing vapor that would have entered the reboiler, the boil-up in the LP column is reduced. The impact of boil-up on oxygen recovery was discussed in Section 3.2.1.1.4, and a curve of oxygen recovery vs. expander flow is shown in Figure 3.17. HP nitrogen production has the same impact as increasing expander flow. As HP N_2 production is increased, the boil-up is reduced in the LP column and oxygen recovery may be reduced. The extent of the loss of recovery is a function of the oxygen purity and the actual production levels of nitrogen. At 95% O_2 purity levels and low expander flows (say, 7% airflow), up to 7% of the airflow can be produced as HP nitrogen with little appreciable reduction in recovery. On the other hand, with 99.5% oxygen

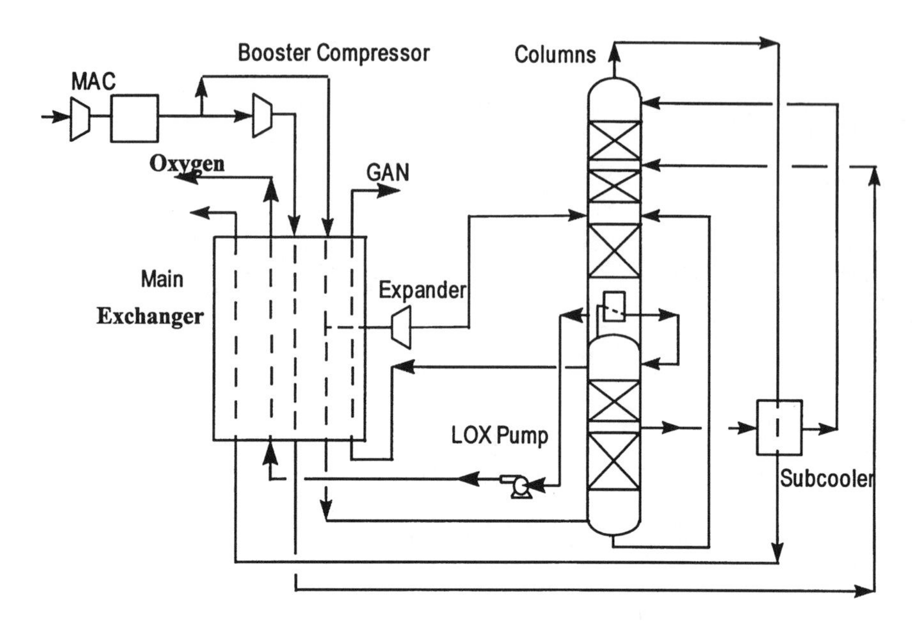

FIGURE 3.22 Nitrogen production from the HP column.

purity, the same HP nitrogen production would significantly impact the oxygen recovery. The production of HP nitrogen is therefore limited.

For large amounts of nitrogen, it is necessary to add significantly more equipment, as illustrated in Figure 3.23. Stages are still needed in the HP column to generate a pure reflux stream which in turn is used to reflux the LP column to produce a pure vapor stream. This requires the addition of a section of stages in the LP column, in addition to extra piping and subcooler and MHE headering and manifolding. Even with this extra equipment, nitrogen production can still be attractive.

The decision to produce nitrogen from the HP column vs. the LP column is an optimization between the power saved in nitrogen compression vs. the power increase in the main air compressor due to loss of oxygen recovery. Typically for lower-purity oxygen plants, nitrogen amounts less than 20% of the airflow can be optimally removed from the HP column. Above that, the nitrogen production is switched to the LP column. For large productions of nitrogen with the standard LP cycle, it is feasible to produce up to 65% of the air as nitrogen. This can be easily demonstrated by referring to the Table 3.7 material balance. A minimum waste flow normally needed to regenerate the adsorption system is 15% of the airflow, which leaves the maximum nitrogen molecules available at 65% of the airflow, assuming distillation capability exists. For the LP cycle, the reflux is available to purify these flows. A pumped LOX design does not have the capability to generate sufficient reflux from the HP column to purify this amount of GAN since about 30% of the air is bypassed around the HP column as a liquid airstream from the pumped LOX system. The vapor flow up the HP column is therefore reduced significantly, as is its ability to generate reflux. The normal maximum production of nitrogen would be approximately 45% of the airflow. In many applications this is not necessarily a

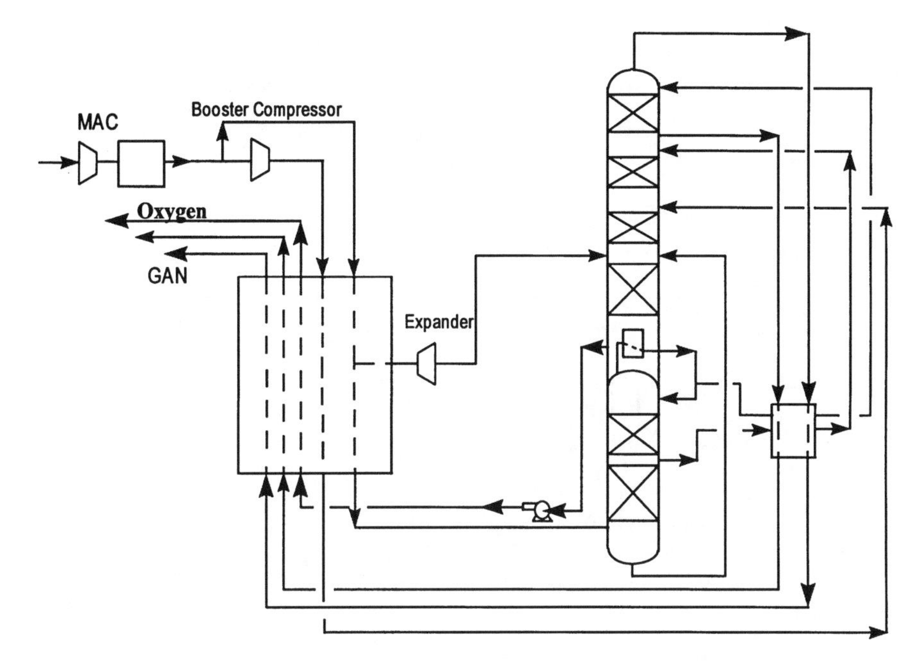

FIGURE 3.23 Production of LP nitrogen.

TABLE 3.7
Material Balance for LP GAN Production

	Pressure psia	Temp. °F	Enthalpy Btu/mca	Btu/lb-mol	Flow (m/mca) N_2	Ar	O_2	Total lb-mol/h
LP Air to coldbox	**100**	**50**	**–211**	–211	0.7812	0.0093	0.2095	1.0000
GOX from coldbox	**20**	45	**–47**	–230	0.0000	0.0000	0.2050	0.2050
LOX from coldbox	**23**	**–300**	**–5.6**	–5598	0.0000	0.0000	0.0010	0.0010
LP GAN from coldbox	**16**	45	**–147**	–229	0.69500	0.0000	0.0000	0.6500
Waste from coldbox (with GOX)	**19**	45	**–33**	–226	0.1312	0.0093	0.0035	0.1440
Heat leak			**20**					
Refrigeration loss			**41**					
Turboexpander			**41**	**600**				**0.069**

problem. A pumped LOX plant producing 1000 t/day of GOX would still be able to produce 2000 t/day of GAN.

3.2.1.2.4 Coproduction of argon

Argon is also a very valuable coproduct of oxygen. The capital requirements to produce argon are not trivial. Figure 3.24 is a flow sheet of the familiar LP cycle with the addition of a crude argon production column. As discussed in Section 3.2.1.1.4, the boiling point of argon is close to that of oxygen but significantly warmer than

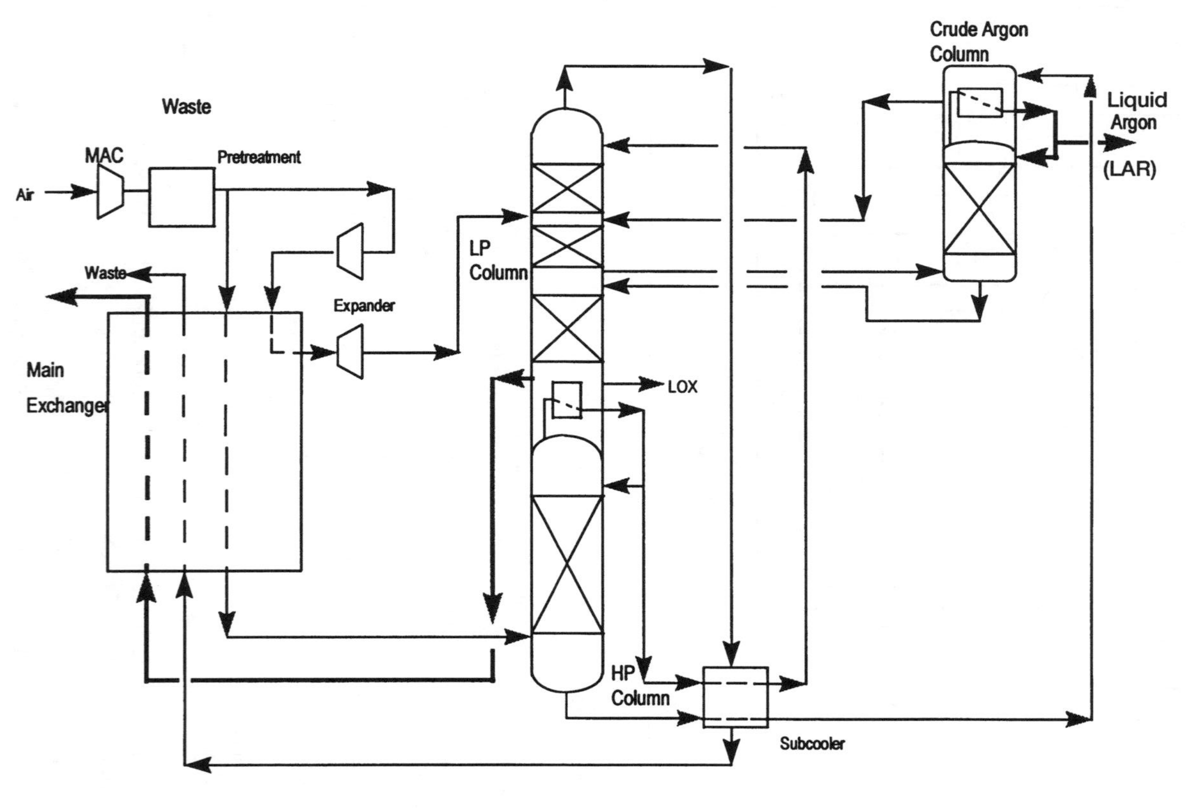

FIGURE 3.24 Production of argon.

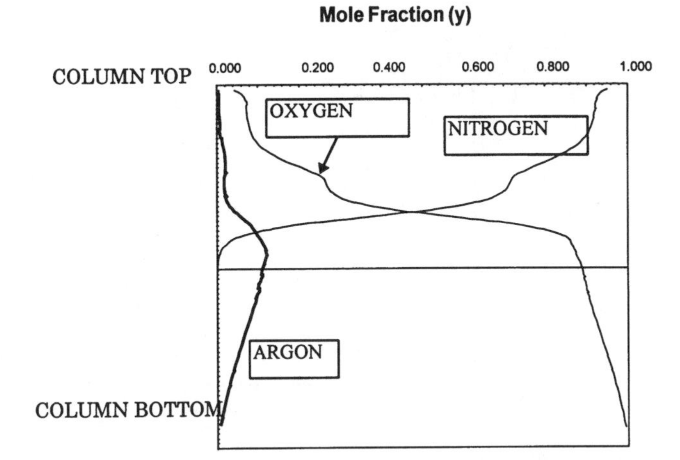

FIGURE 3.25 LP column composition profile.

that of nitrogen. The composition profile of a typical LP column is shown in Figure 3.25. At the top of the column, the composition is almost 100% N_2. At the bottom of the column, the composition is almost 100% O_2. The argon composition is low at both ends, but peaks toward the bottom of the column. This is the optimum feed point location to a side stripping column with maximum argon, but close to zero nitrogen content. The feed to the crude argon column typically has a composition of 90% O_2 and 10% Ar with nitrogen in the low ppm levels.

In the crude argon column, the argon/oxygen vapor feed becomes richer in argon as it passes up the column since argon is the slightly more volatile component. With sufficient stages, the crude argon purity will increase to 97% Ar, or better. The reflux for the column is provided by a condenser at the top of the column which uses the oxygen-rich liquid from the bottom of the HP column. This liquid contains approximately 65% N_2, so at low pressures it is cold enough to condense the argon-rich stream at the top of the column.

One of the concerns in operating a crude argon column is the possibility of significant quantities of nitrogen entering the column as a result of poor operation. The nitrogen, being the more volatile of all the components, will concentrate at the top of the column and form a noncondensable mixture. The result of this is a total disruption of the column operation. It is therefore very important to control the LP column composition profiles to prevent nitrogen ingress.

Commercial-grade argon has maximum allowable O_2 and N_2 levels in the ppm level; therefore, the crude argon needs to be further purified. The most common final purification of the crude argon is performed by a combination of catalytic oxygen removal using hydrogen over a Pt/Pd catalyst and distillation for the argon nitrogen separation. In recent years, column technology has improved with the addition of efficient structured packing to replace trays. This has allowed distillation to be used to remove oxygen down to ppm levels as a replacement for the catalytic systems. Prior to this time, the pressure drop required by distillation trays had made this option not viable since the pressures at the top of the column would have been under vacuum.

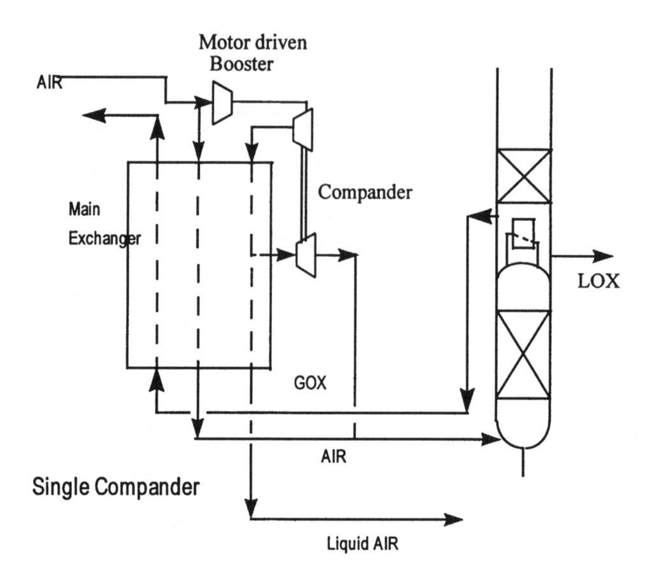

FIGURE 3.26 Simple air booster.

Pure argon production by all distillation technology has the advantage in that it does not require hydrogen, which poses some safety concerns with O_2/H_2 mixtures and incurs a long-term operating cost as a utility. All distillation is, in general, more capital intensive but is attractive when long-term H_2 costs are included.

3.2.1.3 Production of Liquid Oxygen

The section on refrigeration generation showed examples illustrating the impracticality of producing significant quantities of liquid from an ASU with LP expanders. Even before the practical limit of 3% of the airflow as liquid, it is not desirable to run the ASUs at these nonoptimum liquid production rates because of the impact of the high expander flows on column performance. This is especially true when argon production is necessary. Sometimes, however, even though continual operation in this mode is not desirable, it may be necessary to design the plant to do this for short periods of time to fill backup tanks when production demands are down.

If a plant has to be designed for constant operation at high liquid production, another method of refrigeration generation has to be utilized. A simple system is shown in Figure 3.26. A booster compressor has been added to the air compressor which allows refrigeration to be generated by a high pressure expander which can expand down to the HP column feed. In this way there is minimum loss of column boil-up. This approach may be attractive when a booster compressor is already required, for instance, with a pumped LOX system.

For large quantities of liquid, a liquifier is preferred. This is similar to the booster system discussed previously but designed to make a higher percentage of the air as liquid product. The principles of this type of system are outlined in the diagram in Figure 3.27. The basic blocks are very similar to a standard ASU with air compression, heat exchange, and separation by cryogenic distillation. The difference is that

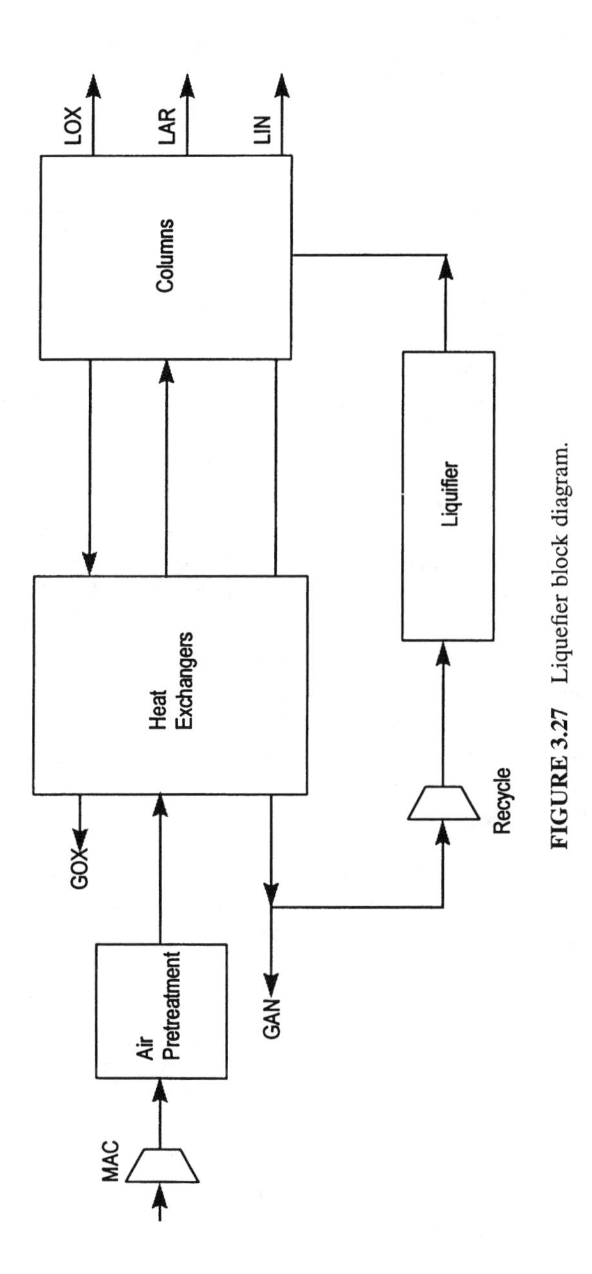

FIGURE 3.27 Liquefier block diagram.

a significant quantity of liquid products is required. The refrigeration to allow liquid production from the column system is provided from the nitrogen liquefier. GAN produced from the ASU is compressed and liquefied in the liquefier. The LIN then provides refrigeration to the column system.

Figure 3.28 shows the typical configuration of a nitrogen liquefier. LP nitrogen from the ASU is compressed in a makeup compressor and fed to a recycle system. The recycle nitrogen is boosted by a large recycle compressor and is further boosted by companders to a pressure which is typically in the 700 to 800 psig (4800 to 5500 kPa) range. Refrigeration is generated at two levels by expanding the HP nitrogen down to the recycle compressor feed pressure. The warm expander provides cooling for the feed to the cold expander and the remaining HP nitrogen stream. The cold expander cools the remaining HP nitrogen stream exiting the exchanger which is flashed to a separator running at the recycle suction pressure. Flash gas is recovered into the recycle system and liquid produced and fed to the ASU.

The ratio of recycle flow to liquid production is on the order of 5 to 7 depending upon liquefier efficiency. Therefore, the production of liquid is very power intensive. It requires about 400 kWh to liquefy 1 ton of LIN.

There are two different types of liquefier plants. The first type is dedicated to only liquid production. This plant would have its own ASU dedicated to making all liquid products for merchant liquid needs. The second type of liquefier is piggybacked onto an ASU that is mainly providing pipeline gas supply. The synergies of this type of "piggyback" liquefier are very attractive. An ASU is already present to provide separation of the air but may need to be larger to provide additional oxygen as LOX. Nitrogen is typically already available. The advantage to gas customers is that a large supply of liquid is available to them for backup purposes.

The advantage of a stand-alone liquefier is that it can be located in an area that is optimum for low power cost and close to markets.

3.2.1.4 Product Backup

ASUs are typically very reliable with plant onstream factors higher than 98% including scheduled outages. This is mainly due to many years of operating experience, a good preventive maintenance program, and good plant design practice. Scheduled maintenance on the compressors generally constitutes the bulk of the plant outages.[6]

Sometimes the processes consuming oxygen cannot afford the impact of a loss of oxygen, even an infrequent one. One example of this is the steel industry. A loss of oxygen to the basic oxygen furnaces (BOF) can have significant cost impact on steel production. In circumstances such as these it is wise to invest in a liquid backup system. The basic system, as illustrated in Figure 3.29, consists of an LP oxygen storage tank of sufficient capacity, a liquid pump, and a vaporizer. On a loss of plant production, the pump and vaporizer system is started. It can take 20 min to start up the system and regain pipeline pressure. If the continuity of product supply is critical, instantaneous backup can be added. This typically consists of an HP liquid tank running at a pressure just above pipeline pressure. On loss of product pipeline pressure, oxygen is automatically introduced into the line from the instantaneous

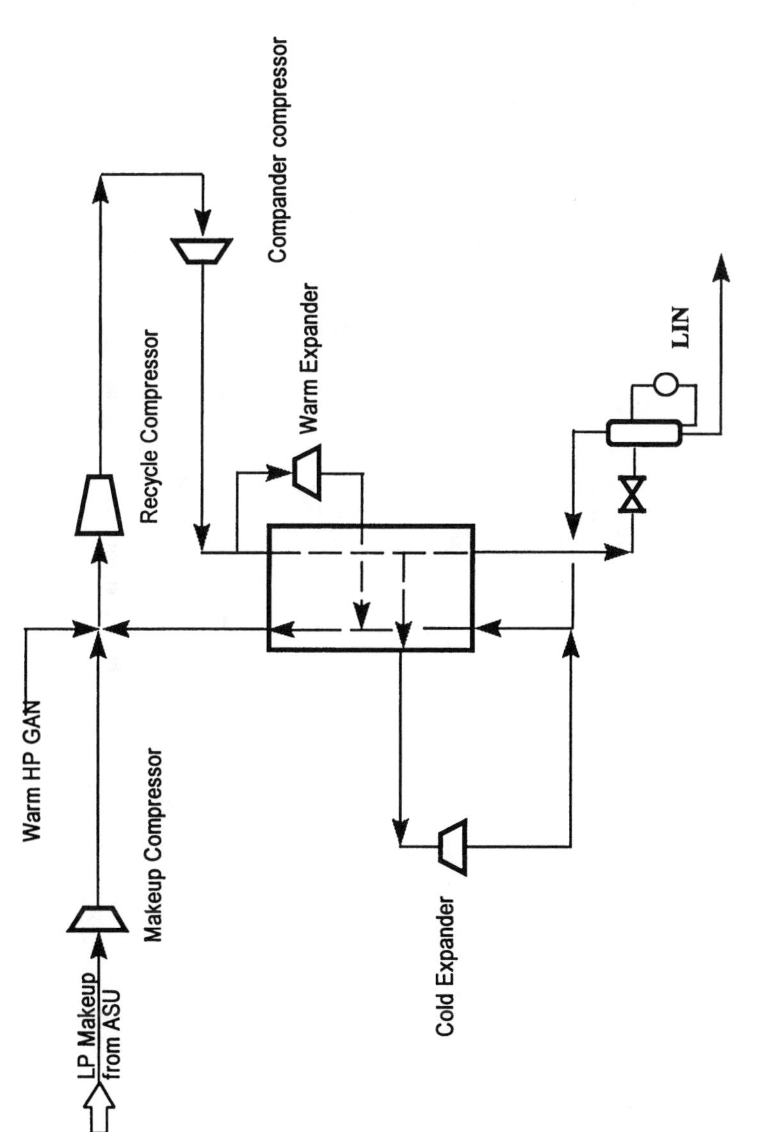

FIGURE 3.28 Simple two-compander liquefier.

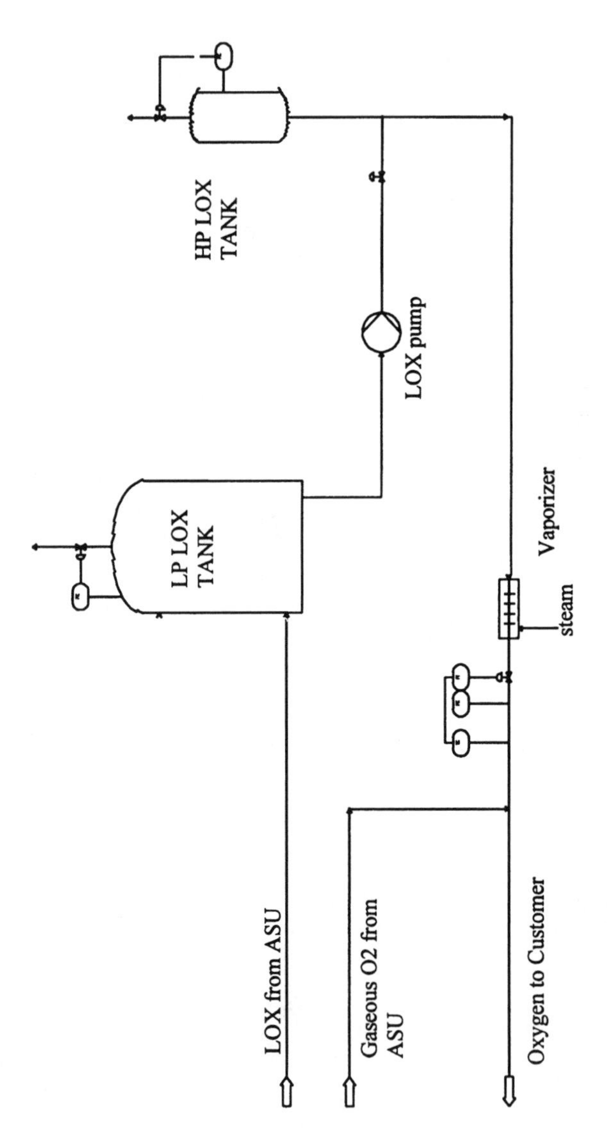

FIGURE 3.29 PFD of storage and backup system.

backup system. Typically, a 30 min to 1 h supply of available liquid is necessary. For a smaller plant, all the liquid backup could be stored at high pressure.

The LP storage tank is typically designed to provide 24 h of product backup. To fill or maintain level will incur a refrigeration penalty. As discussed in Section 3.2.1.1, this refrigeration penalty can be offset by producing liquid at periods of low oxygen demand or by adding external liquefaction capability.

The method of vaporization and the heat source will be a function of what utilities are available. For small quantities of LOX, ambient vaporizer systems are practical. For larger amounts, steam, electric, or natural gas heaters may be used.

3.2.2 Equipment Used in the Cryogenic Production of Oxygen

3.2.2.1 Compressors

3.2.2.1.1 Main air compressor

For most air separation applications the main air compressor can be an integral gear centrifugal compressor. The integral gear design consists of a motor-driven driveshaft connected to a large bull gear. The bull gear, in turn, drives pinions connected to the compressor wheels. The function of the pinion is to increase the wheel speeds closer to optimum aerodynamic values. For a three-stage compressor, two pinions would be required. The first- and second-stage wheels are driven by the first pinion and the third stage by the second pinion. Thus, the first and second wheels run at the same speed. The components of an integral gear compressor are illustrated in Figure 3.30. Increasing the number of compressor stages beyond four requires the addition of a third pinion.

Oil, used for lubricating the gears and bearings, is separated from the compressed airstream by labyrinth seals. Shown in Figure 3.31, the seals consist of two separate components, an air seal and an oil seal. In between the two, the seal is open to atmosphere. Any oil bypassing the oil seal leaks to atmosphere. The air seal is always at a higher pressure so that any leaks are always from air to atmosphere and oil cannot enter the process fluid. Oil is typically provided by a shaft-driven pump with a standby motor-driven pump available.

The operating curve of a typical centrifugal compressor is shown in Figure 3.32. As the discharge pressure of the compressor increases, the flow reduces (point 4 to point 3). If the pressure increases sufficiently, the compressor will not be able to generate flow and will enter a surge in which the air will attempt to flow backward through the compressor (point 2). This condition needs to be avoided to prevent machine damage. This is particularly necessary when feeding a system with a variable feed pressure such as an ASU.

To prevent compressor surge, a surge control system is needed. These control systems generally consist of an automatic controller which actuates a control valve at the machine discharge, venting air to atmosphere to maintain flow through the machine in the event that the process cannot accept it. A surge control system can vary from a pressure control for a fixed discharge pressure system to a sophisticated surge mapping control in which the flow and pressure are compared with a

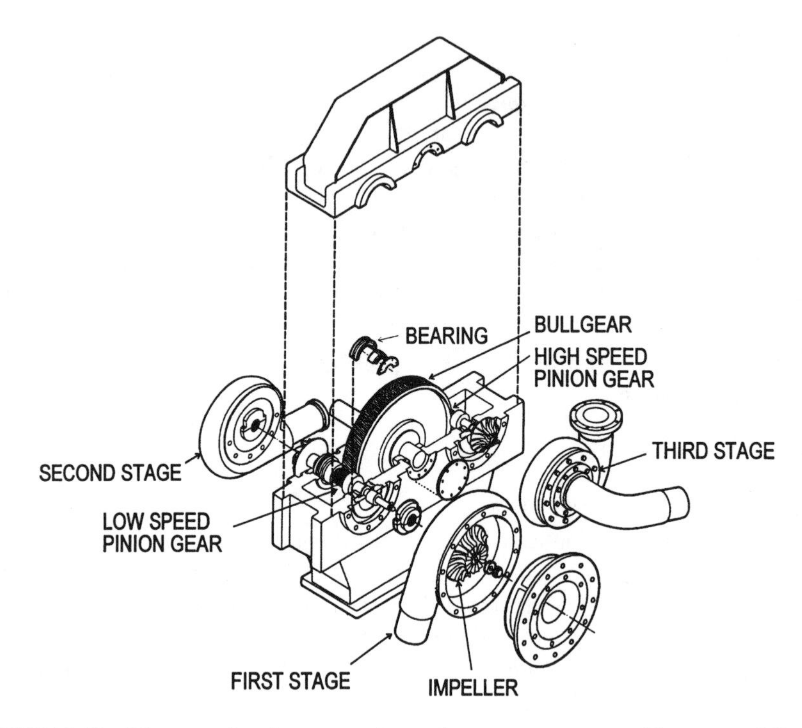

FIGURE 3.30 Blowup of a three-stage integral gear compressor. (Courtesy of Cooper Industries.)

programmed surge curve and adjusted to keep away from surge if the discharge pressure is variable. In Figure 3.32, the antisurge line would be used, as control set points provide a safe distance away from the surge line. At the pressures and flows of point 2, the discharge vent valve would open to prevent operation closer to the surge line.

Integral gear compressors have a very wide operating range from an equivalent of 30 to 3000 t/day oxygen production. Figure 3.33 shows the installation of a large integral gear compressor at an air separation facility. For plants higher than 3000 t/day, axial/centrifugal compressors are generally used, and screw compressors are used for the low end.

3.2.2.1.2 Product oxygen compressors

Depending upon the capacity and pressure requirements, oxygen compressors will either be centrifugal or reciprocating. The selection between the two is determined by discharge volumetric flow on the final stage of compression, which affects wheel size. This is a factor of both mass flow and discharge pressure. A simple rule of thumb to gauge the breakpoint is that the discharge pressure (in psig) cannot be higher than its capacity in tons per day. Thus, a 200-t/day oxygen compressor can be centrifugal below 200 psig. As the technology of the centrifugal compressors changes, this number changes, but it can be used as a first-pass approximation.

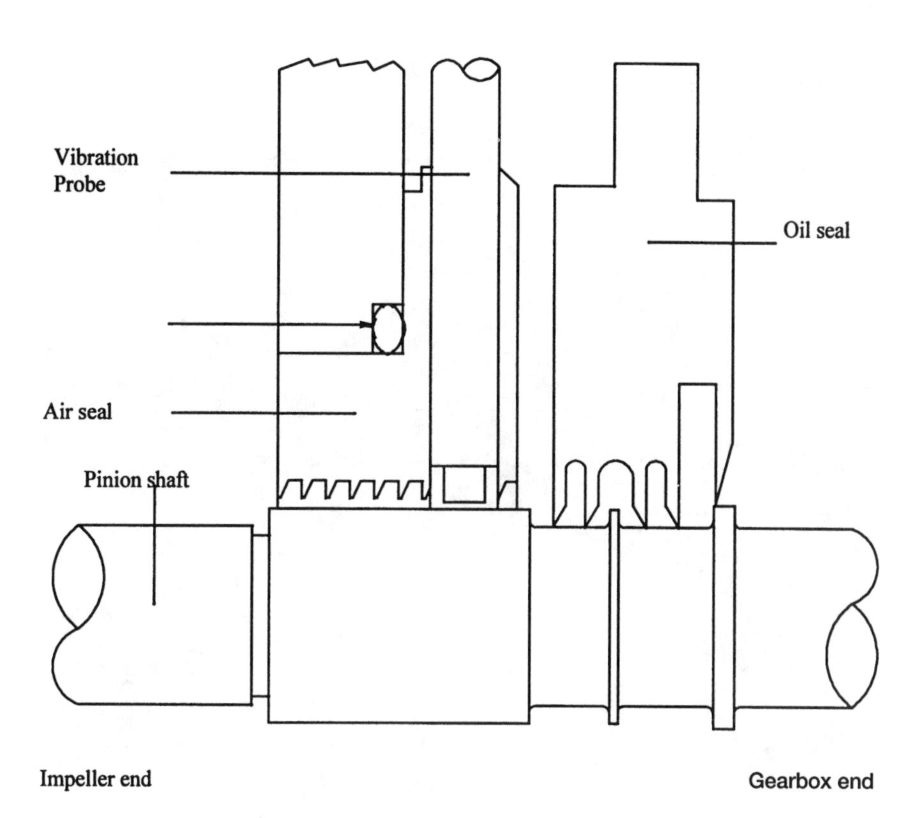

FIGURE 3.31 Labryrinth seal arrangement. (Courtesy of Cooper Industries.)

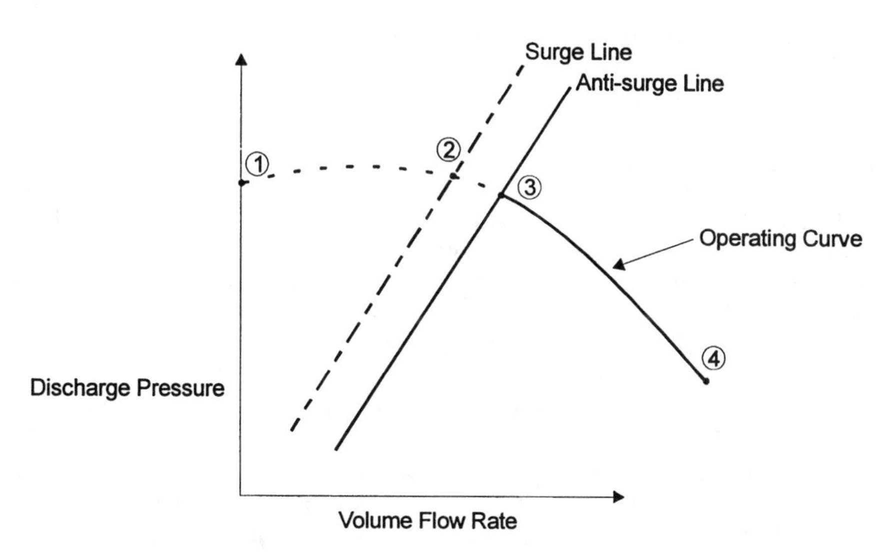

FIGURE 3.32 Centrifugal compressor performance curve.

FIGURE 3.33 Photograph of a compressor installation. (Courtesy of Cooper Industries.)

Oxygen compressors are generally located within a building that contains an external barrier. This barrier provides some containment in the rare event that an oxygen fire takes place.

3.2.2.2 Front-End Adsorption Systems

There is a significant optimization required to select the best system design for a particular customer requirement. The factors that need to be considered are the plant size, TSA regeneration energy type (steam, electric, natural gas), power costs, equipment size breaks, and the availability of waste gas. The DCAC/waste tower combination is generally justifiable for large plants due to the savings in energy consumption and costs of the adsorbers. For a plant needing a high product recovery as a percentage of air feed, a DCAC/waste tower combination is not practical since waste gas flows would be low. In those circumstances, it would be desirable either to have an in-line refrigeration system or to design the adsorbent beds for warm operation. The final selection depends on utility costs and equipment size breaks.

Vessel sizes are determined by the airflow processed. For small plants, two-bed vertical vessel systems are appropriate (see Figure 3.46). As the plant size and airflow increases, the required vessel diameter increases until it reaches a shippable limit at

FIGURE 3.34 Four-bed temperature swing adsorber installation. (Courtesy of Chemical Design Inc.)

about 14 ft diameter. This shipping limit is very much a factor of shop and plant location, but is a good rule of thumb. Above this diameter, choices are to use a multiple vertical bed design, use horizontal beds, or use radial bed designs.

Multiple vertical beds are commonly used in the U.S. Figure 3.34 shows a four vertical bed design. As an alternative to multiple vertical vessels, two horizontal bed designs[11] are competitive. A horizontal vessel has its diameter in the vertical plane, but with the adsorption bed oriented longitudinally. Air flows upward through the bed (Figure 3.35). The cross section of the bed is now a function of diameter and length of the vessel. When the shippable limit of diameter is reached, the vessel can be designed for a larger capacity by increasing the vessel length. The third choice is a radial or concentric bed design in which the bed(s) are contained within concentric screens.[8,11] Air flows into the bottom of the vessel and radially through the beds. This type of bed is good for large capacities, requiring small plot area. Regeneration energy is also more effectively used in this design with less heat loss through vessel walls. The design by nature is complicated and only appropriate for larger plants.

The choice of energy supply for regeneration is a strong function of available utilities. Electrical heaters are generally the lower capital option, but their long-term operating costs are high. If steam is available and dependable, it is a good choice for regeneration heat. The heaters themselves are generally more expensive but justifiable. Another excellent heat source is natural gas which, if available, has a reasonable utility cost and moderate capital investment. Typically, for very small plants, electric heaters are used since the absolute power cost is low and the capital savings are attractive. For most applications, steam or natural gas would be selected

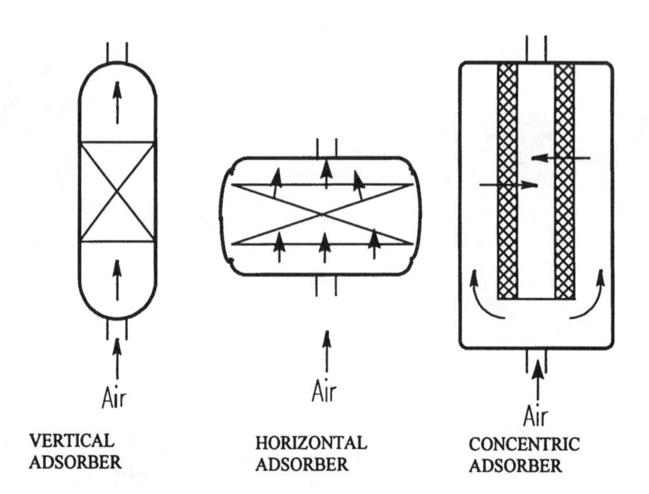

FIGURE 3.35 Adsorber bed options.

if available. Energy levels can be reduced by utilizing heat recovery systems using compressor energy.

In recent years, designs have been used that do not require heat energy and depend only upon pressure swings to desorb water and CO_2. Pressure swing adsorbers (PSAs) require more adsorbent and regeneration flow than TSAs but save on long-term operating costs.

3.2.2.3 Heat Exchangers

3.2.2.3.1 Main heat exchanger

The main heat exchangers, as discussed in Section 3.2.1.1.2, are exclusively brazed aluminum heat exchangers (BAHX) in counterflow service with air cooling against warming product and waste gas streams. In a BAHX the surface area available for heat transfer is greatly increased by the use of finning on all streams. The corrugated fins are sandwiched between plates to form a passage through which the gas flows. These heat exchangers are also called plate–fin heat exchangers or cores. As illustrated in Figure 3.36, the fins are held in place with side bars. A passage is completed with the addition of distributor fins that serve to introduce gas to and from the heat transfer zone. The function of the distributor fin is to spread the gas across the width of the passage to ensure even temperature profiles. The distributor fins are arranged such that they exit or enter the heat exchanger on the sides or top in an orientation that is specific to the stream being processed. In this way, all streams will enter the heat exchanger at a specific and consistent position. For instance, all the air passages will enter the core at the top right-hand corner and the waste all exit at the top left-hand corner. Each completed passage is stacked to form the heat exchanger assembly. The stacking is arranged such that cooling and warming streams are evenly distributed throughout the core. For instance, the airstream will have a warming stream on either side of it (Figure 3.37). When fully assembled, the exchanger is brazed in a saltwater bath or vacuum furnace.

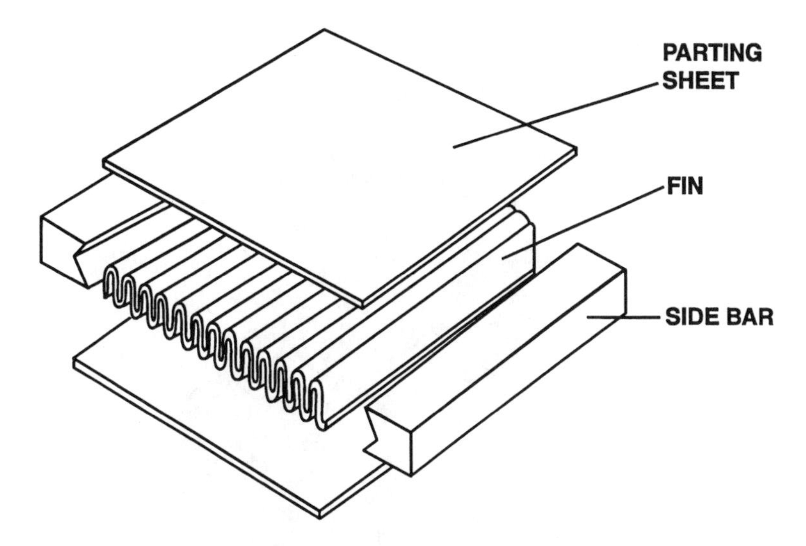

FIGURE 3.36 Plate-and-fin arrangement.

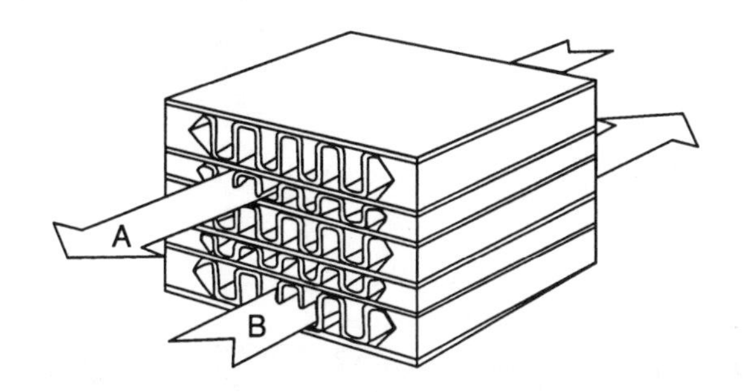

FIGURES 3.37 Passage fin layout.

The heat exchanger core assembly now consists of a block of stacked passages with the distributor openings visible (Figure 3.38). The gases are collected from the passages into a header welded over the openings and piped away. A main exchanger assembly may consist of many individual cores manifolded together. Figure 3.39 shows manifolded core assemblies contained within a coldbox frame.

The maximum size of an individual core is a function of the specific manufacturer's processes but is typically 4 ft wide by 4 ft stack height by 20 ft long (1.2 × 1.2 × 6 m).

3.2.2.4 Expanders

As discussed in the refrigeration section, Section 3.2.1.1.3, an expander is used to create cold temperatures in the process flow stream by the direct removal of gas enthalpy. HP gas is expanded through efficient, aerodynamically contoured nozzles, followed by passage through a rotating radial inflow wheel. Inlet gas pressure energy is converted to kinetic energy in the stationary nozzles. This occurs due to a significant

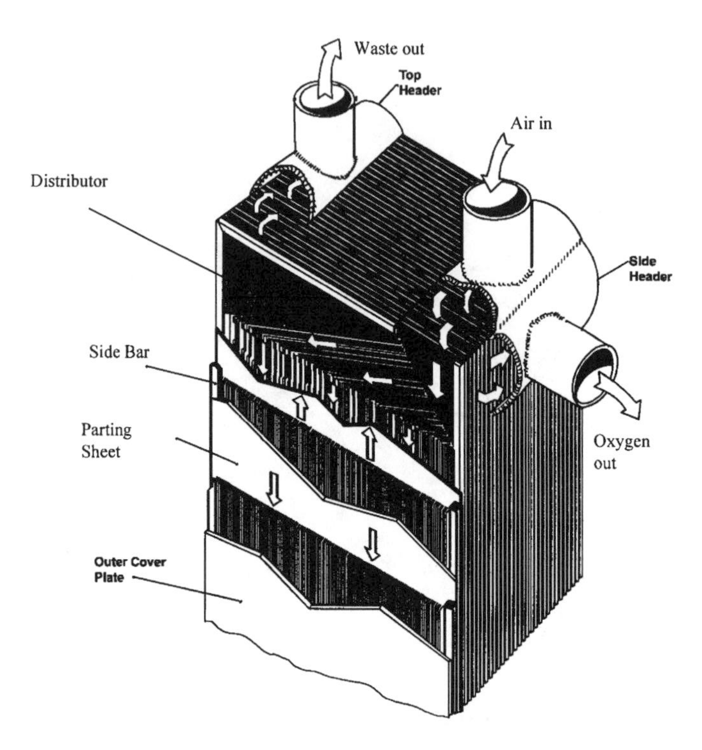

FIGURES 3.38 Core assembly model. (Courtesy of Air Products.)

FIGURE 3.39 Manifolded core assembly. (Courtesy of Air Products.)

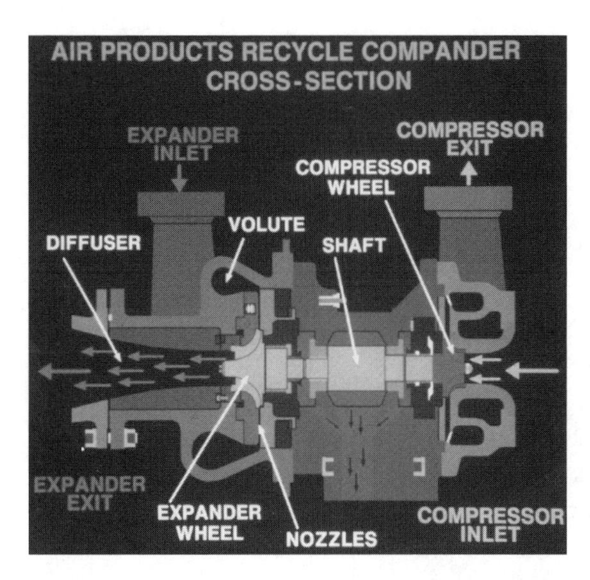

FIGURE 3.40 Compander cross section. (Courtesy of Air Products.)

increase in gas velocity. Gas velocity can be near or at sonic velocity. Because gas velocity exiting from the nozzles is not radial to the wheel axis of rotation, but has a tangential component, this process creates a net angular momentum in the gas flow stream. Entering the outer diameter of the rotating expander wheel, the blades of the wheel reverse the rotational direction of the gas, opposite to that imparted by the nozzles. This removes the angular momentum created by the nozzles. The gas exits the wheel and the machine with no angular rotation or momentum. The redirection of the gas flow in the wheel requires force to be supplied from the shaft to each blade of the wheel. This process creates shaft torque, which as multiplied by the rotational speed of the shaft results in shaft power. Having imparted this power to the shaft, the gas exits the expander at a lower enthalpic condition. This results in both a pressure and temperature reduction, as compared with the expander inlet. The shaft power created by the expanding gas must be transferred to an alternative medium to keep the net power balance of the expander rotor at zero, for no mechanical energy can be stored in the rotor. The alternative medium can directed to use this energy productively or to dispose of it as waste heat.

One method of expander power recovery is to apply a process stream as the alternative medium and compress it as may be required by the overall process cycle. This is extensively used in liquifiers. This *comp*ressor powered by an exp*ander* is sometimes called a *compander.* Figure 3.40 shows the cross section of a compander. HP gas enters the expander volute casing on the left, passes through the nozzles, enters the outer diameter of the expander wheel, and exits to the left. The compressor wheel (on the right side) draws gas into the compressor case and then into the eye of the wheel. The wheel imparts angular momentum to the gas via a kinetic energy increase. The gas discharges outward into the diffuser section and then the voluted compressor casing. The diffuser and volute casing remove angular momentum by

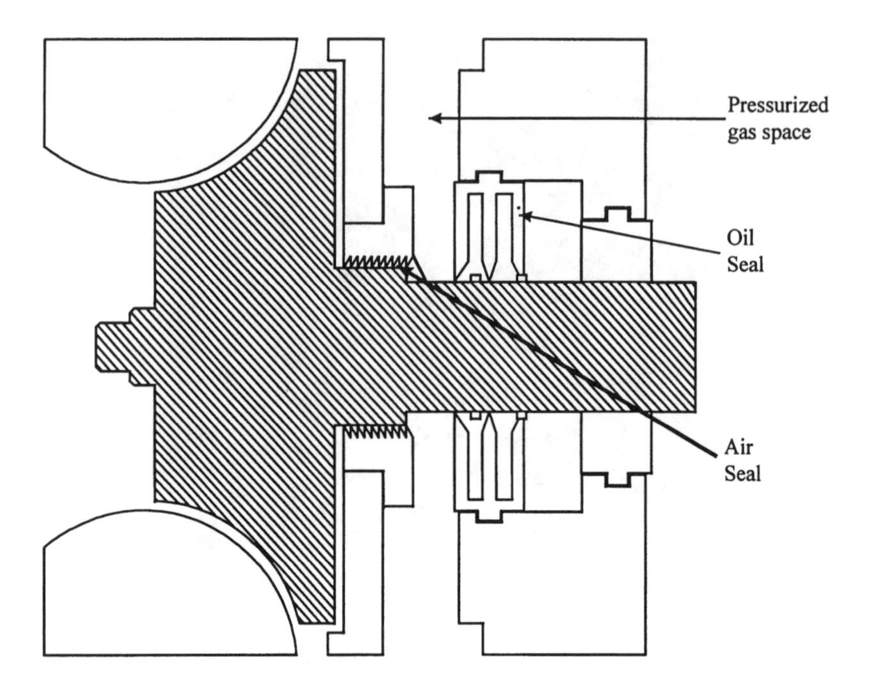

FIGURE 3.41 Seal arrangement. (Courtesy of Air Products.)

decelerating the gas, thus converting kinetic energy into pressure energy. The compressor operates nearly exactly opposite to the expander. A compressor raises the pressure and temperature of a gas flow stream.

Typically, the shaft is supported on oil-lubricated bearings. It is extremely important that oil from these bearings is not allowed to contaminate either the expander or compressor gas streams. On the expander end, oil could freeze in the process causing plugging of the heat exchangers downstream. Ultimately, it could reach the oxygen-rich LP column, creating a considerable safety hazard. Bearing oil is separated from both process streams with multiple group labyrinth seals similar to those described in Section 3.2.2.1. Seal gas injected into these seal groups at standby condition and/or during normal operation assures zero oil migration into the process streams. On the expander end, injection of warm seal gas further assures that cryogenic process gas will not enter the bearing area, causing freezing of oil in the bearings. A typical expander seal system is shown in Figure 3.41.

With advances in wheel, nozzle, casing, and diffuser design, expanders are reaching very high levels of isentropic efficiency. Values greater than 90% have been achieved. To obtain these levels of performance, expanders wheels need to be properly sized for the process duty and operate at the optimum speed. For small wheels, typically 3 in. in diameter, speeds of 70,000 rpm may be necessary. Large wheels operate optimally at lower speeds. Shaft rotational speed is inversely proportional to wheel diameter for similar process enthalpy drops. At these high speeds, bearing and seal designs need to be sophisticated. With improvements in bearing designs, the power transfer to the compressor has reached better than 97%.

FIGURE 3.42 Compander assembly. (Courtesy of Air Products.)

Other methods of removing power from the expander wheel are powering an electric generator, dissipation to an oil brake, or dissipation to an ambient air blower. The latter two methods have the advantage of low capital cost, but the ultimate disposal of expander power to waste heat provides no benefit to reducing overall plant power consumption. For this reason, dissipative methods are used typically for small plants where the level of plant power is low and the additional capital cost of a recovery loading mechanism is not worth the operating economic benefit. Small plant expanders also tend to operate at higher speeds, with less efficiency and a higher percentage of parasitic power loss by bearings and rotor windage. At small sizes, equipment for power recovery, such as gear sets to a generator, tend to be more expensive per unit of power recovered. Higher speeds also tend to reduce mechanical reliability.

Process compressors (companders) and geared generators are the typical means for expander power recovery in larger plants. Economics begin to favor a power recovery scheme at expander powers of 150 hp for generator loading and 75 hp for companders. From an operational perspective, generator-loaded expanders have more reliability and a wider flow range, while companders offer fewer parasitic losses in power recovery and can be tailored to maintain higher efficiencies over changing plant flow conditions. The adjustable speed of a compander can be used to maintain high efficiency in turndown or turnup conditions if the pressure drop across the expander also increases and decreases, respectively. A complete compander assembly is shown in Figure 3.42.

3.2.2.5 Distillation Columns

Distillation in general is a mature technology and trayed distillation columns have changed little in the last 15 years. Due mainly to the high product purity requirements

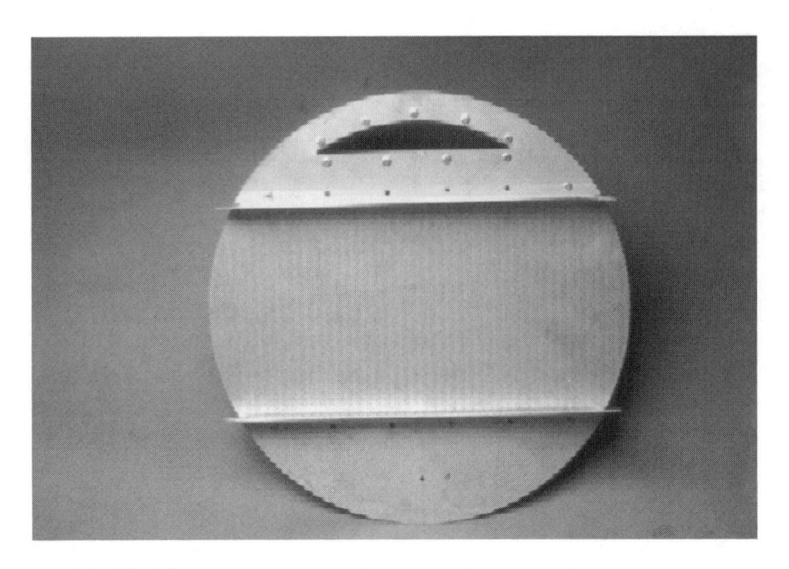

FIGURE 3.43 Simple cross-flow tray. (Courtesy of Air Products.)

and the difficulty of Ar/O_2 separation, the cryogenic distillation of air requires large numbers of distillation stages. This is specifically true when ultrahigh purity (UHP) nitrogen and argon are produced. Columns with over 100 distillation stages are not unreasonable. Due to these large stage requirements, ASU vendors have developed their own proprietary tray designs with the goal of maximizing efficiency and reducing overall costs. Sieve tray designs with many variations have been used,[9] from multipass cross flows, parallel flow, to circular flow. Each design focused on maximizing the distillation efficiency/tray pressure drop ratio. A simple cross-flow tray is shown in Figure 3.43.

Vapor passing through a sieve tray encounters several resistances to flow. The first is the pressure drop needed to pass through the holes. The second is the static head of liquid on the tray and resistances through the two-phase froth. The net result is a substantial pressure loss especially for columns with high stage counts. In the lost work analysis of an ASU presented by Thorogood,[2] the total column exergy loss is worth 20% of the total power of an ASU. Pressure drop is a significant portion of this loss.

The development of structured packing with better efficiency/pressure drop characteristics than trays is making it the preferred distillation medium for a wide range of applications in cryogenic air separation. Structured packing is formed by combining corrugated metal sheets, with the sheets arranged in such a way as to optimize the liquid distribution through the packing. Figure 3.44 shows a sample of structured packing banded together for a small column diameter. In structured packing, the liquid runs as a thin film along the corrugated sheets or rains between sheets. The vapor does not have the inherent resistance it would have on a sieve tray.

In the pressure balance shown in Figure 3.18, a column pressure drop of 2.5 psi (17 kPa) was used. In a packed LP column, the pressure drop would be of the order of 0.5 psi (3 kPa). The 2-psi (14-kPa) saving in pressure drop translates to a 6-psi

FIGURE 3.44 Structured packing photograph. (Courtesy of Sultzer Chemtech.)

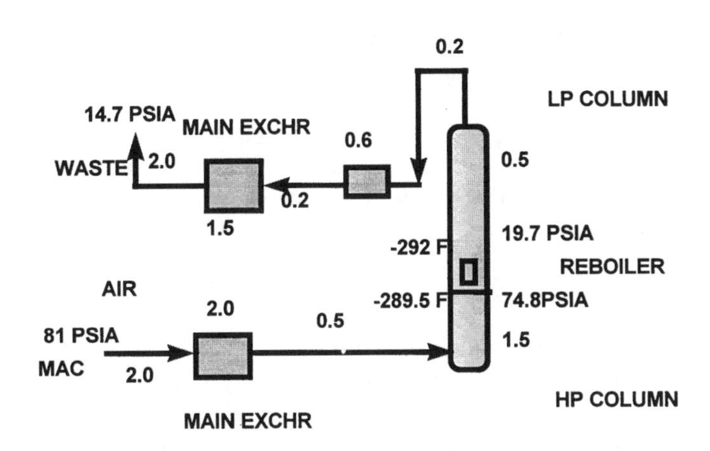

FIGURE 3.45 Pressure balance with packed LP column.

(41-kPa) drop in air compressor discharge, which translates to a 3% saving in power on this compressor (Figure 3.45). For large compressors this power saving can be considerable. For example, for an ASU producing 1000 t/day of oxygen, the air compressor power would be on the order of 10 MW. A 3% power saving would therefore be 0.3 MW. With a power cost of 5¢/kWh, the power savings would be worth $130,000/year. There is therefore considerable incentive to replace trays with packing for large plants.

Typically, distillation columns on an ASU dictate the height of a coldbox. Column heights can reach 150 ft (45 m) for columns making pure products and argon. Column diameters can vary from several inches to 20 ft (6 m) depending on the required capacity of the ASU. For low-purity oxygen, supply column heights are somewhat more modest but still a major cost component of the plant.

3.2.2.6 Coldboxes

As discussed in previous sections, it is important to minimize the heat leak into the cryogenic section of the ASU. This is accomplished by containing the equipment in an insulated enclosure termed a *coldbox*. The coldbox is a structural steel frame which is paneled with metal sheeting and provides structural support for the equipment and also containment for the insulation. There are essentially two different types of insulation material used in ASUs. The first and most popular is perlite. The perlite coldboxes are insulated in the field with perlite blown in from the top of the coldbox. The second insulation material is rock wool. Rock wool has to be packed manually.

The advantage of perlite is that its installation costs are lower since it can be blown in. The perlite is free running so provides a more even insulation. The disadvantage is that in the event of a leak inside the coldbox, the perlite has to be removed from the box to access the leak. Rock wool is more labor intensive to install, but has the advantage that leaks can be repaired by tunneling through the frozen rock wool. In practice, the perlite boxes are the most popular, with areas of potential leakage either encased in a local rock wool enclosure or designed for access from outside the box.

As an alternative to a structural rectangular coldbox, cylindrical cans can also be used as an insulation containment. Figures 3.46 and 3.47 show photographs of coldboxes and cans.

It is extremely important to keep the insulation material dry. Wet insulation loses its insulating properties and will ice up, becoming difficult to remove and possibly damaging lines. To keep it dry, a coldbox is continually pressured with dry nitrogen through a purge system.

3.2.2.7 Storage and Backup

There are two types of cryogenic storage tanks used for ASUs. The first is a vacuum-insulated shop-built tank. These tanks vary considerably in size. Small-capacity tanks in the range of 2000 to 10,000 gal (8 to 40 m^3) are used for customer stations where intermittent or low oxygen flows are needed. For instantaneous backup of product, larger tanks may be required. For instance, to provide one hour of HP backup of 1000 t/day of oxygen at 200 psia (1380 kPa) would require a storage tank of 13,000 gals (50 m^3) operating at a pressure of approximately 230 psia (1590 kPa).

For longer-term backup, larger sizes are needed and generally not maintained at high pressures since pumps can be used to supply the pipeline. The shippable limit of shop-built tanks is in the 50,000 gal (190 m^3) ranges. At low pressure 50,000 gals of LOX is equivalent to 216 tons. For longer-term backup or for storage of volumes generally needed for centralized merchant liquid supply, larger tanks are needed.

The larger tanks are site built with a stainless steel inner tank and a carbon steel outer layer. The annulus between the layers is filled with perlite insulation and purged with dry nitrogen gas, in a similar manner to coldboxes. Due to the large diameters,

FIGURE 3.46 Coldbox with two-vessel adsorber system. (Courtesy of Air Products.)

the maximum allowable working pressures are very low, generally below 5 psig. Figure 3.48 shows two large, flat-bottomed storage tanks for LOX and LIN.

Vaporizer systems are needed to vaporize and warm up the LOX to a temperature close to ambient to maintain supply to the customer. That liquid may be supplied from an HP tank or from pumps. The selection of vaporizer type is a function of the volumes of gas required and the utilities available.

The most reliable vaporizer type is the ambient vaporizer. Ambient vaporizers are essentially bundles of finned tubes with a cryogen on the inside of the tube and ambient air warming the outside fins. The pipe and fins are generally fabricated by extruding aluminum with the fins running the length of the tube and penetrating radially out from the pipe. Figure 3.49 shows a cross section of a typical tube. An

FIGURE 3.47 Insulated cans. (Courtesy of Air Products.)

FIGURE 3.48 Flat-bottom storage tanks. (Courtesy of Air Products.)

FIGURE 3.49 Sketch of ambient vaporizer fin.

FIGURE 3.50 Ambient vaporizers. (Courtesy of Air Products.)

ambient vaporizer is not dependent upon the supply of utilities and, if sized correctly, needs little maintenance.

Ambient vaporizer tubes are set into modular arrangements for easy shipping and construction. The photograph in Figure 3.50 shows an arrangement of many modules. For large capacities the plot area required to situate these modules becomes prohibitive and different types of vaporizers need to be considered. When long-term continuous vaporization is required, ice buildup also becomes a problem and operator intervention or an automatic defrosting system is needed. In locations with very cold ambient temperatures, electric trim heaters may be necessary to keep oxygen temperatures warm enough to prevent problems with carbon steel pipe.

In circumstances when ambient vaporizers are not suitable, steam or natural gas vaporizers are used. The steam is typically directly injected into a water bath maintaining the bath temperature. Tubes are arranged within the water bath. As the cryogen is vaporized, the water temperature drops and steam is automatically injected to maintain bath temperature. An alternative type of steam vaporizer has tubes within a high pressure shell containing steam. The steam is condensed and removed in steam traps. This direct steam vaporizer is generally lower cost but produces a variable product outlet temperature. On loss of steam, there is also an immediate loss of product backup. A water bath has significant ballast. Natural gas is used in a similar manner and maintains the temperature of a water bath. The selection of steam or natural gas is a function of availability.

An alternative method of instantaneous backup for HP oxygen is to use warm gas bottles which operate at pressure higher than line pressure and are let down in pressure to feed the product line when pressure drops. Large gas bottles are also used to provide inventory when the usage requirement for oxygen is very irregular. For short periods of high gas flow, the pressure in the bottles is reduced and rebuilt during low-gas-demand periods. This system is suitable for rapid and short duration changes in flow. For longer-duration changes in flow, the ASU will need to operate at different rates.

3.2.3 Factors in Plant Optimization and Selection

When a requirement for oxygen has been identified, several factors need to be considered. In addition to the required flows and purity needs, the user needs to consider either a purchase of equipment to supply the product or an on-site purchase of gas.

If purchasing equipment with the desire to own and operate the plant, several issues should be decided. What is the average oxygen requirement? What is the peak requirement? Is the usage requirement steady or fluctuating? What purity does my process need? Do I need to generate liquid for backup purposes? What is the level of reliability needed? What is the long-term cost of power? In addition to this, major issues such as land, plot plan, staffing, utility supply, and finance need to considered, among others. If supplying a plant with an intent to own, operate, and supply gas to a customer, the owner needs to also consider the value of coproducts in the local market.

3.2.3.1 Power Rates

Power as a percentage of total oxygen cost varies with plant size, oxygen purity, and the cost of power. Klosek et al.[4] showed the typical impact of plant size on the ratio of power costs to total costs (Figure 3.51), and Scharle and Wilson[6] also showed a breakdown of nonutility costs. As the plant size increases, the capital component increases in a nonlinear way. The standard cost vs. size relationship is $C_n = R^{0.6}C$,[10] where R is the ratio of plant capacities, C_n is the new cost, and C is the old cost, whereas power has an exponent much closer to unity. As plant size increases, power assumes a larger percentage of the product cost. An air separation plant is by nature

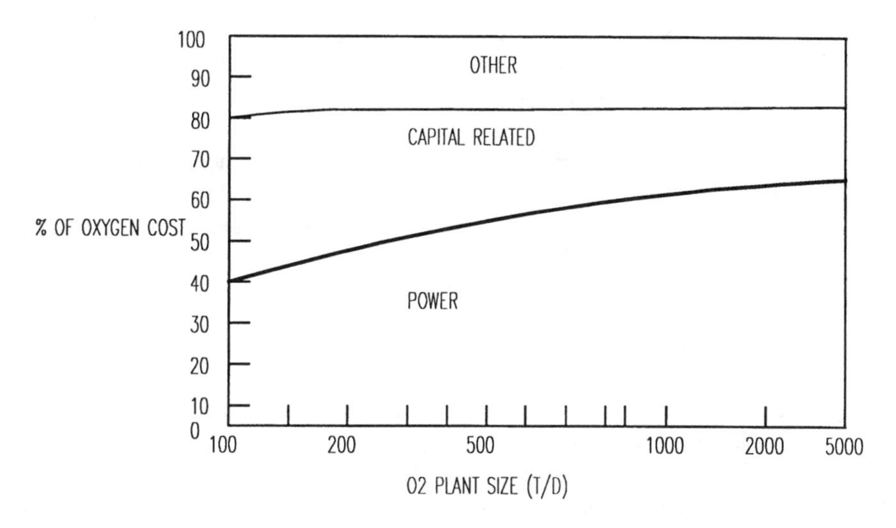

FIGURE 3.51 Components of product cost.[4] (Courtesy of Air Products.)

power intensive. Consideration should be given to the cost of power consumption for all plant sizes, but especially for large plants.

In the large plant range, power is 60% of the total cost. It is extremely important to ensure that the correct optimizations are performed and that trade-offs between capital and long-term power costs are evaluated. For large plants, it is appropriate to invest more in capital to save power. This investment may be in the form of larger equipment with lower pressure drops or, if appropriate, the use of more-sophisticated and complex plant cycle designs with more equipment (Section 3.2.3.3). For smaller plants, when power is 40% of the oxygen cost and at a much lower absolute value, the benefits from saving power may not justify significant capital investment.

Power rate structures vary from area to area. A common rate structure is a variable one with high power costs during high-demand periods and lower costs during the lower-demand periods. In these circumstances, it may be appropriate to shut noncritical equipment down or turn down equipment and supplement from backup liquid. The power rate contract should be understood and the equipment optimized with this in mind.

3.2.3.2 Value of Coproducts

Since coproduction of nitrogen and argon typically do not require significant power consumption increases other than nitrogen compression, they are attractive. The cost to produce high-purity argon is significant but can generally be justified when a pure oxygen product is required. To produce reasonable quantities of argon the product oxygen should be at least 99.5% O_2. At purities less than 99.5% the main contaminant is argon and valuable product is lost. If the required oxygen purity is 95% or less, the cost of argon has to include cost and power adders to bring that purity up to 99.5% O_2. Even with these additions, if the market value of argon is good, the production of argon is viable for large plants. There are also optimizations on capital investment vs. product recoveries. Investment for small amounts of nitrogen is not significant, whereas a large nitrogen requirement substantially changes the ASU (Section 3.2.1.2.3).

The addition of liquid product is another factor that can increase the value of an air separation facility. If the market needs are such that liquid product is required and the investment is attractive, a piggyback liquefier and storage capacity may be added. An analysis of the capital required and the value of the coproducts should be included in an overall facility optimization.

3.2.3.3 Product Purity

Many papers have been written documenting the effect of oxygen purity on power.[2,5,6] Typically, with a pumped LOX cycle or LP cycle (as described in previous sections), the oxygen specific power is improved as purity is reduced. Below 95% O_2, however, this benefit is reasonably constant. This is especially the case if the product pressure requirement is high. The product compression power (or booster power) would now include (or have to vaporize) the impurities in the oxygen.

For 95% O_2 production, more efficient cycle designs are available using multiple reboiler designs that can further improve the specific power of oxygen. Klosek et al.[4] documented potential power savings of 10% below the LP cycle powers for oxygen purity requirements below 95% O_2. These designs would typically require higher equipment costs which would need to be justified by the power savings.

3.3 ADSORPTION

3.3.1 Process Description

In the VSA process (Figure 3.52), air is fed via the air blower (K111) into an adsorbent bed where the nitrogen preferentially adheres to the adsorbent material and oxygen passes through freely. After approximately 37 s on feed, the adsorbent bed becomes saturated with nitrogen. Regeneration of the bed is accomplished by lowering the bed pressure with a vacuum pump (K190/K192) and by purging under vacuum with oxygen-rich gas from another bed. Once regeneration is complete, the bed is repressurized and its cycle is repeated. The process uses two adsorber beds and a GOX buffer tank so that one bed will be in the production stage while the other is at some phase of regeneration. In this way, product may be provided in a continuous stream to the customer.

The fundamental operating principle of the VSA process is that nitrogen adheres to the adsorbent material at higher pressures and is released at lower pressures. In addition, at a fixed pressure, nitrogen will adhere to the adsorbent when its concentration in the gas space is high and it will be released when its concentration in the gas space is low. The VSA process utilizes low bed pressure during evacuation and low nitrogen concentration during the purge step to release nitrogen from the adsorbent material and regenerate the bed

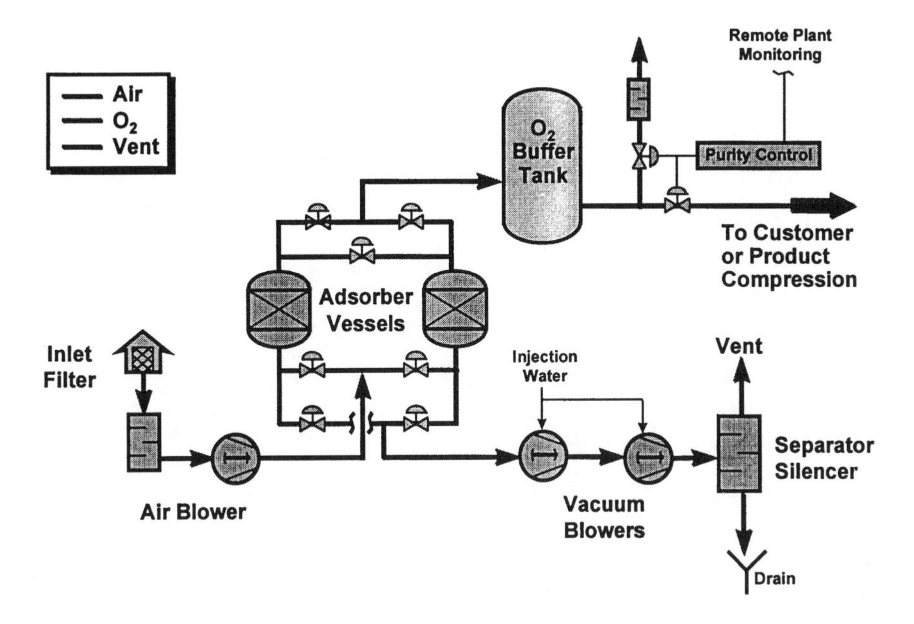

FIGURE 3.52 VSA flow diagram.

The adsorbent beds contain two types of material. At the bottom of the beds is a pretreatment layer for the removal of CO_2 and water from the feed stream. The CO_2 and water will be removed from the pretreatment during evacuation. The remainder of the bed is main sieve material that is responsible for the removal of nitrogen from the feed air.

The VSA process consists of a six-step batch production process. The six steps of the process cycle are described in Section 3.3.2.

3.3.2 Cycle Description

The VSA cycle is broken down into six separate steps. Each step is characterized by a different adsorber bed pressure. Pressures are increased and decreased in the adsorber vessel during the cycle to produce oxygen and regenerate the adsorbent material. Table 3.8 describes the VSA cycle in detail. The pressure variation in the vessels is controlled by valves opening and closing at different times in the cycle.

3.3.3 Equipment Description

3.3.3.1 Vacuum Blowers (K190/K192)

Vacuum blowers (vacuum pumps) are used to draw waste nitrogen, carbon dioxide, and water vapor from the adsorber vessels during the evacuation step. There are two rotary-lobe blowers operating in series. They are positive displacement machines. A timing gear arrangement maintains the close tolerances that are essential for efficient operation.

TABLE 3.8
VSA Cycle Description

Step Number (Length of Step)	**Step Description**	**Step Diagram**
Step 1 (15 s)	**Bed A — Air Repressurization** **Bed B — Evacuation** Air from the air blower is fed into the A bed until the pressure increases to about 1.24 bara. Carbon dioxide and water are removed from the feed air by the pretreatment sieve layer. Nitrogen is removed from the feed stream in the main sieve layer. The vacuum blower evacuates the B bed, removing water vapor, nitrogen, and carries carbon dioxide from the adsorbent. The KV-1928 valve is open until the vacuum level is too deep for one stage to handle. The valve is then closed and the second-stage vacuum blower is no longer bypassed.	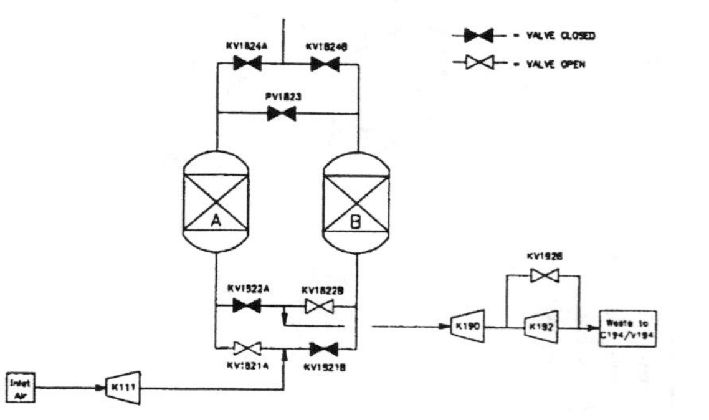

Step 1B (~15 s)

Bed A — Feed and product
Bed B — Evacuation

Air from the air blower continues to feed the A bed. Oxygen-rich gas is removed from the top of the A bed as product. The pressure in the A bed continues to increase to about 1.45 bara. B bed continues to evacuate.

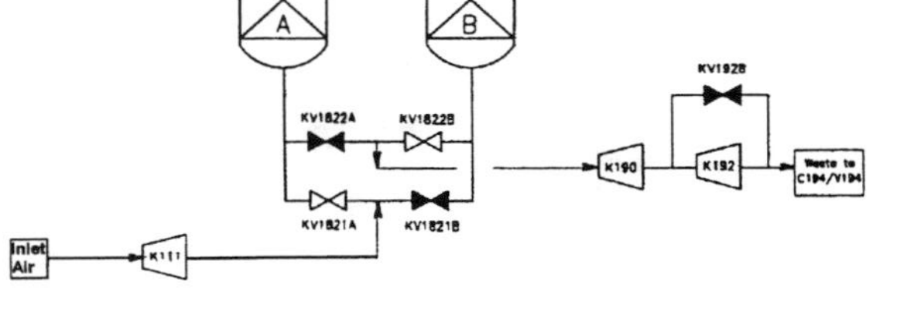

Step 2 (~7 s)

Bed A — Feed and provide purge
Bed B — Receive purge

The A bed continues to be fed by air from the air blower. O_2 from the top of A bed is used to purge the B bed. The downward flow through B bed continues to carry CO_2, water, and nitrogen out of the bed through the vacuum blower. Pressure in A bed decreases to about 1.25 bara.

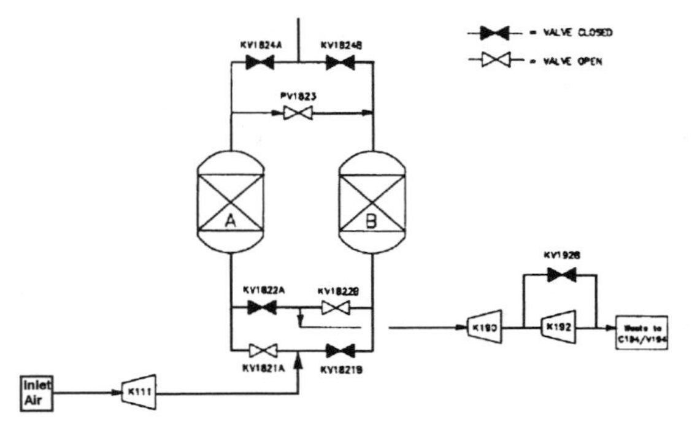

TABLE 3.8 (continued)
VSA Cycle Description

Step Number (Length of Step)	**Step Description**	**Step Diagram**
Step 3 (~4 s)	**Bed A — Feed and provide PE** **Bed B — Receive PE** Oxygen-rich gas from the top of the A bed continues to repressurize the B bed while the KV1822A valve is still open. The pressure in the A bed drops to about 0.862 bara. KV-1821B is opened, and air from the air blower helps to repressurize the B bed.	
Step 4A (~15 s)	**Bed A — Evacuation** **Bed B — Feed repressurization** The vacuum blower continues to evacuate the A bed. The B bed is repressurized with air from the air blower. During this step, the vacuum level will increase causing KV1928 to close.	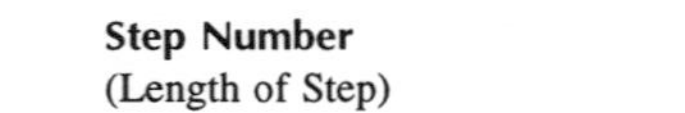

Step 4B
(~15 s)

Bed A — Evacuation
Bed B — Feed and product

The vacuum blower continues to evacuate the A bed until the pressure drops to about 0.310 bara. The B bed continues to be fed with air from the air blower. Product oxygen is taken from the top of the B bed.

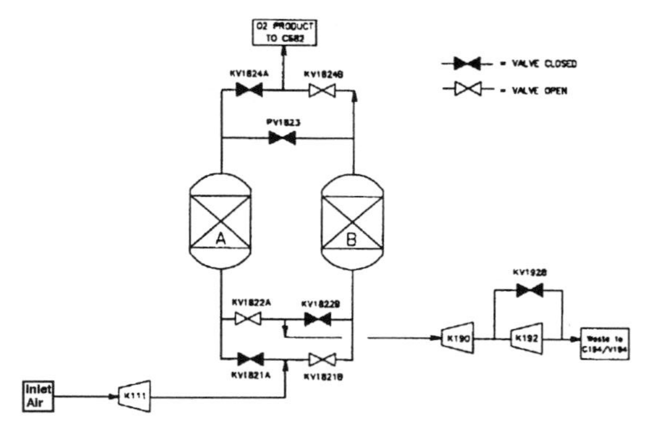

Step 5
(~7 s)

Bed A — Receive purge
Bed B — Feed and provide purge

Evacuation of the A bed continues. The top of the A bed receives purge from the B bed. The vacuum pulls the oxygen-rich gas down through the A bed, removing nitrogen out through the vacuum train. This step lowers the partial pressure of the nitrogen in the A bed, thereby promoting the release of nitrogen from the adsorbent and cleaning the bed. The pressure in the A bed rises to about 0.49 bara. The A bed is fully regenerated after this step.

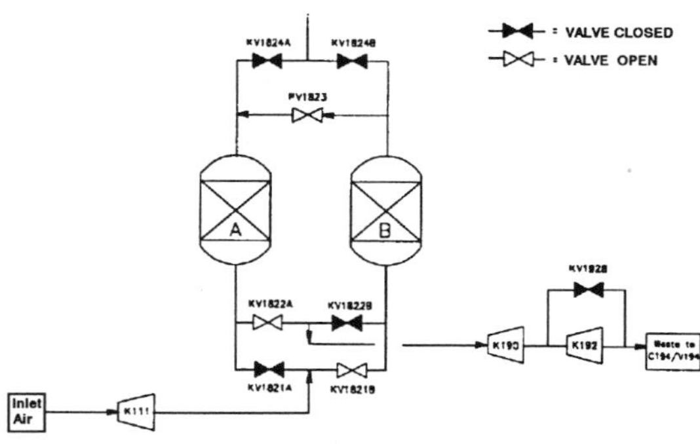

TABLE 3.8 (continued)
VSA Cycle Description

Step Number (Length of Step)	Step Description	Step Diagram
Step 6 (~4 s)	**Bed A — Receive PE** **Bed B — Feed and provide PE** Oxygen-rich gas from the top of the A bed continues to repressurize the B bed while the KV1822B valve is still open. The pressure in the A bed drops to about 0.85 bara. KV1821A is opened and air from the air blower helps to repressurize the B bed.	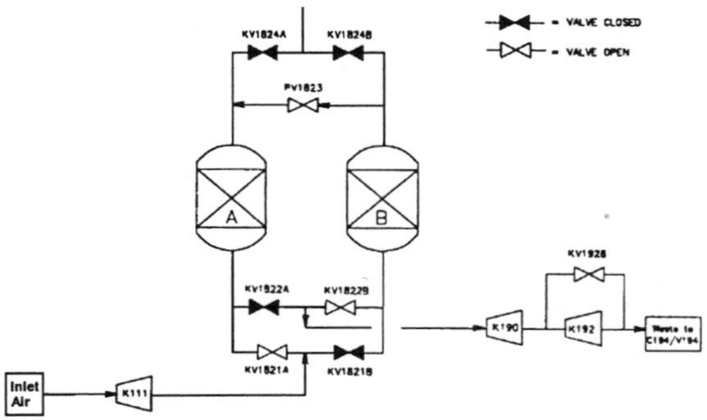

A water injection system sprays water directly into the suction of the machines to seal the rotor clearances and remove the heat of compression. The machines will consume approximately 10 gpm.

Downstream of the vacuum train, the waste gas vents directly to the atmosphere through the discharge separator/silencer. The purpose of this device is to separate any excess seal water which then passes to a drain, and to quiet the waste gas stream while venting to the atmosphere.

3.3.3.2 Air Blower (K111)

The feed air blower, a single rotary-lobe blower, is used to bring air into the plant at a maximum pressure ranging from 0.344 to 0.551 barg. The machine operates similarly to the vacuum blower except that the air blower operates at pressure, not vacuum, and is not water injected.

Air enters the intake of the blower through the inlet air filter, which traps any particulate matter. Without this filtration system, foreign matter could collect in various sections of the plant fouling instrumentation connections, machinery, and, in severe cases, decreasing the efficiency of the motor. Note, however, the filter will not remove gaseous contaminants, such as H_2S, SO_2, chlorides, etc. The air then passes through the suction silencer, which prevents machine noise from going back through the inlet filter.

3.3.3.3 Adsorber Vessels (C182A & B)

The actual separation of air into oxygen and waste streams occurs within the adsorber vessels. The vessels are vertical carbon steel tanks penetrated by piping stubs and a man-way. The adsorber vessels are designed to operate within a pressure range of full vacuum to 1.034 bara. Actual operating range will be 0.344 to 1.44 bara (5 to 21 psia).

Each of the beds contain two layers of molecular sieve. After leaving the air blower skid, the airstream enters the bottom of a vessel and passes through the pretreatment layer of molecular sieve. The pretreatment layer removes water vapor and carbon dioxide from the airstream. Both of these are feed stream contaminants that can damage the main bed permanently if allowed to pass beyond the guard layer.

Above the pretreatment material lies a layer of molecular sieve that attracts or “adsorbs” nitrogen from the air, leaving a 90 to 95% pure oxygen to pass through. Since the sieve cannot separate argon from the oxygen, the resulting product stream contains about 4% argon with the difference being nitrogen.

3.3.3.4 Adsorbent

The molecular sieve adsorbent material is a hard, granular substance that is beadlike in appearance. Molecular sieves are crystalline alumino-silicates.

If it were possible to view a single piece of adsorbent at the molecular level, each pellet would be seen as a cluster of regular geometric forms. Each molecular grouping contains openings that expose the interior of the grouping. This is a very important characteristic of the adsorbent, since the internal surface is extremely large in comparison with the outer dimensions of the structure.

Adsorption is a physical process in which a substance, in this case a gas, is attracted to and held on the surface of another material. Neither material is changed in the process, and it can be easily separated again under the right conditions. In the VSA cycle, adsorption takes place on both the outer and inner surfaces of each structure. Adsorption is a temperature- and pressure-sensitive phenomenon. Both of these variables must be kept in an ideal range for efficient oxygen production to occur.

During the feed step, nitrogen is preferentially adsorbed, thereby producing an oxygen-rich product. In order to remove the nitrogen from the adsorbent, the pressure of the bed is reduced. Reducing the pressure causes each structure in the bed to release the held nitrogen molecules. Evacuating the bed from the bottom removes not only the waste nitrogen but also clean water vapor and carbon dioxide out of the pretreatment layer. A purge gas is used to sweep waste by-products from the bed.

The main adsorbent material is highly sensitive to water and carbon dioxide. For this reason, each of the vessels contains a pretreatment layer at the air inlet end. Unlike nitrogen, which forms a weak physical bond with the main adsorbent, water and carbon dioxide form a very strong physical bond with the adsorbent. This bond is not reversible using pressure swing. Small quantities of water passing through the pretreatment layer will cause localized damage to the main adsorbent bed. This damage will cause a loss of capacity, which results in a loss of production efficiency. If contamination becomes severe enough, the bed material must be replaced.

3.3.4 Buffer Vessel

Due to the discontinuous nature of the VSA process, a buffer vessel is used to provide continuous product. The buffer tank is sized so that the pressure from the VSA does not vary by more than 0.15 bar. If tighter pressure control is required, additional pressure control can be added downstream of the product booster compression.

REFERENCES

1. Isalski, W. H., *Separation of Gases*, Oxford Science Publications, Clarendon Press, Oxford, 1989, Chap. 3.4 and 3.5.
2. Thorogood, R. M., in *Cryogenic Engineering,* Hands, B. A., Ed., Academic Press, London, 1986, Chap. 16.
3. Timmerhaus, K. D. and Flynn, T. M., *Cryogenic Process Engineering*, Plenum Press, New York, 1989.
4. Klosek, J., Smith, R. S., and Solomon, J., The role of oxygen in coal gasification, paper presented at 8th Annual Industrial Energy Technology Conference, Houston, June 1986.
5. Springmann, H., The planning of large oxygen plants for steelworks, Linde Rep., *Sci. Technol.*, 25, 28–33, 1977.

6. Scharle, W. J. and Wilson, K. B., Oxygen facilities for synthetic fuel projects, paper presented at the A.S.M.E. Cryogenic Processes and Equipment Conference, San Francisco, August 1980.
7. Wilson, K. B. and Smith, A. R., Air purification for cryogenic air separation units, *IOMA Broadcaster,* January 1984.
8. Grenier, M., Lehman, J. Y., Petit, P., and Eyre, D. V., Adsorption purification for air separation units, paper presented at ASME Cryogenic Processes and Equipment Conference, December 1984.
9. Lockett, M. J., *Distillation Tray Fundamentals*, Cambridge University Press, Cambridge, 1986, Sect. 1.
10. Weaver, J. B. and Bauman, C. H., in Cost and profitability estimation, Perry, R. H. and Chilton, C. H., Eds., *Chemical Engineers Handbook*, 5th ed., McGraw-Hill, New York, 1973, Sect. 25.
11. Kerry, F. G., Front-ends for air separation plants — the cold facts, *Chem. Eng. Prog.*, August 1991.

4 Heat Transfer

Xianming Li and Vladimir Y. Gershtein

CONTENTS

4.1 INTRODUCTION

Heat transfer occurs whenever there is a temperature difference. It is a common phenomenon that affects everyday life. The very nature of combustion — the process that liberates the chemical energy in the fuel — dictates an intimate relationship with heat transfer. Indeed, heat transfer is not only the objective of any technologically significant combustion process to deliver thermal energy to the load, it is also the key step that makes the combustion process itself possible. Heat transfer within the flame is responsible for ignition, extinction, and a host of other flame properties. In fact, heat transfer determines the shape and behavior of a flame. Understanding the mechanisms of heat transfer and the solution techniques of heat transfer problems can lead to optimization of the combustion process, improvement in heat transfer efficiency, and, ultimately, improvement in process efficiency. Because combustion is the result of the interaction among fluid mechanics, chemistry, mass transfer, and heat transfer processes, understanding heat transfer starts from the interrelationship among these processes.

0-8493-1695-2/98/$0.00+$.50

This chapter begins with a general discussion of the physical phenomena that govern combustion. Then, the mechanisms of heat transfer are discussed. The governing equations of fluid flow, heat transfer, and mass transfer are summarized next, along with important constitutive relations such as turbulence closure models. Finally, solution techniques for heat transfer problems are discussed. The intent here is to have only brief overviews of the fundamental issues, but to highlight the differences in heat transfer in an oxygen-enhanced combustion (OEC) as compared with a conventional air/fuel system. The objective of this chapter is to serve as the information clearinghouse for heat transfer–related theories behind the applications discussed in this book. However, no attempt is made to present a comprehensive treatment of the subject of heat transfer, or even to resemble a handbook with extensive data. Numerous publications are available for that purpose; see, for example, References 1 through 8. Finally, except calculus-level mathematics, no prior knowledge on the topics covered is assumed.

4.2 KEY PROCESSES IN COMBUSTION

4.2.1 CHEMISTRY

Chemistry is the characteristic process in combustion, as compared with other fluid mechanical phenomena. In fact, a sufficiently exothermic reaction is a prerequisite to form a flame. Because chemistry changes the composition of the system, mass transfer process follows. Because of heat release, heat transfer becomes inevitable, and density changes due to heat release as well as reaction stoichiometry subsequently change the fluid flow pattern. Therefore, chemistry is the root cause of a series of changes in a flame.

In general, a series of chemical reactions are involved in a flame. The exact reaction pathway depends on the fuel, oxidizer, and heat transfer environment. For example, simple fuels such as methane and propane involve a different reaction pathway than more-complex fuels such as coal and heavy fuel oils. The reaction pathway will be different for OEC and air/fuel combustion because the flame temperatures will be significantly different. Even if the fuel and oxidizer are fixed, greater heat loss from the flame can result in temperature differences that alter the reaction pathways. The exact reaction mechanisms are usually complicated and in general not well known. However, for engineering purposes, it is often sufficient to conceptualize the combustion process with a few reaction steps, even a single-step global reaction. For example, the single-step global reaction for methane combustion can be defined as

$$CH_4 + 2O_2 \rightarrow CO_2 + 2H_2O \tag{4.1}$$

Although the single-step global reaction approach is valuable in conceptualizing the combustion process, its predictions are only qualitative. First-order quantities, such as oxygen requirement, mean flue gas temperature and gross heat transfer rates, can be estimated approximately, but predictions on distributions, such as velocity, temperature, species concentration, and heat transfer rates, are unreliable. The literature[9] suggests that it is necessary to include at least CO as an intermediate step for reasonable temperature estimates. In that case, the reaction mechanism can be approximated as

$$CH_4 + 3/2O_2 \rightarrow CO + 2H_2O \tag{4.2}$$

$$CO + ½O_2 \Leftrightarrow CO_2 \tag{4.3}$$

Both forward and backward rates are considered for the CO reaction. Other elementary steps that are significant at moderate and high temperatures include the water–gas shift reaction and water vapor dissociation:

$$CO + H_2O \Leftrightarrow CO_2 + H_2 \tag{4.4}$$

$$H_2O \Leftrightarrow H_2 + ½O_2 \tag{4.5}$$

$$H_2O \Leftrightarrow ½H_2 + OH \tag{4.6}$$

Generally speaking, the complexity of the reaction mechanism should be related to the desired output accuracy and the available analysis resources. The additional dissociation steps above form the basis for attempting to predict second-order quantities, such as NOx and soot. In OEC or preheated air/fuel flames, flame temperatures are extremely high. Under such conditions, additional dissociation steps can be important. It is clear that the task of analysis becomes increasingly difficult as the reaction mechanism gets more detailed. For more-complex fuels, such as heavy hydrocarbons or coals, additional steps exist in the process of thermal decomposition, as complex fuel elements break down to simpler but more reactive species. Because such fuels consist of many component species and impurities, each of which may follow a separate path, approximations are necessary.

4.2.2 Mass Transfer

Sustained combustion requires a continuous supply of fresh reactants and a continuous removal of reaction products. This process is loosely known as mass transfer. Specifically, mass transfer is a consequence of three possible modes: bulk fluid motion, molecular and turbulent diffusion, and reaction sources and sinks. Mass transfer due to bulk fluid motion is generally known as convection. It is similar to the convection heat transfer process. Mathematically, the rate of change for species i per unit volume, ρY_i, via convection can be described as $\partial(\rho u_j Y_i)/\partial x_j$, where ρ is fluid density, Y_i is the mass fraction of species i, u_j is the j-component of the fluid velocity.

Mass transfer by molecular diffusion is a result of molecular collisions on the microscopic scale. Fick's law states that mass flux due to molecular diffusion is proportional to the gradient of concentration:

$$J_{i,j} = -\rho D \frac{\partial Y_i}{\partial x_j} \tag{4.7}$$

where $J_{i,j}$ is mass flux of species i in coordinate direction j and D is the molecular diffusion coefficient.

Mass transfer due to reaction refers to the consumption of reactant species (sinks) and the generation of product species (sources). As reactions proceed, reactant species disappear and product species are created. The rate at which mass transfer takes place is generally called the mass source due to reaction. Consider a general system of elementary reactions as follows:

$$\sum_i \nu_{i,k} F_{i,k} = 0, \quad k = 1, 2, \ldots$$

where $\nu_{i,k}$ is the molar stoichiometric coefficient and $F_{i,k}$ is the molecular formula, both for species i in reaction k. According to general convention, $\nu_{i,k}$ is positive for reactants and negative for products. The reaction rate for species i in reaction k follows from the Arrhenius law of mass action:

$$\dot{R}_{i,k} = -\nu_{i,k} M_i T^{\beta_k} A_k \prod_j C_j^{\nu_{j,k}} \exp\left(-E_k / R_u T\right) \tag{4.8}$$

where M_i is the molecular weight of species i, T the mixture temperature, and R_u the universal gas constant. β_k, A_k, and E_k are the temperature exponent, Arrhenius preexponential factor, and activation energy, respectively, for reaction k. C_j is the molar concentration of species j. The overall rate of mass source for species i is the sum of all such sources over all reactions:

$$\dot{R}_i = \sum_k \dot{R}_{i,k} \tag{4.9}$$

4.2.3 Heat Transfer

Heat transfer occurs as long as there is a temperature gradient, and heat release as required by a combustion reaction guarantees a temperature gradient. The heat transfer process controls the stability of the flame. It is responsible for ignition, extinction, and other flame properties. Heat transfer is responsible for flame temperature distribution, and ultimately it is responsible for delivering the energy to the process of interest. Because of its importance and its relevance to the topic of discussion, heat transfer is addressed in detail in Section 4.3.

4.2.4 Fluid Mechanics

Fluid mechanics is the foundation of combustion. Mixing, heat transfer, or chemistry all depend on the underlying flow distribution. In turn, they affect the flow pattern. All four processes are intimately linked and interdependent.

The equations of fluid mechanics originate from the momentum and mass conservation principles. The overall mass conservation or continuity equation for laminar flows is

$$\frac{\partial \rho}{\partial t} + \frac{\partial}{\partial x_j}\left(\rho u_j\right) = 0 \tag{4.10}$$

This equation is general, and is valid for incompressible as well as compressible flows. By far, the majority of combustion applications can be treated as incompressible in the sense that the fluid density varies with temperature and composition, but not with pressure.

The k-component momentum conservation equation for laminar flows is

$$\frac{\partial}{\partial t}\left(\rho u_k\right) + \frac{\partial}{\partial x_j}\left(\rho u_j u_k\right) = -\frac{\partial p}{\partial x_k} + \frac{\partial \tau_{jk}}{\partial x_j} + \rho g_k + F_k \tag{4.11}$$

where g_k is the k-component of the gravitational acceleration, F_k is the k-component of any applied force such as an electromagnetic force for example, p is the fluid pressure, and τ_{jk} is the shear stress tensor. Assuming a Newtonian fluid, the stress tensor can be written as

$$\tau_{jk} = \mu\left(\frac{\partial u_j}{\partial x_k} + \frac{\partial u_k}{\partial x_j}\right) - \frac{2}{3}\mu\frac{\partial u_l}{\partial x_l}\delta_{jk} \tag{4.12}$$

where μ is the molecular viscosity and δ_{jk} is the Kronecker delta.

Most combustion applications of engineering importance involve turbulent flows. Turbulent eddies introduce a fluctuating component for every variable such as velocity, temperature, and concentration:

$$\phi = \bar{\phi} + \phi' \tag{4.13}$$

where ϕ denotes any variable and $\bar{\phi}$ and ϕ' are the average and fluctuation of ϕ, respectively, defined as:

$$\bar{\phi} = \frac{1}{\tau}\int_0^\tau \phi dt, \quad \int_0^\tau \phi' dt = 0 \tag{4.14}$$

where the integration length τ is assumed to be much greater than the characteristic time period of fluctuation and the integration starting point has no impact on the averaging process. Although most combustion systems involve incompressible flows, the fluid density does vary with temperature and composition. In such situations, it is convenient to introduce a density-weighted average, $\tilde{\phi}$, such that

$$\phi = \tilde{\phi} + \phi'', \quad \tilde{\phi} \equiv \frac{\overline{\rho\phi}}{\bar{\rho}}, \quad \overline{\rho\phi''} = 0 \tag{4.15}$$

where the overbar denotes conventional time averaging defined above. It is clear that

$$\tilde{\phi} - \bar{\phi} = \frac{\overline{\rho'\phi''}}{\bar{\rho}} \tag{4.16}$$

Although the concept of time averaging is straightforward, experimentally measured quantities are generally closer to density-weighted averages rather than time averages except for pressure, stresses, and heat flux. It also eliminates certain terms introduced by conventional time averaging, e.g., $\partial\left(\overline{\rho' u_j'}\right)/\partial x_j$ in the continuity equation, which simplifies the final form of the governing equations. According to Favre,[10] density, stresses, and heat fluxes are to be time averaged, whereas all other quantities are density-weighted averaged. Then the continuity equation for the mean flow becomes

$$\frac{\partial \bar{\rho}}{\partial t} + \frac{\partial}{\partial x_j}\left(\bar{\rho}\tilde{u}_j\right) = 0 \tag{4.17}$$

and the mean momentum equation becomes

$$\frac{\partial}{\partial t}\left(\bar{\rho}\tilde{u}_k\right) + \frac{\partial}{\partial x_j}\left(\bar{\rho}\tilde{u}_j\tilde{u}_k\right) = -\frac{\partial \bar{p}}{\partial x_k} + \frac{\partial}{\partial x_j}\left(\bar{\tau}_{jk} - \overline{\rho u_j'' u_k''}\right) \tag{4.18}$$

Note that for clarity, gravity and additional force terms are excluded from the above equation. The last term, $\overline{\rho u_j'' u_k''}$, is a result of the averaging procedure. It represents the turbulent stresses, an additional source of momentum diffusion due to turbulence. A constitutive relationship, or modeling, must be introduced to relate the turbulent stresses to mean flow quantities, an area known as turbulence modeling.

Although numerous turbulence models are reported in the literature,[11–13] by far the most popular is the two-equation $k - \varepsilon$ model, first proposed by Jones and Launder.[14] In this model, the turbulent stresses are recast in a form similar to the molecular stress tensor with mean velocity gradients, an assumption generally known as the Boussinesq hypothesis:

$$-\overline{\rho u_i'' u_j''} = \mu_t\left(\frac{\partial \tilde{u}_j}{\partial x_j} + \frac{\partial \tilde{u}_j}{\partial x_j}\right) - \frac{2}{3}\left(\bar{\rho}k + \mu_t \frac{\partial \tilde{u}_l}{\partial x_l}\right)\delta_{jk} \tag{4.19}$$

where μ_t is the turbulent viscosity, a term analogous to its molecular counterpart, and k is the turbulence kinetic energy, defined as $k \equiv \frac{1}{2}\overline{u_i' u_i'}$. The rate of turbulence energy dissipation is defined as

$$\varepsilon \equiv \nu \overline{\frac{\partial u_i'}{\partial x_j} \frac{\partial u_i'}{\partial x_j}} \tag{4.20}$$

which is the product of molecular kinematic viscosity and the mean square of the strain rate of turbulence. In the $k-\varepsilon$ model, the transport equations for k and ε are, respectively,

$$\frac{\partial}{\partial x_i}\left(\bar{\rho}\tilde{u}_i k\right) = \frac{\partial}{\partial x_i}\left(\frac{\mu_t}{\sigma_k}\frac{\partial k}{\partial x_i}\right) + G_k - \bar{\rho}\varepsilon \tag{4.21}$$

$$\frac{\partial}{\partial x_i}\left(\bar{\rho}\tilde{u}_i \varepsilon\right) = \frac{\partial}{\partial x_i}\left(\frac{\mu_t}{\sigma_\varepsilon}\frac{\partial \varepsilon}{\partial x_i}\right) + C_{1\varepsilon}\frac{\varepsilon}{k}G_k - C_{2\varepsilon}\bar{\rho}\frac{\varepsilon^2}{k} \tag{4.22}$$

Here $C_{1\varepsilon}$, $C_{2\varepsilon}$, σ_k, and σ_ε are model constants (1.44, 1.92, 1.0, and 1.3, respectively). G_k is the rate of production of turbulence kinetic energy by means of shear rate in the mean flow:

$$G_k = \mu_t \frac{\partial \tilde{u}_j}{\partial x_i}\left(\frac{\partial \tilde{u}_j}{\partial x_i} + \frac{\partial \tilde{u}_i}{\partial x_j}\right) \tag{4.23}$$

Turbulence kinetic energy k and turbulence energy dissipation rate ε are then used to evaluate a velocity scale and a length scale:

$$v = \sqrt{k}, \quad l = k^{3/2}/\varepsilon \tag{4.24}$$

By analogy to molecular viscosity, the turbulence viscosity can be computed as

$$\mu_t \propto \rho v l, \text{ so that } \mu_t = \rho C_\mu \frac{k^2}{\varepsilon} \tag{4.25}$$

where C_μ is a model constant ($C_\mu = 0.09$). With a turbulence model, proper closure is obtained for the equations of fluid mechanics for a turbulent flow. Note that the $k-\varepsilon$ model is not the only turbulence closure model available; in fact, it is not even the most accurate model available, but it is probably the one most commonly used. It is used here for illustrative purposes only.

4.3 MECHANISMS OF HEAT TRANSFER

Three mechanisms of heat transfer are known as conduction, convection, and radiation. On the microscopic scale, all three mechanisms share the aspect of energy exchange via interparticle collisions. Yet the macroscopic behavior and analysis methods have distinct characters. In this section, each of the aforementioned mechanisms are discussed separately, providing the emphases on the importance of each heat transfer mechanism in air/fuel and oxy/fuel combustion systems.

Heat transfer or thermal energy exchange occurs if and only if there is a temperature difference. Moreover, thermal energy can only be transferred from a system or substance with a higher temperature to a system or substance with a lower temperature. The phenomenological laws will be discussed here to provide a quantitative relation of a heat flux, as a measure of energy transfer, with a system temperature gradient. Such a relation will be discussed for conductive, convective, and radiative heat transfer mechanisms.

The thermal state of a system can be characterized by its temperature field, i.e., by the temperature at every point of the system at a given time. The temperature field or temperature distribution inside the system depends on its properties. Before the discussion on the mechanisms of heat transfer, the system and some of its properties are defined.

A system generally consists of a collection of substances or subregions with different thermophysical properties and with explicit boundaries. A system can be classified as homogeneous or nonhomogeneous, isotropic or anisotropic, steady or transient, one dimensional or multidimensional, and so on, for the purpose of analysis. In general, a system is nonhomogeneous. That is, material properties are different at different points of the system. However, a particular part of a nonhomogeneous system can be homogeneous if material properties are constant within that subregion. A homogeneous body or subregion can be isotropic or anisotropic depending on whether or not thermophysical properties change in different directions. A system is considered to be at a steady state if its properties do not change with time, or it is at a transient state. All systems in the real world are three dimensional, but can be idealized as one or two dimensional depending on whether the properties change significantly, for practical purposes, along one or two axes in the chosen coordinate system.

4.3.1 Conduction

Heat can be conducted through solids, liquids, and gases. Conduction in solids is the most illustrative since it is the most common heat transfer mechanism in that type of medium. Conduction is the energy transfer between adjacent molecules or atomic particles at motion. The nature of the motion depends on the system and on the molecular and particle state. The motion can range from vibration of atoms in a crystal lattice of solids to the chaotic fluctuations of gas molecules. In metallic solids, movement of free electrons contributes to heat conduction.

As mentioned previously, heat flux is a measure of the rate of heat transfer, defined as the amount of energy exchange per unit area and per unit time in a specified direction. At this point, it is convenient to introduce the concept of an isothermal surface. An isothermal surface is a geometric surface defined by a collection of points with the same temperature. By definition, temperature changes in any direction except along an isotherm. Moreover, the largest temperature change per unit length occurs along the direction normal to the isotherm. A temperature gradient refers to the temperature increase between two isotherms along the normal direction, defined as the ratio of the temperature change ΔT to the distance between the isotherms Δn:

$$\lim_{\Delta n \to 0} \frac{\Delta T}{\Delta n} = \nabla T = \frac{\partial T}{\partial x_i} \vec{e}_i \tag{4.26}$$

where $\vec{e}_i$ is the unit vector in the ith coordinate direction. Note that a summation on index i is implied in the last relationship, consistent with the conventions of Cartesian tensor notation. Fourier's law of conduction states that heat flux is proportional to the temperature gradient in magnitude, but opposite in direction, as follows

$$q = -k\nabla T = -k \frac{\partial T}{\partial x_i} \vec{e}_i \tag{4.27}$$

where the parameter k is known as thermal conductivity.

In practical combustion systems, the predominant mode of heat transfer is usually not molecular conduction, but turbulent diffusion, except at the boundaries and the flame front. Conduction is the only mode of heat transfer through refractory walls, and it determines ignition and extinction behaviors of the flame. Turbulent diffusion, an apparent or pseudo conduction mechanism arising from turbulent eddy motions, will be discussed in Section 4.4. The relations from the theory of conduction heat transfer[15–17] can be used to evaluate heat losses through furnace walls and load zones, and through the pipe walls inside boilers and heat exchangers, etc.

4.3.2 Convection

Heat transfer by convection implies energy transfer due to bulk fluid motion and mixing of macroscopic elements of liquid or gas. Since fluid motion is involved, convective heat transfer is practically governed by the laws of fluid mechanics. One can distintinguish between *natural* and *forced convection*. Natural convection is induced by density differences inside a fluid volume under the influence of gravity. A density difference can be the result of the differences in temperature and/or composition. On the other hand, forced convection usually involves an applied pressure gradient imposed by devices such as pumps and compressors. The distinction between natural convection and forced convection is intuitive rather than strict, and it depends on the extent of the analysis domain. For example, atmospheric motions are related to natural convection. But to an observer next to a building, wind currents are an applied force on the building, thus heat transfer involves forced convection. In either case, convection heat transfer involves energy exchange due to bulk fluid motion, or mathematically, a term $\partial(\rho u_i c_p T)/\partial x_i$ in the governing equation. Note that c_p is the specific heat of the fluid.

In engineering analysis, however, convection heat transfer is more readily associated with the following relationship:

$$Q = Ah(T - T_w) \tag{4.28}$$

where h is known as the heat transfer coefficient, T_w is the temperature of a solid wall in direct contact with the fluid, and A is the heat transfer area. This correlation is known as the Newton–Rihman equation of convection.

The heat transfer coefficient characterizes the heat flux through a unit surface area at a unit temperature difference between the fluid and the solid. The complexity of convective heat transfer lies in the evaluation of the heat transfer coefficient which is a complex function of many parameters. Note that the concept of heat transfer coefficient is an engineering convenience. Some argue that it is the heat flux rather than the heat transfer coefficient that is fundamental. The fact is, however, the heat transfer coefficient has been widely used and accepted, and it is intimately related to heat flux.

For natural convection, the heat transfer coefficient h has the following general functional dependence:

$$h = f\left(T, T_w, \beta, k, \nu, \alpha, g, \Phi\right) \tag{4.29}$$

where β is the volumetric expansion coefficient of the fluid, k the thermal conductivity, ν the kinematic viscosity, α the thermal diffusivity, g gravity, and Φ represents parameters characterizing the solid wall (dimensions, shape, material structure, etc.).

The dependence of natural convection heat transfer on the aforementioned parameters can be established based on the physics of the process. Let us assume that a vertical wall is in contact with a fluid. The wall temperature T_w is higher than the fluid temperature T. When a unit volume of fluid contacts the hot wall, the fluid receives energy from the wall due to molecular collisions. The fluid molecules begin to move with a higher velocity. The initial fluid volume expands. From this description one can conclude that energy transfer should depend on the parameters T, T_w, β, and c_p.

The expanded fluid volume has a lower density, so it rises. The fluid motion is the result of several applied forces such as gravity, internal friction, etc. Thus, the parameters g and ν appear in Equation 4.29. It is clear that the fluid motion is connected to its density. Nevertheless, density is not explicitly taken into consideration in Equation 4.29. It is possible to show that the density, ρ, is present in Equation 4.29 implicitly through the relation of parameters α, c_p, and k:

$$\rho \alpha c_p = k \tag{4.30}$$

So far, we showed that convective heat transfer is characterized by fluid motion. Yet fluid motion is affected by the characteristics of the solid boundary, such as dimensions, orientations, roughness, etc. Therefore, the parameter Φ is present in Equation 4.29.

Very often fluid motion near the solid is not only the result of the temperature gradient, but also the result of some outside forces. In such cases, heat transfer is by forced convection. Here, a pressure gradient appears as a result of an external force exerted by a pump, for example. Fluid mixing takes place as a result of the applied pressure gradient. The heat transfer coefficient in forced convection has the following general functional dependence:

$$h = f(U, k, \nu, \alpha, \Phi) \tag{4.31}$$

where U is the average fluid velocity as a result of the applied pressure gradient. In general, the influence of temperature gradient on fluid motion is not as strong as in the case of natural convection. Thus, Equation 4.31 does not explicitly include temperature, but it implicitly relates to temperature through property variations.

In some cases both natural convection and forced convection are of comparable importance. Then Equations 4.29 and 4.31 should be combined to obtain the relationship for the heat transfer coefficient.

The functional dependence expressed in Equations 4.29 and 4.31 governs the behavior of convective heat transfer. In some cases the functionality can be determined analytically, but in most cases it can be determined only as a statistical correlation of experimental data. Dimensionless groups are used to generalize empirical correlations for convective heat transfer. These groups can be determined from the parameters in Equations 4.29 and 4.31. They are the Reynolds, Nusselt, Grashof, and Prandtl numbers, respectively, defined as:

$$\mathrm{Re} = \frac{UL}{\nu} \tag{4.32}$$

$$\mathrm{Nu} = \frac{hL}{k} \tag{4.33}$$

$$\mathrm{Gr} = \beta g \frac{L^3}{\nu^2} \left(T - T_w\right) \tag{4.34}$$

$$\mathrm{Pr} = \frac{\nu}{\alpha} \tag{4.35}$$

where L is a characteristic length scale representing the influence of solid boundaries, previously denoted as Φ. Note that L is a subset of Φ in the parameter space. Now Equations 4.29 and 4.31 can be rewritten in these dimensionless numbers, respectively, as

$$\mathrm{Nu} = f(\mathrm{Gr}, \mathrm{Pr}) \tag{4.36}$$

$$\mathrm{Nu} = f(\mathrm{Re}, \mathrm{Pr}) \tag{4.37}$$

These expressions are much more compact and general than their dimensional counterparts. Numerous empirical formulas in such forms are available in the literature; see, e.g., Reference 1.

4.3.3 Radiation

Energy transfer by radiation is quite different in nature than conduction or convection. It is the only mechanism of energy exchange through vacuum. Radiation can be conceptualized as energy exchange via electromagnetic waves or via photon particles. The wave theory construes a continuous nature, while the photon theory relates to discrete energy bundles. Both theories contribute to the development of radiation as we know it today; neither alone can explain all experimental observations.

The unique nature of radiation and its relation to conduction can be understood with a description of the process of radiation energy exchange in terms of the photon theory. An atom or a molecule becomes excited when it gains some excess energy. A particle at an excited state is unstable. At the first opportunity, it releases some of its excess energy by discharging a photon at a particular frequency and returns to its undisturbed state. The photon travels away from the host particle, and eventually collides with another particle. By absorbing the photon, the new particle gets excited and repeats the sequence. As a result, energy exchange occurs across space. These photon collisions and exchanges bear some resemblance to molecular collisions in conduction. Indeed, in the limit of an optically thick medium where a large population of excited particles exists and the mean free path of the photons is short, radiation heat transfer can be approximated with conduction. However, under conditions of engineering importance, the population of particles in an excited state comprises only a tiny fraction of the total population of particles, unlike particles available for conduction which involve the total population. As a result, phenomena such as reflection, absorption, and scattering are important in radiation. The temperature at a given point depends not only on its immediate neighbors as in conduction, but also on locations everywhere. This global character makes the governing equation for radiation heat transfer, as seen later, an integrodifferential equation, rather than a differential equation for conduction. Therefore, a completely different mathematical approach is required for radiation.

In combustion applications, a particular type of radiation called thermal radiation is of interest. Because a particle can become excited through different approaches, the frequency of the discharged photon can be different. Radiation at different frequencies can be classified as cosmic rays, γ-rays, X-rays, thermal radiation, microwaves, radio waves, etc. Note that wavelength can also be used to classify radiation, because wavelength λ and frequency ν are related to the speed of light c as $\lambda\nu = c$. In terms of wavelength, thermal radiation generally refers to a range from about 0.1 to about 1000 μm, covering the long-wave fringe of the ultraviolet, visible light, near infrared, and far infrared spectra. In this wavelength range, vibration, rotation, and electronic transitions are mainly responsible for photon generation in gases; molecular and lattice vibrations and bound electron transitions are the responsible mechanisms in solids; and molecular vibrations are the mechanisms for liquids. Because these mechanisms are associated with the temperature state of the particle, such radiation is thus known as thermal radiation. In this book thermal radiation is the subject of interest.

Some fundamental definitions are now introduced to simplify further discussion. Assume incident energy is approaching a body. This incident energy can be transmitted,

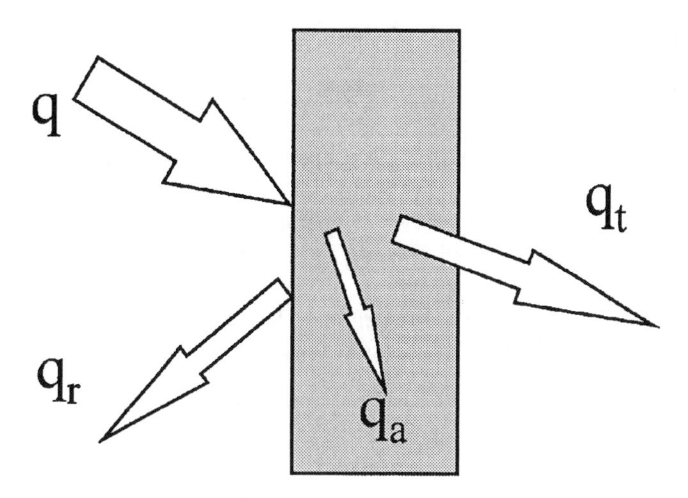

FIGURE 4.1 Incident radiation is divided into reflection, absorption, and transmission.

absorbed, and reflected by the body, as shown in Figure 4.1. The following relation can then be written:

$$q = q_t + q_a + q_r \tag{4.38}$$

where q is the incident energy, q_t transmitted energy, q_a absorbed energy, and q_r reflected energy. If we divide both sides of Equation 4.38 by q, we have

$$1 = \frac{q_t}{q} + \frac{q_a}{q} + \frac{q_r}{q} = \tau + \alpha + \rho \tag{4.39}$$

where τ is known as the transmissivity, α the absorptivity, and ρ the reflectivity of the body. It is obvious that these characteristics are related, and they vary from 0 to 1. Three special cases are of interest:

1. $\alpha = 1$, $\rho = 0$, $\tau = 0$. The incident radiant energy is completely absorbed by a body. Such a body is known as a *blackbody.*
2. $\alpha = 0$, $\rho = 1$, $\tau = 0$. The incident radiant energy is completely reflected by a body. In this case the body is known as a *white body.*
3. $\alpha = 0$, $\rho = 0$, $\tau = 1$. The incident radiant energy is completely transmitted through a body. Such a body is known as a *completely transparent body.*

Evaluation of these special cases helps establish the radiation laws for real bodies. One can differentiate between monochromatic and total radiation. If radiation is emitted in the narrow wavelength from λ to $\lambda + d\lambda$, we call it monochromatic radiation. If the emitted radiation is taken over the whole wavelength spectrum, from $\lambda = 0$ to $\lambda = \infty$, then we call it total radiation.

Absorption usually does not occur equally for all wavelengths. Monochromatic or selective radiation is especially significant in gases, which are transparent to some wavelengths of radiation but highly absorbing for others. Since the spectral distribution of energy is a function of the temperature of the emitter, the amount of energy absorbed by the receiver depends on the temperature of the emitter. The absorptivity of a material also depends on its own temperature. Thus, the specification of absorptivity requires the designation of two temperatures. If the absorptivity of a body does not depend on the wavelength and is equal to any constant lower than 1, the body is defined as a *gray body.*

For a blackbody, the spectral distribution of hemispherical emissive power (integrated over the hemispherical solid angle) follows Plank's law:

$$e_{\lambda b} = \frac{2\pi C_1}{\lambda^5\left(e^{C_2/\lambda T} - 1\right)} \tag{4.40}$$

where C_1 and C_2 are constants. The wavelength at which the blackbody emissive power is a maximum at a given temperature follows Wien's displacement law:

$$\lambda_{max} T = C_3 \tag{4.41}$$

where C_3 = 5216.0 μm·°R (2897.8 μm·K). This relationship follows readily from Plank's law by setting to zero its derivative with respect to wavelength. The implication is that, at a higher temperature, the peak emissive power occurs at a shorter wavelength. If Plank's law is integrated over the entire wavelength spectrum, we have the fourth-power temperature dependence law of radiation known as Stefan–Boltzmann's law:

$$e_b = \sigma n^2 T^4 \tag{4.42}$$

In this equation, σ is Stefan–Boltzmann constant — 0.1712×10^{-8} Btu/(hr·ft^2·°R^4) or 5.67×10^{-8} W/(m^{2}K^4) — and n is the refractive index. Most media have refractive indexes close to 1, but glass, for example, has a refractive index close to 1.5.

A concept closely related to emissive power is emissivity, which describes how "black" the surface is. Recall that a blackbody is a perfect absorber; it is also a perfect emitter. By comparing with a blackbody, a unique datum can be defined. The emissivity of a surface is the ratio of its emissive power to the emissive power of the blackbody at the same temperature:

$$\varepsilon_\lambda = \frac{e_\lambda}{e_{\lambda b}} \tag{4.43}$$

For blackbody, ε = 1 by definition. In general, emissivity is related to absorptivity by Kirchhoff's law, loosely described as

$$\varepsilon = \alpha \tag{4.44}$$

Note that this equality carries underlying qualifications. To enumerate the details of these qualifications, however, is beyond the scope of this book. The interested reader is encouraged to consult any radiation textbook, such as that by Siegel and Howell.[2]

In the analysis of participating medium, one often encounters the concept of the absorption coefficient, a. This quantity describes the amount of attenuation of incoming radiation by the medium via absorption per unit length so that it has the dimension of reciprocal length. Analogously, the scattering coefficient, s, describes the amount of attenuation of incoming radiation by the medium via scattering. It also has the dimension of reciprocal length. The sum of scattering and absorption coefficients is generally known as the extinction coefficient, k:

$$k_\lambda = a_\lambda + s_\lambda \tag{4.45}$$

where subscript λ reflects the dependence of wavelength. The extinction coefficient is related to absorptivity as follows:

$$\alpha = 1 - e^{-kS} \tag{4.46}$$

Here, the extinction coefficient is assumed to be constant over the path length S. If Kirchhoff's law holds, then emissivity of the medium over length S is

$$\varepsilon = 1 - e^{-kS} \tag{4.47}$$

If a body at temperature T_1 and emissivity ε emits into a large black enclosure with temperature T_2, the heat flux between the body and the enclosure can be calculated as

$$q = \varepsilon\sigma\left(T_1^4 - T_2^4\right) \tag{4.48}$$

This relationship is true for gray surfaces in terms of total hemispherical radiation and that the emitting body "sees" the entire enclosure. In engineering, this relationship can be used to estimate the amount of radiative heat loss from a refractory wall to the surroundings, for example.

Radiative intensity I refers to the amount of radiation energy per unit time, per unit area, and per unit solid angle normal to the area. From the definition, the intensity is a function of the direction of radiation. Such a direction can be defined by polar and azimuth angles θ and φ, and the cross-sectional area of the pencil ray at an element dA. The differential heat flux dq is related to radiative intensity as

$$dq = I(\theta,\varphi)d\Omega \cos\theta \tag{4.49}$$

where q is heat flux, θ is polar angle, φ is azimuth angle, $I(\theta, \varphi)$ is radiation intensity in the direction of θ and φ, and Ω is the solid angle related to θ and φ. For one-way flux, the total radiative energy rate is

$$dQ = I(\theta, \varphi) d\Omega \cos \theta dA \tag{4.50}$$

Here, dA is the area of a differential element.

Radiative intensity is related to the blackbody emissive power. To illustrate, consider an isotropic system where the intensity is independent of angle. The total energy flux at dA throughout the hemisphere of 2π steradians is an integration of Equation 4.49. It must also be the total emissive power of the element. Therefore,

$$e_b = I \int_{2\pi} d\Omega \cos \theta = \pi I \tag{4.51}$$

Now the radiative transfer equation can be derived. Consider a ray of radiation passing through a participating medium. This ray will be attenuated along the path S. The amount of attenuation or change is related to the starting intensity and the distance it travels:

$$dI_\lambda = -k_\lambda I_\lambda dS \tag{4.52}$$

This expression is known as the Bouguer–Lambert Law. Here, k_λ is the extinction coefficient along the direction S.

If the radiation along the ray is in local thermodynamic equilibrium, the amount of radiation added to the ray by spontaneous emission is related to the local absorption coefficient and blackbody intensity as follows:

$$dI_\lambda = a_\lambda I_{\lambda b} dS \tag{4.53}$$

The overall change of radiative intensity as it passes through a participating medium is the net effect of attenuation by absorption and outscattering, augmentation by emission and in-scattering. Mathematically, this relationship is

$$\frac{dI}{dS} = -kI + aI_b + \frac{s}{4\pi} \int_{\Omega'=4\pi} I(S, \Omega') \Phi(\lambda, \Omega, \Omega') d\Omega' \tag{4.54}$$

where Φ is known as the scattering phase function. Note that the explicit wavelength dependence has been dropped for simplicity. This is the radiative transfer equation, an integrodifferential equation.

The concept of view factors is quite convenient in the analysis of diffuse and gray radiation exchanges. Under these assumptions, the view factor, $F_{1\text{-}2}$, is purely a geometric quantity. Physically, it means the fraction of radiative energy leaving surface 1 that reaches surface 2. In other words, it describes how much surface 1 "sees" surface 2, thus the name view factor. Due to the restricted nature of this chapter, the expression for the view factors will not be derived here. Instead, the expression will be given here and the reader will be referred to a more-detailed discussion in References 2, 18, and 19. Mathematically, the view factor is defined as

$$F_{1-2} = \int_{A_2} \int_{A_1} \frac{(dA_1 \cos \theta_1)(dA_2 \cos \theta_2)}{\pi r^2} \tag{4.55}$$

In this equation A_1 and A_2 are the areas of the appropriate surfaces 1 and 2. Angles θ_1 and θ_2 are the angles defined by the shortest distance r between the surfaces and the appropriate normal vector to each surface. There are many different ways of solving the view factor equation. The solution depends on the shape and orientation of the surfaces. Comprehensive listings of view factors for commonly encountered configurations can be found in References 2 and 18 through 20. The application of view factors can be illustrated with a simple example. Suppose we have surface 1 with temperature T_1 and emissivity ε and a black surface 2 ($\varepsilon = 1$) with temperature T_2; then the total radiant heat flux from surface 1 to surface 2 is

$$q = \varepsilon \sigma F_{1-2} \left(T_1^4 - T_2^4\right) \tag{4.56}$$

Now, after all three mechanisms of heat transfer have been briefly discussed, attention can be focused on the differences in heat transfer between air/fuel combustion and OEC.

4.3.4 Heat Transfer with Air/Fuel and Oxy/Fuel Firing

High-temperature systems under air/fuel firing mode and OEC are extensively employed in high-temperature heating and melting industries. Such systems can be industrial furnaces, boilers, heaters, engines, etc. Traditionally, people are used to air/fuel combustion when air is used as the oxidizer. In the last decade, more and more attention has been paid to oxy/fuel firing, where high-purity oxygen or oxygen-enriched air is used as the oxidizer. For example, a large portion of the glass industry has converted, or is in the process of converting, from air/fuel combustion to oxy/fuel combustion. The aluminum industry is also extensively using the oxygen-enriched firing mode. There are similarities and significant differences between the two firing modes. These similarities and differences are the subject of interest in this section.

Let us first look at the similarities of the two firing modes. In both cases, fuel and oxidizer are introduced into the system. A diffusion flame is formed (in this section premixed flames are not a subject of interest). Both flames are practically governed by a mixing process which indicates the importance of the flow regime of the gases. An exothermic chemical reaction takes place with an energy release that then is transferred to the object of interest. The hot gases, the products of combustion, are exhausted from the system.

Now, let us list the major differences of air/fuel and oxy/fuel firing. Air/fuel firing includes nitrogen in the combustion zone. The following relation can be offered to calculate the volume difference in the case of methane combustion for air/fuel and oxy/fuel firing modes:

$$G = CH_4 + 2\left(O_2 + \frac{1-x}{x} N_2\right) \tag{4.57}$$

In the case of air-firing mode $x = 0.21$ and $G = 10.52$. In the case of oxy-firing mode $x = 1$ and $G = 3$. Thus, considering the same firing rate, the gas volume entering the burners for the air methane system is 73% greater than that of oxy/methane system.

During combustion, nitrogen heats up and consumes energy otherwise available for useful purpose. Due to the high nitrogen content in the oxidizer, the volume of oxidizer needed for stoichiometric firing is large so that convection plays a more significant role as a heat transfer mechanism. The adiabatic flame temperature of air methane combustion is approximately 3500°F (2200 K) with no air preheat.

In the oxy/methane firing mode, considering pure oxygen as an oxidizer, the volume of oxidizer needed for stoichiometric firing reduces by 73% as shown above. There is no extra gas to heat up, and the adiabatic flame temperature is considerably higher, around 4940°F (3000 K). At such a temperature dissociation and recombination processes are taking place at the flame region. Additional species such as OH, H, O, etc. now are important and should be taken into account along with the major species H_2O, CO, and CO_2. The concentration of these combustion products in oxy/methane firing is significantly higher than in air/methane firing. Indeed, the combustion products are not diluted with a large amount of nitrogen. Thus, the absorptivity of the furnace gas is higher since all these species are radiatively participating. In some literature the average absorption coefficient of the furnace gas for oxy/methane firing furnace is mentioned to increase up to 0.06 to 0.09 ft^{-1} (0.2 to 0.3 m^{-1}) compared with its value of 0.012 to 0.015 ft^{-1} (0.04 to 0.05 m^{-1}) for the air/methane firing environment. More details for absorptivity of CO_2, H_2O, etc. can be found in Reference 18. Here, we only want to emphasize that the oxy/fuel gas medium is more radiative than air/fuel.

The emissivity of the gas media is a function of many parameters including gas pressure, temperature, partial pressures of radiatively participating species, and optical path length or characteristic dimension. Thus, if the concentration of the absorbing/emitting species is increased, the emissivity of the media increases as well. If the optical thickness of a medium tends to infinity, then the emissivity of such a medium tends to 1, which corresponds to the blackbody limit. At this limit, radiation becomes a totally diffusive process.

The role of convection is diminishing and the role of radiation is increasing in oxy/fuel firing compared with air/fuel firing, due to considerably lower volume of the gases involved and significantly higher temperature. To illustrate the relative importance of convection, let us look, for example, at the average gas velocity inside an industrial furnace. The average velocity of the furnace gas inside a typical air/methane-fired glass or aluminum furnace is about 10 to 15 ft/s (3 to 5 m/s). This velocity reduces to 1 to 5 ft/s (0.3 to 1.5 m/s) for oxy/methane firing. Both a reduction in furnace gas velocity and an increase in flame temperature lead to the conclusion that radiation is the main heat transfer mechanism inside an oxy/fuel firing furnace. An evaluation of some industrial oxy/fuel firing glass furnaces showed that up to 98% of energy is transferred via radiation.

Because gas radiation comes in discrete bands, the increased contribution from CO_2 and H_2O makes radiation in an oxy/fuel system more spectral. From Wien's displacement law, we know that shorter-wavelength radiation becomes more important due to higher gas temperature. More importantly, increased levels of CO, OH, and CH radicals help increase shorter-wavelength radiation. This might be important for some types of materials being heated in the furnace, such as clear glass, so that more radiation can penetrate the material to provide even heating.

4.4 GOVERNING EQUATIONS

In Section 4.2.4, the governing equations of fluid mechanics for a turbulent flow are derived. Similarly, the governing equations for heat transfer and mass transfer can be derived from the principles of energy and mass conservation. In fact, the species conservation equation is an extension of the overall mass conservation (or the continuity) equation. For species i, it has the following form:

$$\frac{\partial}{\partial t}\left(\bar{\rho}\tilde{Y}_i\right)+\frac{\partial}{\partial x_j}\left(\bar{\rho}\tilde{u}_j\tilde{Y}_i\right)=\frac{\partial}{\partial x_j}\left(\bar{\rho}D\,\frac{\partial \tilde{Y}_i}{\partial x_j}-\overline{\rho u_j''Y_i''}\right)+\overline{\dot{R}_i} \tag{4.58}$$

where Fick's law of diffusion, as introduced in Equation 4.7, is substituted into the general conservation equation, and proper averaging has been applied. The additional term inside the parentheses on the right-hand side represents mass fluxes due to the fluctuating velocities in a turbulent flow. A turbulence closure model must be used. Consistent with the $k-\varepsilon$ model, a gradient diffusion assumption is commonly used to provide the closure:

$$\overline{\rho u_j''\phi''}=-\frac{\mu_t}{\sigma_t}\frac{\partial\tilde{\phi}}{\partial x_j} \tag{4.59}$$

where σ_t is the turbulent Prandtl/Schmidt number for ϕ. The species conservation equation now becomes

$$\frac{\partial}{\partial t}\left(\bar{\rho}\tilde{Y}_i\right)+\frac{\partial}{\partial x_j}\left(\bar{\rho}\tilde{u}_j\tilde{Y}_i\right)=\frac{\partial}{\partial x_j}\left(\bar{\rho}D+\frac{\mu_t}{\sigma_s}\right)\frac{\partial \tilde{Y}_i}{\partial x_j}+\overline{\dot{R}_i} \tag{4.60}$$

The mean species source term, $\overline{\dot{R}_i}$, requires further examination. Because this term is usually highly nonlinear as illustrated in Equation 4.8 and its value directly controls the reaction progress, a proper closure model is necessary. Indeed, various methods are available from the literature.[21,22] For the purpose of discussion, we consider the eddy breakup model of Magnussen and Hjertager.[23] Note that this model is chosen for its simplicity rather than accuracy, much as the $k-\varepsilon$ model was selected for turbulence closure. In this model, a turbulent reaction rate is computed which is then compared with the kinetic rate. The smaller of the two is used as the reaction rate because it limits the reaction progress. The kinetic rate is simply the laminar reaction rate evaluated at the mean temperature, pressure, and concentrations:

$$\overline{\dot{R}_i^k} = \dot{R}_i\left(\tilde{T}, \tilde{Y}_l, \ldots\right) \tag{4.61}$$

where superscript k indicates the kinetic rate.

To determine the turbulent reaction rate, conceptualize the reaction as a single-step process where the fuel and oxidant are converted to a product. Furthermore, visualize a turbulent flame as a collection of burning eddies. For a reaction to take place, the fuel and oxidant must be simultaneously available, and the reaction product must be removed properly. That is, the reaction can be limited by the availability of fuel, oxidant, or by the removal of product. Once the fuel and oxidant are mixed, combustion takes place, on average, within one eddy turnover time, k/ε, so that the reaction rate can be written as

$$\overline{\dot{R}_{i,j}^t} = \min\left\{\tilde{c}_F, \tilde{c}_o, B\tilde{c}_P\right\} A \nu_{ij} M_i \bar{\rho} \frac{\varepsilon}{k} \tag{4.62}$$

where superscript t indicates turbulent rate. $\tilde{c}_F$, $\tilde{c}_O$, and $\tilde{c}_P$ are the mean molar concentrations of fuel, oxidant, and product, respectively, and A and B are model constants. Note that the turbulent rate has been generalized to a multistep system, so that the overall mass source for species i due to turbulence is

$$\overline{\dot{R}_i^t} = \sum_j \overline{\dot{R}_{i,j}^t} \tag{4.63}$$

and the mass source term used in the species transport equation is

$$\overline{\dot{R}_i} = \min\left\{\overline{\dot{R}_i^t}, \overline{\dot{R}_i^k}\right\} \tag{4.64}$$

The energy conservation equation for laminar flows can be written in terms of enthalpy as

$$\frac{\partial}{\partial t}(\rho h) + \frac{\partial}{\partial x_j}\left(\rho u_j h\right) = -\frac{\partial q_j}{\partial x_j} + S \tag{4.65}$$

where the mixture enthalpy is related to component enthalpies as

$$h = \sum_i h_i Y_i \tag{4.66}$$

and the enthalpy for species i is a combination of sensible enthalpy and the formation enthalpy:

$$h_i = \Delta h_{f,i}^{\circ} + \int_{T_0}^{T} c_{p,i}\, dT \tag{4.67}$$

Note that the standard temperature (77°F or T_0 = 298.15 K) is used in this definition. $c_{p,i}$ is the specific heat and $\Delta h^\circ_{f,i}$ is the enthalpy of formation at the standard state, both for species i. The heat flux, q_j, includes contributions from conduction, radiation, differential diffusion among component species, and concentration gradient-driven Dufour effect. For combustion applications, the most important contributions come from conduction and radiation. As discussed in Section 4.3, conduction heat flux follows Fourier's law (Equation 4.27) and radiation heat flux is related to the local intensity as

$$-\frac{\partial q_{r,j}}{\partial x_j} = a\int_{4\pi} I d\omega - 4a\sigma T^4 \quad (4.68)$$

where a is absorption coefficient of the mixture and σ is the Stefan–Boltzmann constant. The radiative transfer equation for intensity I is described in Equation 4.54.

The source term S in the energy conservation equation in general includes contributions from flow work ($u_j \partial p/\partial x_j$), viscous dissipation ($\tau_{ij}\partial u_i/\partial x_j$), and any external heat source. For simplicity, only conduction and radiation are included in further discussions.

In turbulent flows, the energy equation becomes

$$\frac{\partial}{\partial t}\left(\bar{\rho}\tilde{h}\right) + \frac{\partial}{\partial x_j}\left(\bar{\rho}\tilde{u}_j\tilde{h}\right) = \frac{\partial}{\partial x_j}\left(k\frac{\partial \tilde{T}}{\partial x_j} - \overline{\rho u_j'' h''} - \bar{q}_{r,j}\right) \quad (4.69)$$

Again, the gradient diffusion assumption can be used to describe the turbulent heat flux term:

$$\overline{\rho u_j'' h''} = -\frac{c_p \mu_t}{\sigma_h}\frac{\partial \tilde{T}}{\partial x_j} \quad (4.70)$$

so that the energy equation becomes

$$\frac{\partial}{\partial t}\left(\bar{\rho}\tilde{h}\right) + \frac{\partial}{\partial x_j}\left(\bar{\rho}\tilde{u}_j\tilde{h}\right) = \frac{\partial}{\partial x_j}\left(k + \frac{c_p \mu_t}{\sigma_h}\right)\frac{\partial \tilde{T}}{\partial x_j} - \frac{\partial \bar{q}_{r,j}}{\partial x_j} \quad (4.71)$$

The governing equations share a common form (notations for averaging dropped):

$$\frac{\partial}{\partial t}(\rho\phi) + \frac{\partial}{\partial x_j}\left(\rho u_j \phi\right) = \frac{\partial}{\partial x_j}\left(\Gamma\frac{\partial \phi}{\partial x_j}\right) + S, \quad \phi = 1, u_i, Y_i, h, k, \varepsilon \ldots \quad (4.72)$$

Physically, these terms mean time variation, convection, diffusion, and source, respectively. By substituting the appropriate transport variable for ϕ, we obtain the corresponding transport equation for that variable. With a common form, the solution procedure becomes similar.

4.5 SOLUTION TECHNIQUES

Combustion is a rather complex phenomenon that involves fluid flow, mixing, chemical reaction, and heat transfer all at once. Different solution techniques can be used to solve the problem of interest. For some problems, one feature might be dominating while others might be secondary. For other problems, several features might be important simultaneously. Thus, the choice of a solution technique depends on the task and the problem definition. Generally, analytical, empirical, and numerical methods can be identified.

Analytical and empirical methods are usually used for fairly simple geometries, well-studied phenomena with experimental data. Flow inside a circular pipe or around a sphere are two examples. Analytical or empirical solutions are difficult to obtain for more-complex geometries or more-complex physics. In such a case, numerical methods can be used.

Several numerical methods, such as finite volume, finite difference, finite element, spectral methods, etc., are widely used for solving the complex set of partial differential equations. The latest computer technology allows us to obtain solutions with a mesh resolution on the order of millions of nodes. More-detailed discussion on numerical methodology is provided later.

4.5.1 SCALE ANALYSIS AND GENERAL PROCEDURE OF PROBLEM SOLVING

Regardless of what solution method one would use, the first step is always focused on the simplification of a specified problem. To do that, an engineer or scientist usually uses the first-order evaluation of the defined processes. The processes and the system of interest should be carefully evaluated and the phenomena of greater influence should be identified. All the identified phenomena then should be simplified, if possible, within reasonable accuracy of the process description. Boundary conditions and system-simplifying assumptions then should be applied.

4.5.2 ANALYTICAL AND EMPIRICAL SOLUTIONS

Suppose one would like to evaluate a furnace combustion space bounded with refractory material. The composite walls have several layers of different refractory material and thickness. To estimate the losses through the furnace walls, one can obtain an analytical solution for temperature change in each layer of the refractory separately (if needed) and in the wall as a whole. If the heat transfer through the wall is considered to be one dimensional, the solution for the energy loss through the furnace wall can be obtained as

$$Q = hA(T_h - T_a) \tag{4.73}$$

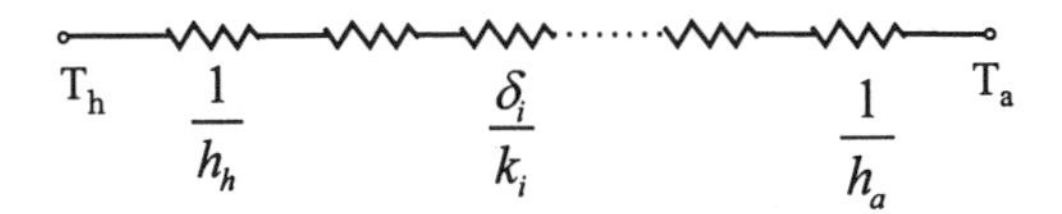

FIGURE 4.2 Electric resistance analogy for a multilayer refractory wall.

where Q is energy loss through the refractory, h is the overall heat transfer coefficient, T_h and T_a are internal wall and ambient temperatures, respectively, and A is the wall surface area.

Let us assume that we know the hot surface refractory temperature which is equal to T_h, thermal conductivity k_i, and thickness δ_i of each refractory layer and the ambient space temperature T_a. Using an electrical resistance analogy as illustrated in Figure 4.2, we find h as follows:

$$\frac{1}{h} = \frac{1}{h_h} + \sum_i \frac{\delta_i}{k_i} + \frac{1}{h_a} \tag{4.74}$$

As a result, the overall heat transfer through the composite refractory wall is known. The hot face and cold face heat transfer coefficients can be calculated from known expressions for forced and free convection near a flat plate. These expressions have the same structure but different empirical constants and can be found in, for example, Reference 20.

If the system geometry is too complex, for example, a detailed car engine, or a number of phenomena are important simultaneously, we may have to resort to numerical techniques. At present, the development and application of numerical methods have led to a new technology known as computational fluid dynamics (CFD).

4.5.3 Numerical Solutions

With ever more powerful computers and increasingly sophisticated software, numerical solution of the governing equations of turbulent reacting flows becomes practical, at least for some classes of problems. A new technology has established itself to be known as computational fluid dynamics or CFD, and it has been regarded as the third prong in the "tripartite science" of combustion: theory, experiment, and computation.[24] By far, numerical solutions through CFD provide the most-detailed information. They also can include far more physical phenomena in the analysis, thus providing insight into the interdependence of various factors. Currently, models up to millions of grid points can be analyzed, full geometry details with feature sizes varying over 1000 times or more can be treated, and detailed reaction mechanisms involving hundreds of reaction steps and species can be solved, although generally not all of these aspects can be included simultaneously. Certainly, direct numerical simulation which fully resolves the length scale and timescale of turbulence and does away with turbulence modeling is not yet possible for turbulent combustion flows of practical importance.

In this section, rather than delve into the details of numerical solution procedures and provide endless examples, we intend to describe the general steps only, and highlight what to expect and what to be cautious about in each step. The analogy of numerical solutions and laboratory experiments is emphasized. This approach acknowledges the fact that today most numerical solutions are obtained through the use of CFD packages. Understanding the physics of the problem and the general solution procedure is more important than a mastery of numerical details inside the CFD packages.

Numerical solutions are experiments in the virtual world of computers. As such, they share many similarities with laboratory experiments in the real world. Each solution is a data point much like a data point in a laboratory setting, and certain parameters, such as flow rate, temperatures, and geometry, are allowed to change and new solutions obtained so that a trend of variation can be identified. Just as laboratory experiments involve preparation, data collection, and data analysis, numerical solutions involve preprocessing, solution control, and postprocessing, three common steps to be discussed next.

Preprocessing for numerical solutions refers to geometry representation, domain discretization, boundary condition specification, and physical model selection. For computers to analyze the combustion system, we must first represent the real-world geometry in the computer. Depending on the analysis objectives, various levels of geometry details need to be included. For example, if the objective is to understand the performance of a low-NOx burner, the details of the burner must be included in the analysis, and some level of simplification and idealization must be employed for the furnace. On the other hand, if the objective is to understand how such burners interact with each other in a multiburner furnace, it may not be necessary to include all the details of the burner in the model. Instead, details of the furnace should be included as much as is practical. The analysis objectives also determine where to begin the domain of analysis and where to end it, and what type of boundary conditions can be applied at those boundaries with reasonable accuracy. Once the geometry is identified, it must be discretized into small control volumes, or grid points, for which the governing equations in differential form can be approximated algebraically. The grid arrangement can follow a global structure, a local structure, or no structure at all, resulting in a single-block-structured grid, a block-structured grid, and an unstructured grid, respectively. The shapes of control volumes range from quadrilaterals and triangles in two dimensions to hexahedrons, tetrahedrons, pyramids, prisms, wedges, and other chopped-off derivatives of these shapes in three dimensions. Geometry representation and domain discretization used to be the most time-consuming and labor-intensive tasks of numerical analysis just a few years ago. At the time of this writing, tools were available that automated most of the tedious steps. Complex models that took months to build can now be completed in days or hours. It is expected that before long complete automation is possible for domain discretization or mesh generation, and seamless interfaces allow two-way exchanges of geometry and analysis data between designers and analysts. For numerical solutions to impact equipment designs, to provide guidance for product optimization and real-time process control, increasingly sophisticated tools are required that make the preprocessing tasks intuitive and transparent.

Boundary conditions and physical models are specified next. These tasks are usually simple, but they have an enormous impact on the validity and accuracy of the results. After all, the solution is at best only as accurate as the input data and the selected physical models. It is therefore important that the correct boundary conditions be applied, and that the appropriate physical models be selected. Today, most commercial CFD packages include a host of physical models that can be turned on and off with the click of a mouse. It is unwise to oversimplify the problem by omitting certain important features; it is equally unwise to turn on every available option in the package.

Implicit in the preprocessing step is the method of equation discretization in which the differential governing equations are transformed into algebraic equation sets. Today, commercial CFD packages complete this task automatically. However, it used to be a major undertaking; in fact, many existing texts on numerical analysis follow the theme of equation discretization methods. Specifically, finite element and finite volume methods are the two main themes, each with its own myriad variations, merits and shortcomings, believers and critics. Suffice it to say there has been enough cross-pollination between the two methods that it is irrelevant to argue which is the preferred. It just so happens that, historically, the finite element method has been more common for low-speed laminar flows and the finite volume method is more prevalent in turbulent combustion applications.

Analogous to the data collection and actually running the experiments in the laboratory, solution monitoring and control refer to the process of obtaining a solution. Today, various options are available to speed up the solution, maintain solution stability, and revise the mesh based on intermediate results. Parallel computing is available in most commercial CFD packages to take advantages of multiple processors in a computer or a network of computers of the same type or different mixture. Massively parallel packages that utilize thousands of processors are emerging. The combination of hardware advancements and software improvements, such as parallelization, code vectorization, and optimization, have reduced solution time from months to days or hours, even minutes. More opportunities are certainly made available as a result.

Solution adaptive mesh refinement is another powerful feature that enables faster and more-accurate results. This method allows the user to modify the grid interactively based on intermediate solutions so that grid points can be put where they are required and can make the most contribution to the solution accuracy. Although not new, solution adaptive mesh refinement has become increasingly easy to use due to more-sophisticated user interfaces and the ability to handle hybrid meshes in the new generation of CFD packages.

Once a solution is obtained, analysis of the data in the postprocessing step is similar to data analysis in a laboratory experiment program. Today, most packages provide contours, vectors, line plots, and alphanumeric reports for almost any quantity and on almost any surface desired. More-sophisticated features such as animation are also available. Flexible user interfaces also allow custom programming to extract additional quantities of interest. However, data exchange back to the geometry side or down to other analyses such as stress still requires significant effort.

Like real-world experiments, the most valuable application of numerical solutions lies in identifying the underlying relationships in the combustion flow system through parametric studies. That means a large number of solutions must be obtained in a timely and accurate way. It is thus important that each of the three solution steps be automatic as much as possible. In the aerospace industry, where CFD was first applied and which is by far the leader in this technology, proprietary front-end programs directly compute aircraft geometry based on performance specifications using CFD as a solution engine, rather than the other way around in a traditional application. New commercial airliners roll out from assembly lines without ever building a physical prototype, thanks in part to highly sophisticated analysis software including CFD. In the automotive industry, where CFD has been experiencing a recent boom, short product cycle and intense competition push the limits of both hardware and software in numerical analysis. Judging from the recent developments in computer technology and software engineering, we can fairly say that CFD has become a viable technology of practical importance.

Nevertheless, we must realize the limits of numerical solutions. Physical models are only as good as our understanding of the real phenomenon. Unfortunately, many aspects involved in turbulent reacting flows are currently not well understood, such as turbulence, chemistry, and turbulence–chemistry interactions, to name a few. While we have reason to be excited about the advancements of numerical solutions and the opportunities made available by those advancements, we should be keenly aware that a lot more is required. Numerical analysis will not replace experiments; it is a complementary tool at our disposal. In the final analysis, nature provides the best solutions.

REFERENCES

1. Kakac, S., Shah, R. K., and Aung, W., *Handbook of Single-Phase Convective Heat Transfer*, John Wiley & Sons, New York, 1987.
2. Siegel, R. and Howell, J. R., *Thermal Radiation Heat Transfer*, 3rd ed., McGraw-Hill, New York, 1992.
3. Modest, M. M., *Radiative Heat Transfer*, McGraw-Hill, New York, 1993.
4. Arpaci, V. S., *Conduction Heat Transfer*, Addison-Wesley, Reading, MA, 1966.
5. White, F. M., *Viscous Fluid Flow*, McGraw-Hill, New York, 1974.
6. Kays, W. M. and Crawford, M. E., *Convective Heat and Mass Transfer*, 2nd ed., McGraw-Hill, New York, 1980.
7. Bejan, A., *Convection Heat Transfer*, John Wiley & Sons, New York, 1984.
8. Kuo, K. K., *Principles of Combustion*, John Wiley & Sons, New York, 1986.
9. Dryer, F. L., The phenomenology of modeling combustion chemistry, in *Fossil Fuel Combustion — A Source Book*, Bartok W. and Sarofim, A. F., Eds., John Wiley & Sons, New York, 1991, Chap. 3.
10. Favre, A., Statistical equations of turbulent gases, in *Problems of Hydrodynamics and Continuum Mechanics* (Sedov 60th birthday volume), SIAM, Philadelphia, 1969.
11. Launder, B. E., Reece, G. J., and Rodi, W., Progress in the development of a Reynolds-stress turbulence closure, *J. Fluid Mech.*, 68(3), 537–566, 1975.
12. Rodi, W., *Turbulence Models and Their Application in Hydraulics*, 2nd ed., International Association for Hydraulic Research, Delft, The Netherlands, 1984.

13. Yakhot, V. and Orszag, S., Renormalization group analysis of turbulence. I. Basic theory, *J. Sci. Comput.*, 1(1), 1–51, 1986.
14. Jones, W. P. and Launder, B. E., The prediction of laminarisation with a two-equation turbulence model, *Int. J. Heat Mass Transfer*, 15, 301, 1972.
15. Carslaw, H. S. and Jaeger, J. C., *Conduction of Heat in Solids*, 2nd ed., Oxford University Press, New York, 1992.
16. Bird, R. B., Stewart, W. E., and Lightfoot, E. N., *Transport Phenomena*, John Wiley & Sons, New York, 1960.
17. Bennett, C. O. and Myers, J. E., *Momentum, Heat, and Mass Transfer*, 2nd ed., McGraw-Hill, New York, 1974.
18. Hottel, H. C. and Sarofim, A. F., *Radiative Transfer*, McGraw-Hill, New York, 1967.
19. Sparrow, E. M. and Cess, R. D., *Radiation Heat Transfer*, Brooks/Cole, Belmont, CA, 1970.
20. Lenenberger, H. and Pearson, R. A., Compilation of Radiation Shape Factors for Cylindrical Assemblies, Paper #56-A-144, ASME, New York, 1956.
21. Sivathan, Y. R. and Faeth, G. M., Generalized state relationships for scalar properties in non-premixed hydrocarbon/air flames, *Combust. Flame*, 82, 211–230, 1990.
22. Bilger, R. W., Turbulent flows with nonpremixed reactants, in *Turbulent Reacting Flows*, Libby, P. A. and Williams, F. A., Eds., Springer-Verlag, Berlin, 1980.
23. Magnussen, B. F. and Hjertager, B. H., On mathematical models of turbulent combustion with special emphasis on soot formation and combustion, in *Proc. 16th Int. Symp. on Combustion*, Cambridge, MA, August 15–20, 1976.
24. Williams, F. A., The role of theory in combustion science, Hottel Plenary Lecture, in *Proc. Twenty-Fourth Symp. on Combustion*, The Combustion Institute, Pittsburgh, PA, 1992, 1–17.

5 Ferrous Metals

Marie DeGregorio Kistler and J. Scott Becker

CONTENTS

5.1 INTRODUCTION

The history of iron and steelmaking has been substantially influenced by the availability of oxidants for the various metallurgical processes. Iron making in North America dates back to the 1600s. The process utilized charcoal and local iron ores in a blast furnace that looked more like a modern cupola into which cold air was blown, powered by a bellows. The primary product was cast iron (containing 2 to 4% carbon). Steel (0.1 to 0.8% carbon) could only be made by a very low quality, labor-intensive process known as puddling.

Nothing much changed in iron or steelmaking for the next 200 years until the Bessemer process was introduced in the mid 1800s. In the Bessemer process, molten iron from the blast furnace was transferred to a separate refractory-lined vessel into which large quantities of cold air were blown to oxidize and remove the carbon. This was made possible by developments in air compression equipment. The modern steel age was born. The availability of higher-quality, low-cost steel drove rapid expansion of the market for iron and steel and led to a rapid development of several new technologies.

The next major improvement was the development of the Siemens or open hearth furnace. Unlike the Bessemer furnace, which relied solely upon the oxidation of various elements in the bath for heat, the open hearth furnace was "fired" by burners, which allowed for external fuel to add considerable melting power to the process. The most important aspect of the open hearth process was its use of preheated air. Without air preheating, the metal would have melted very slowly, if at all. The air preheating occurred in refractory brick "checkers" (or checker-work) located under

0-8493-1695-2/98/$0.00+$.50

the furnace. The checkers were heated by countercurrent heat exchange from the hot exhaust gases from the furnace. The hot checker-works then preheated the incoming air to the furnace. With two separate checkers alternately being heated by the exhaust and then preheating the air, the open hearth was one of the first effective industrial uses of heat recovery. Ultimately, the open hearth would come to utilize substantial amounts of oxygen.

The advent of the open hearth dramatically lowered the cost of steel and fueled unprecedented growth in the steel markets. Meanwhile, the blast furnace was still producing iron in much the same way it had for many years. The charcoal had been replaced with coke, the air was being preheated in stoves, and the furnace size was increased, but the process was similar. With increased steel demand there was great interest in enhancing the productivity of the blast furnace. In the mid 1950s the concept of injecting oxygen into the blast air emerged, and from then on blast furnaces have been using ever increasing volumes of pure oxygen mixed into the hot blast air. This is done to improve both process efficiency and productivity. Injection of supplemental fuels also began to develop, and before long blast furnace productivity once again came to the forefront of steelmaking.

The open hearth furnace benefited from two new developments in oxygen-based combustion which substantially improved its productivity. The first was called undershot enrichment. Essentially this is a technique in which pure oxygen is injected into the underside of a conventional air/fuel flame. In effect, this serves to concentrate the enrichment in the part of the flame closest to the molten bath. While some mills also tried conventional enrichment of the hot combustion air, this was generally found to be substantially less efficient. The second development was the water-cooled oxy/fuel burner developed and patented by Air Products and Chemicals, Inc.[1] Because these burners were water-cooled, they could be installed into the roof of the open hearth (through the refractory and actually protruding into the furnace space) and could concentrate considerable energy input on the charge. This was the first significant use of conventional oxy/fuel combustion in iron and steelmaking. When equipped with both undershot and oxy/fuel burners, the productivity of the open hearth furnace was impressive, but not high enough to compete with the newly developed basic oxygen furnace (BOF).

The development of the BOF process (also called LD for Linz–Donowitz) probably began in Germany or Austria before World War II. Modern application of the process did not occur until the mid 1960s. Basically, the BOF was very similar to the Bessemer furnace except that the air was replaced with pure oxygen and it was blown into the molten iron bath from above through a water-cooled lance. The BOF became and remains today the standard for mass production of high-quality steel derived primarily from molten iron. The oxygen injected into the iron oxidizes (or burns) out the dissolved elements in iron (primarily silicon, manganese, carbon, and phosphorus). Since the whole process relies on oxygen, supplemental oxygen use is unnecessary, although split injection of the oxygen to enhance the post combustion of the CO evolved from the bath was developed and will be discussed later in this chapter.

Even before the arrival of the BOF, a major alternative to conventional ore-based steelmaking had begun to gain in acceptance. The electric arc furnace (EAF) was conceived as a unit to melt scrap steel and recycle the iron units back to useful service. It did not rely on molten iron from a blast furnace, but rather solid scrap was melted by energy input from electrical arcs passed between three carbon electrodes. Originally this furnace was conceived as an all-electric melter. Freedom from the blast furnace and comparatively cheap electric power fueled the growth of the EAF, particularly in what came to be known as minimills.

As the EAF grew in popularity it also came under continued cost pressure to lower operating cost as well as increase specific productivity. Thus began the development of a series of oxygen uses that have caused the average oxygen used in EAF melting to grow from nearly zero to almost 2000 SCF/ton (a BOF also uses about 2000 SCF/ton by comparison). The first use of oxygen in EAFs was to react chemically with the dissolved carbon (in this case from the scrap and electrodes) present in the molten bath and to lower it to desired levels based upon the steel grade produced. Direct injection of oxygen through carbon steel pipes proved much faster and easier to control than the previous practice of charging mill scale (iron oxide) to promote its reduction with carbon. Once oxygen was being injected into the furnace, operators discovered that it could be used to help the furnace melt faster. Oxygen cutting soon became the norm in most shops and productivity soared, but at the expense of yield loss (the oxygen was in effect being used to burn scrap). Since scrap is a very expensive fuel, additional carbon was also frequently charged into the furnace. This practice was difficult to control and led to considerable process variability.

The stage was clearly set for installing oxy/fuel burners in the EAF. Beginning in Japan in the early 1970s and North America in the early 1980s, the growth of oxy/fuel-assisted EAF melting has been remarkable. Oxy/fuel burners are now almost standard for EAFs. They are used to meet one or more of the following needs: to eliminate cold spots thereby improving melting efficiency and speed, to increase dramatically the total energy input rate, to increase production where the existing electrical power delivery system is limited, to replace electrical energy where it is relatively more expensive than oxy/fuel-derived energy, or selectively to avoid peak electricity demand charges. Specific power density (kW/metric-ton) has been increasing over time. Stoichiometric combustion of natural gas and oxygen provides 5.7 kWh/Nm^3 of oxygen. Literature indicates that 0.15 MW of burner rating should be supplied per metric ton of furnace capacity.

5.2 OXY/FUEL BURNERS IN THE EAF

Initially, firing through the EAF slag door, oxy/fuel burners mounted on a boom or a carriage were introduced to increase scrap melting. Later, to target the cold spots efficiently in the EAF, the burner location was moved to either the roof or sidewalls, to aim at the cold spots associated with the spaces between the electrodes, with one to four burners in a furnace, providing supplemental energy. In addition to productivity improvements of 5 to 20%, the burners provide economical energy for melting scrap at a lower cost. Figures 5.4 and 5.5 show scatter plots of published data on the impact of specific burner capacity (MW_{burner}/metric-ton of steel) on EAF productivity and electrical power savings (kWh/metric-ton).[2] Depending on the region

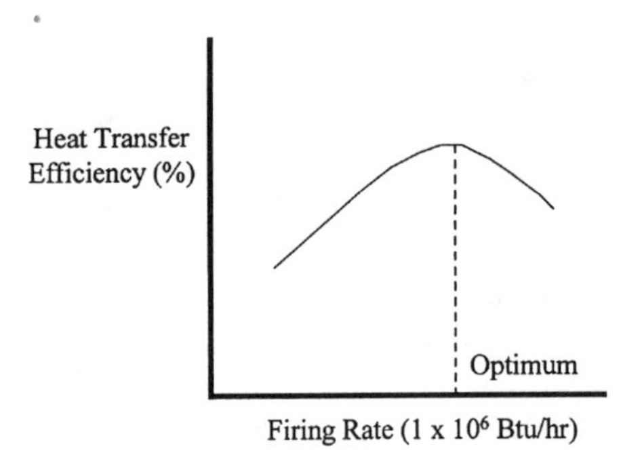

FIGURE 5.1 Optimum firing rate is at maximum heat transfer efficiency.

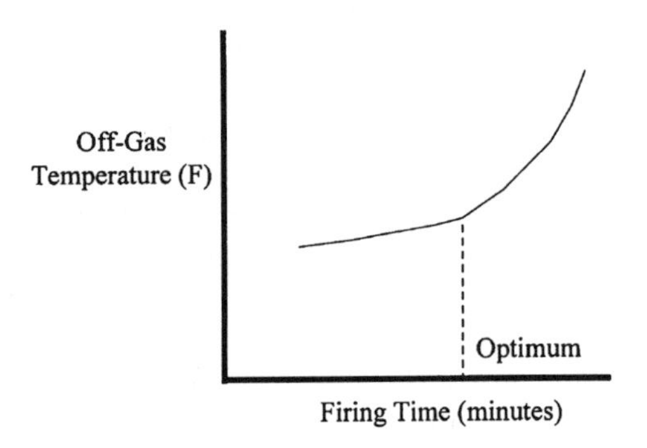

FIGURE 5.2 Optimum firing time is determined by plotting off-gas temperature vs. time for several heats.

or country, the energy cost savings benefit becomes more attractive as the unit cost of electricity and high-quality electrodes rise, whereas that of oxygen and natural gas remain constant or actually fall.

Oxy/fuel burners are most efficient during cold scrap heating and meltdown. Typically, burner effectiveness decreases during the latter half of the meltdown period from an initial 60 to 80% to less than 20%, when firing on a flat molten bath. Excessive use of burners with low efficiency can result in yield loss and potential furnace damage due to higher off-gas temperatures.

For smaller furnaces, a single slag door burner is convenient and adequate since the operator can effectively reach all cold spots by aiming the burner as necessary. For larger furnaces, generally three sidewall-mounted burners are optimal (Table 5.1). If a foamy slag practice is employed, roof-mounted burners can reduce maintenance requirements. However, in many cases, roof mounting has led to considerable installation and maintenance costs.

Some early burner designs were maintenance nightmares, largely due to mounting or retraction mechanisms that were constantly becoming jammed with slag and metal. Today's state-of-the-art burners are easily mounted, nonretracting, and relatively low maintenance. The number and design of the burner selected for a particular

TABLE 5.1
Typical EAF Oxy/Fuel Results

	Company A	Company B	Company C
Transformer, MVA	19	64	100
Transformer, MVA	7.5	25	28
Mounting location	Slag door	Roof	Sidewall
Number of burners	1	3	3
Firing rate, MW	2.6	11.4	21.1
Results			
Production increase, %	28	15	15
Electrode savings, kg/metric-ton	0.7		0.2
Power savings, kWh/metric-ton	88	27	44

operation is determined by three primary considerations: (1) choice of mounting location, (2) available oxygen and natural gas supply pressures, and (3) furnace size and configuration.

With regard to design considerations, it is important to maximize flame velocity for maximum heat transfer efficiency (i.e., deeper scrap penetration, higher convective heat flux). By using the smallest-diameter burner for the available oxygen and natural gas supply pressures and desired firing rate, the highest flame velocity is achieved. Converging/diverging nozzle designs have also been successfully used in a number of burner designs.

Regardless of mounting location, water-cooling is necessary to ensure burner longevity. In the case of door or roof mounting, it is most practical to cool the burner via a water-cooled jacket. Sidewall-mounted burners may be cooled by either a water-cooled jacket or a separate water-cooled mounting block. A mounting block approach permits burner replacements without disconnecting the water supply and return hoses. In this case, the burners are somewhat lighter and therefore easier to handle and maintain.

Sidewall mounting has traditionally utilized three burners with one mounted in each conventional "cold spot." A recent development is the Split-Fire™ burner concept, in which two flames are fired from a single location.[3] This burner is ideal for small arc furnaces which typically exhibit only two cold spots.

An efficient oxy/fuel practice typically supplies 25% of the total energy required to melt the steel. This may be higher or lower depending on the desired results and characteristics of the furnace and operation in question. The optimum burner firing rate is limited by how quickly the scrap can absorb the heat supplied by the burners. In other words, the optimum firing rate is at the point where the heat transfer efficiency to the scrap is at a maximum (Figure 5.1). Experience and energy-balance-based computer models are often used as aids in predicting the optimum firing rate. In practice, the firing rate is further optimized during burner start-up by experimenting with different firing rates while monitoring melting rate, yield, and off-gas temperatures.

TABLE 5.2
Comparison of Firing Ratios

	Substoichiometric Ratio (Fuel-Rich)	Stoichiometric Ratio (No Excess Fuel or O_2)	Superstoichiometric Ratio (O_2 Rich)
Advantages	Minimal carbon and alloy oxidation	Highest flame temperature Maximum operating efficiency	Scrap lancing Better penetration of dense charges
Disadvantages	Reduced flame temperature Increased off-gas temperature		Carbon and alloy losses Increased electrode consumption Reduced chemistry control

TABLE 5.3
Air-Fuel vs. Oxy-Fuel Ladle Preheating

	Air Fuel		Oxy/Fuel				
	Average Firing Rate (MM/Btu)	Time to Preheat (h)	Average Firing Rate (MM/Btu)	Time to Preheat (h)	Net Natural Gas Savings	Preheat Time Decrease (%)	Ladle Size (ton)
Company A	6.0	1.5	1.77	1.5	70.5	N/A	60
Company B	5.8	2.5	2.75	1.5	71.6	40.0	50
Company C	6.2	1.3	2.40	0.8	76.1	38.5	55
Company D	10	14	3.5	5.0	87.5	64	90
Company E	9.2	2	4.4	0.75	56	62.5	292

The desired firing time can also be estimated based upon experience and computer models; however, this too is often further optimized during start-up. A practical method for optimizing firing time is to plot the off-gas temperature vs. the firing time (Figure 5.2). Initially, the heat transfer efficiency of the oxy/fuel flame is high. However, as the scrap melts, the surface area for heat transfer decreases, and consequently the heat transfer efficiency is reduced. Eventually, as the charge approaches the flat bath condition, most of the heat reflects off the bath surface and is lost to the off-gas and furnace walls. Therefore, the burners should be fired until the heat transfer efficiency drops substantially as indicated by an increase in off-gas temperature. This normally occurs when 60 to 80% of the scrap is below the bath surface.

Burners should have the flexibility of operating in either the fuel-rich or fuel-lean mode (Table 5.2). They can be fired fuel-rich with a bushy, luminous flame early in the heat to preheat scrap without excessive bridging or welding of the scrap. They are then switched to an oxygen-rich mode to cut away scrap and assist with uniform melt-in. Operating in an oxygen-rich mode is beneficial for slag door installations to assist with cutting away the scrap to allow earlier use of an oxygen/carbon lance manipulator to produce foamy slag.

TABLE 5.4
Typical Cost Savings with Oxy-Fuel Ladle Preheating

	Air/Fuel	Oxy/Fuel
Average fuel consumption (scfh)	6000	1770
Oxygen consumption (scfh)	0	3540
Preheater costs ($/operating hour)	$20.40	$13.50
Operating hours/year	6300	6300
Preheater costs/year	$128,520	$85,050
Savings/year[a]/preheater	$0	$43,470

Assumptions:Oxygen:natural gas ratio = 2:1; ladle operating hour/elapsed hour = 0.75.

[a] *Note:* The operating costs do not take into account the electricity costs of the systems that use an air blower for combustion air, nor do they reflect the savings in fire wall life and flue maintenance, which may exceed $20,000/year with oxy/fuel.

Oxy/fuel burners can also be operated with propane and various fuel oils. Compared with natural gas, propane has better combustion characteristics, while oil, especially the high-viscosity type, is relatively more difficult to combust efficiently and generates additional emissions. The oil burner equipment is also more complex and expensive.

The economic benefit of employing burners is greatly influenced by the relative costs of fossil fuel and electrical energy. Figure 5.3 shows the reduction in specific electrical power as more oxygen is used in the EAF. Figure 5.4 shows increased production ranging from 5 to 34% using oxygen. Figure 5.5 shows how much electrical power can be saved using oxy/fuel burners in a state-of-the-art EAF.

5.3 OXYGEN INJECTION IN THE EAF

5.3.1 Slag Foaming in the EAF

While use of oxy/fuel burners in the EAF grew rapidly, oxygen injection did not decrease, but also grew as improved methods of injection were developed. Especially important was the development of the combined injection of solid carbon and oxygen for the purpose of both increased energy input and the formation of a "foamy" slag which added additional operating benefits. As scrap melting proceeds in a typical EAF, the electrodes are no longer surrounded by scrap, and the heat transfer from the arc to the scrap and bath drops off with a corresponding increase of radiation to the furnace sidewalls. The efficiency of electrical energy input significantly increases when the arc is covered with slags. Therefore, a more recent trend has been to start injecting both carbon and oxygen to form CO in the slag. This is done comparatively early in

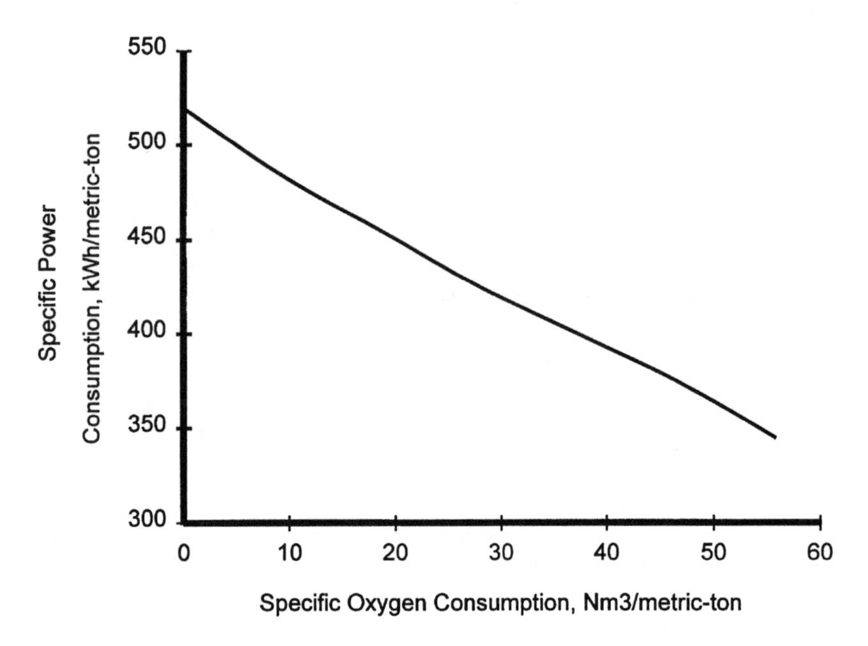

FIGURE 5.3 Power savings as a function of oxygen usage.

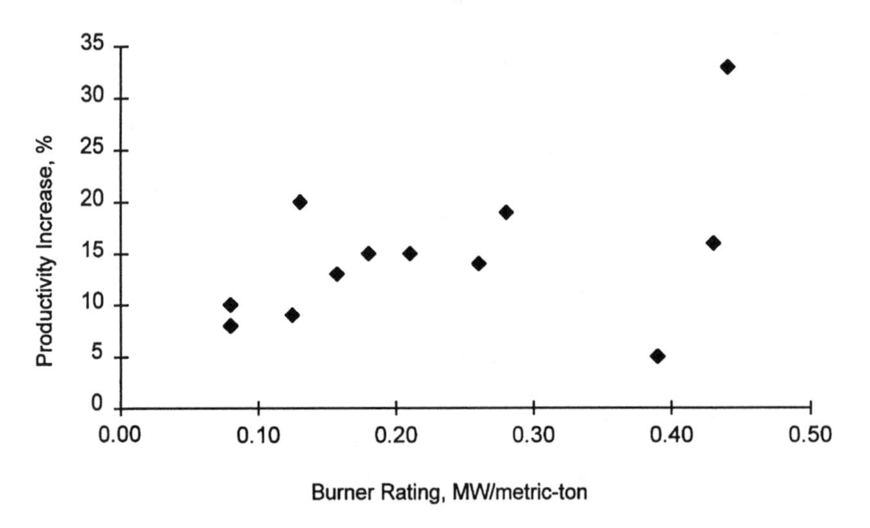

FIGURE 5.4 Productivity increase as a function of burner rating.

the melting process. The formation of CO in the slag produces an emulsified or foamy slag which can completely cover the arc, resulting in higher electrical energy utilization and shielding of the sidewalls from the arcs. It has been reported that at least 0.3% carbon should be oxidized to form good foamy slag. Reported improvements in electrical energy utilization range from an efficiency of 40% without slag foaming to 60 to 90% with slag foaming. Also, with a deeper layer of foamy slag, the arc voltage

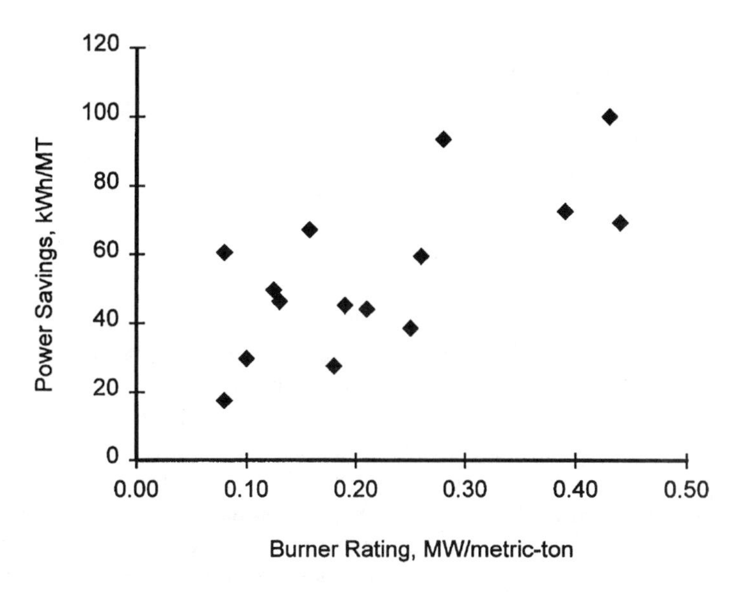

FIGURE 5.5 Power savings as a function of burner rating.

can be increased, allowing a higher power input rate. This can be especially beneficial when DRI (direct reduced iron) is being used as a scrap substitute, since DRI requires additional energy to melt. Early foamy slag formation is especially important in direct current (DC) and ultrahigh-power (UHP) alternating current (AC) furnaces.

When utilizing a foamy slag practice, it is also important to control closely slag basicity with simultaneous injection/addition of lime. In addition, the generation of higher CO concentrations and an associated increase in off-gas volume and heat content must be adequately addressed, preferably by some form of post combustion.

5.3.2 Post Combustion

The most recent application of oxygen in EAF melting is for post combustion. The constant drive for increased productivity in EAF melting has resulted in increased use of oxygen, natural gas, and carbon in the EAF industry as a whole, with a resulting increase in carbon monoxide and hydrogen in the furnace off-gases, the combustion of which is a potential additional heat source (effectively free fuel). Consequently, post combustion has become a high-priority opportunity and is currently being developed and optimized for electric arc furnace steelmaking. To date, many inconsistent claims have been made regarding the results and efficiencies obtained in implementation of the technology. It appears that the technology and its application to different furnace configurations and operating practices are far from optimized across the industry. It should also be remembered that post combustion is only one element of an overall low-cost, energy-efficient EAF steelmaking process which may also include the other technology elements (e.g., foamy slag, oxy/fuel burners, submerged gas injection, and alternate iron units) in addition to innovative process control concepts. Therefore, it must be integrated effectively with the overall process operation and should not be treated as an independent operation.

In basic terms, carbon and hydrogen react with oxygen in the EAF as illustrated in the following primary reactions:

$$2C + O_2 \rightarrow 2CO \quad (5.1)$$

$$2CO + O_2 \rightarrow 2CO_2 \quad (5.2)$$

$$2H_2 + O_2 \rightarrow 2H_2O \quad (5.3)$$

The conversion of carbon to carbon monoxide (Reaction 5.1) generates 2.4 kWh/Nm^3 of oxygen. The post combustion conversion of carbon monoxide to carbon dioxide (Reaction 5.2) generates approximately 6.3 kWh/Nm^3 of oxygen, nearly three times that of the first reaction, depending upon the temperature of the reactants. The post combustion of hydrogen, a product of natural gas decomposition, to water (Reaction 5.3) generates 6.2 kWh/Nm^3 of oxygen. These reactions illustrate the potential energy available by converting the products of decarburization and partial combustion — carbon monoxide and hydrogen to carbon dioxide and water, respectively, within the EAF. Due to the many sources of carbon utilized in the EAF process, the majority of the post combustion will involve CO.

During decarburization, 2.5 Nm^3 of CO may result from each 1 Nm^3 of oxygen injected. Typical ranges reported for CO evolution are between 26 and 66 Nm^3/min. Actual analyses of off-gases clearly indicate that carbon monoxide exits in the furnace in large volumes. When a furnace is not equipped for post combustion, "post combustion" of CO with excess air takes place in the ductwork, thereby increasing the temperature and loading on the baghouse system and essentially wasting the associated energy value. Also, up to 10% of the CO leaves the ductwork uncombusted and is introduced into the secondary fume-capture system. This represents a significant environmental concern.

Post combustion occurs in many forms, making it difficult to promote and control efficiently. The various CO sources and evolution rates (i.e., decarburization, burner combustion products, electrode oxidation, and slag-foaming carbon) need to be balanced with the means of post combustion oxygen introduction (i.e., oxygen flowrate, injection angle, and velocity) with due consideration to cost issues related to each carbon source. In addition, an understanding of the limitations of various heat transfer mechanisms and efficiencies is a key consideration in developing an effective process since the energy is wasted if not effictively transferred to the charge.

For instance, post combustion in the headspace of the furnace (i.e., heat transfer to both scrap and flat bath surfaces) and in the slag layer (heat transfer to slag and slag–metal interaction) must be understood and addressed for specific equipment and furnace configurations. Furthermore, the process control mechanisms (vis-à-vis response times and oxygen utilization) and equipment utilized to effect the post combustion are critical to its viability.

Post combustion is not a new concept. Post combustion techniques in the BOF were actually developed long before post combustion in the EAF. While various approaches have been applied, all basically involve the injection of oxygen high up in the BOF vessel to combust the CO being liberated in the bath by oxidation of dissolved carbon in the molten iron. In BOF post combustion the intention is that the heat released will be radiated back to the molten bath either directly from the flame or indirectly from reradiation from the hotter upper walls of the BOF vessel. Generally, the critical issue in designing the oxygen injection is to obtain very rapid mixing so as to achieve uniform and rapid combustion of the CO as it is being rapidly exhausted from the vessel.

5.4 LADLE PREHEATING

Another major use for oxy/fuel combustion in the melt shops of the steel industry is for ladle preheating. Today's refractories and steelmaking processes often require ladle preheat temperatures above 2000°F (1400 K), which can be difficult to obtain with the relatively low temperatures generated by conventional air/fuel flames. Because of its higher flame temperature, oxy/fuel provides the energy required for ladle preheating in a form that can reach the desired temperature in a shorter period of time with minimum flue losses and maximum fuel efficiency. Table 5.3 shows comparative operating data from five ladle preheaters converted from air-fuel to oxy-fuel. Table 5.4 provides comparative economics of air-fuel vs. oxy-fuel preheating.

Basically, a steel ladle is a relatively small, refractory-lined vessel that during preheating behaves very much like a small furnace. Heat transfer from the preheater to the ladle is primarily by convection and radiation. The convection can be described using

$$Q_c = h_c A \Delta T \tag{5.4}$$

where Q_c = convective heat flow
h_c = convective heat flow coefficient
A = hot-face surface area of the ladle (a constant)
ΔT = temperature difference between the hot face of the ladle and the combustion gases

To maximize heat flow (Q_c), the preheater must maximize h_c and ΔT. The convective heat flow coefficient (h_c) is a function of the combustion gas velocity, which is a function of the burner firing rate and design. To maximize h_c, a preheater should incorporate a "high-fire/low-fire" logic. This approach allows the burner to fire at high fire whenever the ladle requires energy so that maximum combustion gas velocity is maintained. This approach enables the preheater to achieve maximum heat flow at the most efficient times.

Since the hot-face temperature cycle is the same whether the ladle is heated with air/fuel or oxy/fuel, ΔT is a function only of the combustion gas temperature, which is a direct function of the flame temperature. Therefore, to maximize ΔT, the flame temperature has to be maximized. As described above, higher flame temperatures are intrinsic to oxy/fuel (see Chapter 1).

Equation 5.5 depicts radiative heat transfer:

$$Q_r = 0.1713 F_e F_a 10^{-8} \left(T_1^4 - T_2^4\right) \tag{5.5}$$

where Q_r = radiative heat flow
F_e = emissivity factor
F_a = radiation coefficient due to geometric arrangement (a constant)
T_1 = absolute temperature of the preheater flame
T_2 = absolute temperature of the hot face of the ladle

To optimize radiant heat flow to the ladle (Q_r), thus making the most efficient use of the energy input, F_e and T_1 must be maximized. T_2 cannot be manipulated because the hot-face temperature cycle is the same, regardless of the heating method.

T_1 is maximized when the preheater has an oxy/fuel flame at a 2:1 (oxygen:natural gas) firing ratio. F_e, the emissivity factor, can be optimized by producing a luminous flame. Several companies have developed such burners including the Air Products proprietary K-Tech® burner. This burner has an outer annulus for natural gas flow (see Figure 2.6), which serves two purposes. First, it cools the burner tip via endothermic cracking of natural gas (the K-Tech burner is not water-cooled). Second, it produces a very luminous outer flame envelope, which complements the high temperature of the oxy/fuel core. The outer annulus maximizes flame luminosity, increases F_e, and also allows for flame length adjustment.

Oxy/fuel ladle preheating installations provide the following benefits: faster heating times, hotter ladle bottoms, fuel savings of over 70%, decreased off-gas of 90%, reduced maintenance, and no water-cooling requirements. Although fuel savings are a major economic benefit, the other advantages associated with the oxy/fuel ladle preheating system are just as important. The fire wall life also improves dramatically. In one reported example, the fire wall in an oxy/fuel system lasted more than 70% longer than a typical air/fuel fire-wall in this particular shop. This result is due to the lower amount of off-gas volume flowing across the firewall.

A good oxy/fuel ladle preheater incorporates several safety features. These include double block and bleed, flame supervision, low and high flow interlocks on oxygen and natural gas, a high-temperature interlock and manual reset (see Chapters 9 and 10). The entire system must conform to current NFPA (National Fire Protection Association) safety standards.

The Air Products state-of-the-art ladle preheating system also incorporates several new features. Ignition is provided by a spark ignitor rod mounted in the center tube of the K-Tech burner. This feature provides reliable ignition and is simple to replace. System controls are PLC (programmable logic controller) based with redundant hard-wired relays. The PLC provides system sequence for the oxygen and natural gas flow control valves, monitors all interlocks, and has primary control over the oxy/fuel system shutdown. The system also includes a digital display of oxygen and natural gas flow rates to the burner, with flow totalization capabilities.

5.5 OXYGEN IN BLAST FURNACES AND CUPOLAS

Enriching the cupola blast with oxygen has been in practice since the early 1930s.[4] Use of oxygen to enrich blast air in iron blast furnaces has an even longer history. More recently, lead and mineral wool cupolas also have been equipped with oxygen enrichment. Due to the comparatively high cost, the use of oxygen was not economically attractive for many cupolas until the 1970s when decreasing oxygen cost and increasing coke and pig iron costs led to widespread use of oxygen.

An increase in melt rate and/or reduction in charge material cost are the prime driving forces to use oxygen in both blast furnaces and cupolas. Furnace production can be increased either intermittently or permanently with oxygen enrichment often at a lower cost per ton of iron than achievable with alternative methods. Oxygen can also be used as a cost-saving tool because it allows decreases in charged coke, partial substitution of lower-cost coal for more-expensive metallurgical coke, replacing pig iron with less-expensive charge metallics in cupolas, and decreasing ferroalloy requirements in foundries.

The ability to vary oxygen enrichment levels instantaneously also provides the cupola or blast furnace operator with an extremely flexible device to control furnace operations better in spite of unavoidable changes in coke and metallic charge quality, blast humidity and temperature, and cast house demand pattern. Today, hundreds of cupola foundries worldwide use oxygen to increase cupola capacity and/or reduce unit operating cost, and most modern blast furnaces are equipped with oxygen injection.

Blast air (wind) is introduced through the cupola or blast furnace tuyeres (injector-like pipes). Dry air at sea level contains by volume 20.99% oxygen, 78.03% nitrogen, 0.94% argon, and 0.04% other gases, mainly carbon dioxide. Since both nitrogen and argon are inert in most processes, the analysis of air is usually taken to be 21% O_2, 79% N_2. Oxygen enrichment simply increases the oxygen content of the blast air from 21% to a higher level. This is normally achieved by adding pure oxygen to the air. When the oxygen content is increased to, say, 23%, the oxygen-enrichment level is typically referred to as 2.0% (23% – 21%). By increasing the oxygen content by 2%, the nitrogen content is reduced from 79% (in atmospheric air) to 77% (in enriched air). Generally, the enrichment levels practiced in cupolas range from 1% to 5% and in blast furnaces range from 0.5% to 8%.

The major benefits of oxygen enrichment are derived from the increased flame temperature in the combustion zone. The increase in flame temperature is a result of several factors, including decreased nitrogen content. The nitrogen supplied per unit of fuel or energy is reduced; therefore, less energy is needed for heating the nitrogen to the flame temperature. This energy is distributed among the remaining constituents of combustion products, bringing them to a higher flame temperature. Since heat transfer efficiency (especially radiation) increases with increasing flame temperature, the higher flame temperature resulting from oxygen enrichment increases heat transfer to the metallic charge in the furnace.

The rate of heat transfer by radiation from a heated surface, such as combusting coke in the cupola or blast furnace, increases with the fourth power of the absolute temperature. Therefore, the effective heat transfer by radiation increases at a much faster rate with increasing flame temperature than does the effective heat transfer by conduction or convection. Figure 5.6 illustrates how net radiated power increases with increases in temperature.[5]

Methods used to introduce oxygen into the cupola or blast furnace can be classified into two main categories: diffuser enrichment and tuyere injection

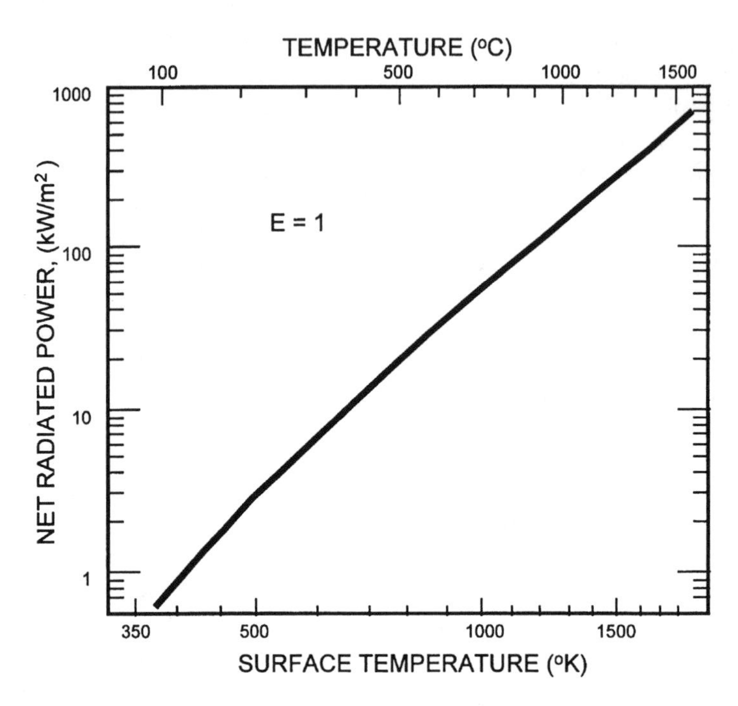

FIGURE 5.6 Net power radiated per unit of black surface to 77°F (298 K) surroundings.

(Figure 5.7). In the diffuser enrichment mode, oxygen is introduced into the wind main downstream of any recuperator through a diffuser installed in the wind main to increase the oxygen concentration of the resulting blast prior to entering into the furnace tuyeres. In tuyere injection pure oxygen is introduced directly into the cupola or blast furnace through additional injectors, located inside the tuyeres. Here, the oxygen introduction is independent of the blast stream. Most tuyere injectors are sized so that optimum operating pressures range from 15 to 100 psig (100 to 670 kPa). Oxygen velocities through these injectors typically range between 650 and 1700 ft/s (200 to 500 m/s).

Oxygen enrichment increases the operating efficiency of the cupola or blast furnace by either providing increased productivity (melt rate) for a given amount of energy input or reducing energy consumption per unit of molten iron produced. A 1976 investigation[6] on energy savings with oxygen enrichment of a cold blast cupola demonstrated that total energy required to produce 1 ton of molten iron could be reduced by about 11% (including the energy required for producing, transporting, and vaporizing oxygen) when 2% enrichment was used compared with no enrichment. The iron-to-coke ratio changed from 7.5:1 (no enrichment) to 9.2:1 (enrichment), an 18% coke savings with enrichment. The major contribution of oxygen in conserving energy in the cupola is the substantial reduction in coke requirements. When the total energy equivalent of coke including the energy utilized in its manufacture and delivery is balanced against the energy required to produce and deliver oxygen to the cupola, net energy savings up to 25% have been experienced.

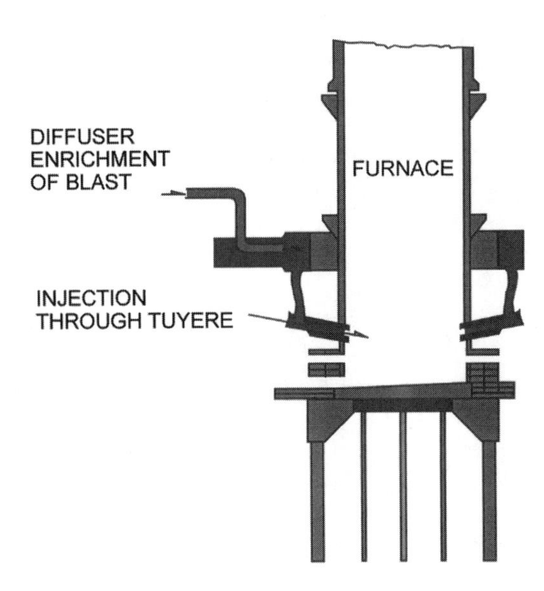

FIGURE 5.7 Methods of introducing oxygen into a cupola or blast furnace.

5.6 CONCLUSION

With the growth of EAF melting and the expansion of the minimill, driven primarily through the use of additional oxygen in oxy-fuel burners, slag foaming, post combustion, and ladle preheating, the older blast furnace/BOF-based integrated steel mills came under great cost/productivity pressure. Contrary to popular belief, the integrated mills did not just roll over and die, but have, in fact, begun to adopt many of the same oxygen-based combustion techniques discussed above. In particular, the integrated mills have moved to adopt oxy/fuel ladle preheating and also post combustion of CO in the BOF, as well as continuing to inject ever increasing volumes of fuel and oxygen into the blast furnace. Today, both integrated and mini-mills are increasingly utilizing oxygen to meet ever more stringent environmental regulations.

Applications for the oxygen-based combustion in the iron and steel industries are not limited to the processing of molten metal. Increased application of oxygen can be found in many metal-heating applications in the rolling facilities as well. In these applications, oxygen is utilized to enrich or even replace the air used in conventional air/fuel burners.

REFERENCES

1. Jones, R. D. and Miller, K. A., Variable Flame Oxy-Fuel Burner, U.S. Patent 3,578,793, May 18, 1971.
2. Wells, M. B. and Vonesh, F. A., Oxy-fuel burner technology for electric arc furnaces, *Iron Steelmaker*, November, 1986.

3. Becker, J. S., Kidd, T. C., McAfee, A. J., Mullane, J. C., and Wells, M. B., Oxy-Fuel Burner for Use in a Metallurgical Furnace, U.S. Patent D312869, December 11, 1990.
4. Morawe, F., Enriching the blast of the cupola furnace with oxygen, *Giesserie*, 74, 132–155, 1930.
5. Landefeld, C. F., Energy savings with covered pouring ladles and cupola troughs, *AFS Trans.*, 86, 187–192, 1978.
6. Rolseth, H. C., The effect of oxygen enrichment on energy consumption in iron cupola, *AFS Trans.*, 84, Paper No. 76-42, 295–299, 1976.

6 Nonferrous Metals

Debabrata Saha and Charles E. Baukal, Jr.

CONTENTS

0-8493-1695-2/98/$0.00+$.50

6.1 INTRODUCTION

The use of oxygen to speed up decarburization of pig iron was recognized by Bessemer in patents around 1855, but the first commercial use of oxygen did not happen until the production of low-cost, oxygen was available and used in steelmaking processes (see Chapter 5). While the first use of oxygen was reported in the Bessemer converter in 1931, the development of LD (Linz–Donawitz) steelmaking, commonly known as the basic oxygen furnace, or BOF, actually opened the floodgate of tonnage oxygen use in metallurgical industries. Cominco in Canada was the first to use oxygen enrichment in zinc and lead pyrometallurgy in the 1930s. INCO in Canada developed flash smelting using pure oxygen in the early 1960s. In hydrometallurgy, Cominco commissioned the first oxygen pressure leaching system in 1981. Although oxygen usage in melting and enrichment in the nonferrous industries is now commonplace, it was not so even until the 1970s. Today, more than half of the world production of copper, lead, and nickel is made with oxygen enrichment in the smelters.

The rapid growth of oxygen usage in the nonferrous industry since the 1970s was prompted by higher fuel prices in the early 1970s, global competitiveness requiring productivity improvements and the ever increasing demand to meet environmental regulations. As we move to the next century, the intensity of oxygen usage in the nonferrous industry will continue to increase at a rapid rate. All major nonferrous smelters will most likely have their own oxygen plants. The secondary nonferrous plants will find it hard to remain competitive without oxygen.

6.2 MELTING

6.2.1 Rotary Furnace

A typical rotary melting furnace is shown in Figure 6.1. Radiation and convection from the hot combustion products are transferred to both the furnace walls and the charge materials. The heat in the furnace wall is then transferred to the charge material as the furnace rotates on its horizontal axis. Usually the charge door for adding the feed materials is located on the burner side of the furnace. The exhaust is located at the opposite end of the furnace, except for a double-pass rotary furnace where the charge door, burner, and exhaust are all on the same end.

6.2.1.1 Aluminum

A typical charge for an aluminum rotary furnace contains mixed scrap, such as sheets, foils, castings, turnings, and dross. The aluminum producer would like to recover as much of that aluminum as possible to maximize the yield. There has

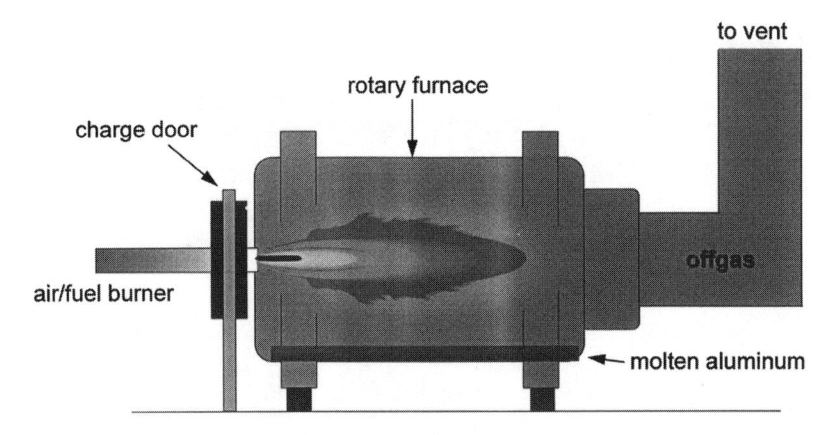

FIGURE 6.1 Conventional air/fuel, single-pass, rotary aluminum melting furnace.

always been a concern in the aluminum industry about using oxygen-enriched combustion air or oxy/fuel burners for melting because of the propensity of aluminum to oxidize and form undesirable oxides. This is especially true for light-gauge scrap and high-magnesium alloys. Some unpleasant memories still linger from experiences using oxygen-assisted combustion for aluminum melting in the late 1960s and 1970s. Most of these experiences occurred when oxy/fuel burners were installed in direct charge and open-well reverberatory furnaces.

Aluminum has an even higher affinity for hydrogen and carbon dioxide. Therefore, it is not enough just to minimize the free oxygen inside the rotary furnace. The aluminum can pick up oxygen from carbon dioxide and water which are in the products of fuel combustion. The main factor that can be influenced is the reaction kinetics. Faster melting minimizes the aluminum oxidation and therefore maximizes the yield.

6.2.1.1.1 Air–oxy/fuel

The rotary furnace lends itself very well to the use of oxygen-assisted combustion. The reduced flue gas volume provides for a longer residence time of the hot combustion gases in the furnace thus allowing more heat to be transferred to the charge material. Also, because of the reduced flue gas velocity, less fines are carried into the dust collection system.

The concept of using an oxy/fuel burner within an air/fuel burner provides the aluminum melter with the flexibility of using either burner, or a combination of both burners (see Figure 6.2). This innovative approach, developed by Air Products under the trade name of EZ-Fire™, has proved to be very successful in the aluminum industry.[1] This differs considerably from both conventional oxygen-enriched techniques and the oxy/fuel burners typically used for melting in the metals industries. This technology is retrofittable on existing air/fuel combustion systems. As shown in Figure 6.3, an oxy/fuel burner is installed in the center of the conventional air/fuel burner. This approach marries the existing air/fuel burner with a retrofitted non-water-cooled oxy/fuel burner, providing the more efficient heat transfer and available heat benefits of oxy/fuel, while maintaining the external flame characteristics of

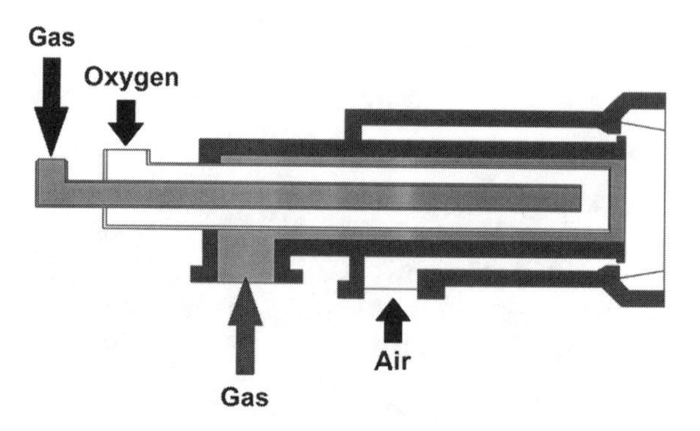

FIGURE 6.2 One variation of the EZ-Fire burner.

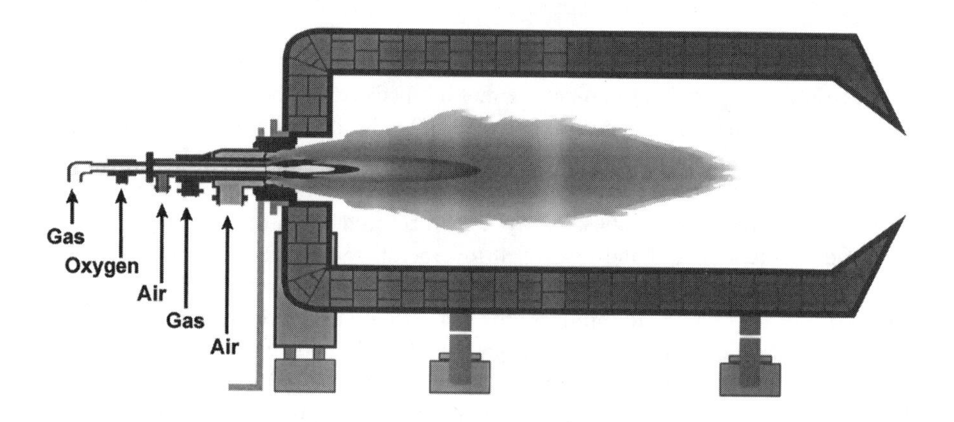

FIGURE 6.3 Another variation of the EZ-Fire burner in a rotary furnace.

air/fuel. The result is a low-capital-cost, air–oxy/fuel combustion system that improves the furnace efficiency and lowers production costs.

During meltdown, the oxy/fuel burner is used along with the air/fuel burner to provide maximum melt rates and minimum flue gas volumes, while during holding and casting only the air/fuel burner is used. For the melting of aluminum, a combination of air–oxy/fuel combustion provides the best results. The outer air/fuel flame envelope provides a shielding effect for the hot oxy/fuel inner flame. By firing the oxy/fuel inner burner on a fuel rich mixture, flame luminosity is increased as combustion radicals and gas particles are heated to incandescence by the hot oxy/fuel flame. This in turn improves the radiative heat transfer of the flame to the charge material and furnace wall while preventing excess oxygen from contacting the load and causing metal oxidation.

This air–oxy/fuel burner concept was first tried in 1989 at a Canadian aluminum dross reclamation plant where drosses and skimmings are smelted.[2] The dross, with 10% salt, is charged through the burner end of a 25,000 lb (11,000 kg) rotary furnace, fired by a dual-fuel burner of 20×10^6 Btu/h (6 MW) capacity for melting. The rotary furnace flue exhaust as well as dross crushers and cooling tables are all connected to the same baghouse. When the dross reclamation unit was used at the same time the rotary furnace was firing, fume collection efficiency was reduced and fugitive emissions became a problem.

At the start of the week, when the furnace is cold, the air/fuel burner can be fired at 14×10^6 Btu/h (4 MW) while still keeping baghouse temperatures below 250°F (120°C). As the furnace becomes hotter during the week, the firing rate has to be reduced to 9.5×10^6 Btu/h (2.8 MW) in order to maintain baghouse temperatures below 250°F (120°C) and prevent burning of bags. With an average tap-to-tap time of 3 to 4 hours, the average gas consumption was 3.5 ft^3/lb (0.22 m^3/kg) of metal produced. Furnace and overall shop efficiencies were 28.3 and 13.5%, respectively.

Because of the draft limitations on the baghouse, the melt furnace firing rate had to be maintained at a reduced rate, which limited metal output. Time constraints and economics precluded upgrading the baghouse system to accommodate the combined dust and fume loads from the melting furnace and dross reclamation systems.

The most logical approach was to address the melt furnace, which generated the largest amount of hot gases and fines processed by the baghouse. The use of oxygen-assisted combustion to alleviate the baghouse problems while simultaneously increasing furnace productivity became the preferred choice. By substituting air with pure oxygen, flue gas output from the furnace would be reduced and combustion efficiency would increase.

The oil tube/atomizer section was removed from the dual fuel burner and an oxy/fuel burner was inserted in its place. The burner retrofit package including the natural gas and oxygen flow control trains were prefabricated to prevent furnace downtime. The oxy/fuel flow controls and safety switches were interlocked to the conventional air/fuel controls, thus creating a hybrid air–oxy/fuel combustion system.

The oxy/fuel burner retrofit was designed to fire up to 15×10^6 Btu/h (4.4 MW). However, the best results were achieved when a combination of air–oxy/fuel combustion was used. By firing the oxy/fuel burner at 9×10^6 Btu/h (2.6 MW) and the air/fuel burner at 4×10^6 Btu/h (1.2 MW), the combined air/oxygen/fuel input of 13×10^6 Btu/h (3.8 MW) provided a meltdown reduction of 50 to 65% from 3 to 4 h/heat to 1 to 2 h/heat, depending upon charge material.

The average gas consumption was 2.5 ft^3/lb (0.13 m^3/kg). Furnace and overall shop efficiencies were 52.2 and 18.95%, respectively. Baghouse temperatures ranged from 215 to 230°F (102 to 110°C) during high fire. Furnace exhaust gases to the flue were 68,500 ft^3/h (1940 m^3/h). The inside furnace temperature, as measured by optical pyrometer, never exceeded 2100°F (1150°C).

Table 6.1 shows the comparative results of the conventional air/fuel and retrofitted air–oxygen/fuel systems. There has been no adverse affect on refractory life, and metal yield was unchanged. The retrofitted air–oxygen/fuel burner proved to be a cost effective and economic method of increasing productivity from a rotary furnace while reducing flue gas volume and baghouse temperature.

6.2.1.1.2 Oxy/fuel

Used beverage cans (UBCs) and other recycled aluminum scrap are typically melted in a rotary furnace. A new process has been developed which reduces the melting

TABLE 6.1
Comparative Results of Air/Fuel and Air–Oxy/Fuel Operations of Aluminum Rotary Furnace

	Combustion System	
	Air/Fuel	**Air–Oxy/Fuel**
Firing rate (10^6Btu/h)	9.5–14.0	12.0–14.0
Heat cycle (h, tap to tap)	3–4	1–2
Furnace efficiency (%)	28.3	52.2
Melting energy consumption (Btu/lb)	1,662	900
Energy savings (%)	—	45.8
Average production (lb, 5 days/week)	193,563	319,497
Production increase (%)	—	65.0
Overall energy consumption (Btu/lb)	3,500	2,500
Overall shop efficiency (%)	13.57	18.95
Gas savings (%)	—	28.4
Average baghouse temperature (°F)	250–270	215–230
Flue gas volume (ft^3/h)	132,000	67,500
Flue gas reduction (%)	—	49.0
Flue dust reduction (%)	—	42.0

costs and pollution emissions from the conventional rotary melter. It is referred to as the LEAM® process, or *low emission aluminum melting* process (patent pending).[3]

The aluminum processor needs an efficient and flexible melter to handle a wide range of scrap materials. An ideal melting process would be able to handle a wide range of scrap and automatically compensate for the contaminants. Scrap aluminum cans commonly have paints and lacquers that must be removed. In some cases, this is done in a pretreatment furnace where the coatings are burned off. Scrap turnings from machining operations typically contain significant quantities of lubrication oils. These oils produce large amounts of smoke during heating and melting in conventional rotary furnaces. This smoke must be removed before the exhaust gases exit the stack.

Stricter regulations are increasing the environmental compliance costs. This includes the costs to clean up contaminated air and water resulting from the melting process, as well as the costs to dispose of any waste products, which is particularly important for hazardous wastes.

A slag layer covers the top of the molten aluminum in the melter to minimize aluminum oxidation. The slag is produced by adding salt that contains 10 to 20% potassium chloride and 80 to 90% sodium chloride. The slag is skimmed off before tapping the furnace. This slag, in certain geographic regions, is classified as a hazardous waste. It is expensive either to reprocess the slag to recover even more aluminum or to dispose of the slag, usually in a landfill.

The exhaust gases from the rotary furnace are usually ducted to a baghouse which removes any particulate matter. In some cases, this filter dust may also be classified as hazardous waste, which again increases the disposal costs.

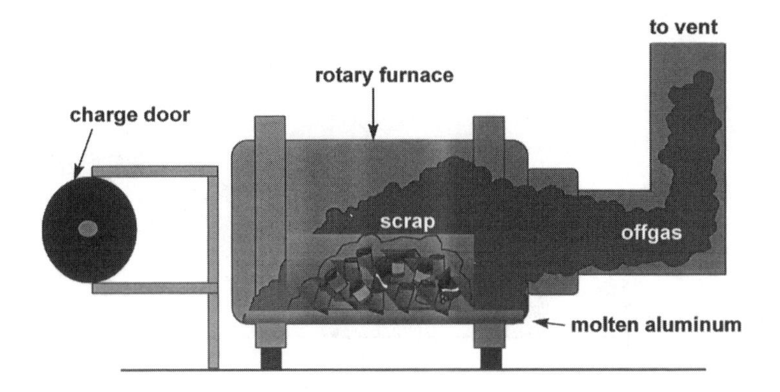

FIGURE 6.4 Hydrocarbon emissions from a new charge of dirty scrap.

Another important factor is the hydrocarbon emissions. In a conventional rotary furnace, with the burner mounted on the charging door, the burner must be turned off when the door is opened for charging. Fumes are generated by vaporization of any combustibles in the charge, as shown in Figure 6.4. Because of the reduced furnace temperature and insufficient free oxygen to burn the newly added combustibles, large quantities of hydrocarbon emissions can be generated. The problem is exacerbated when the door is closed and the burner restarted. Even less free oxygen is available when the door is closed, and more combustibles are vaporized by the additional heat from the burner. An afterburner could be used to incinerate these fumes. However, this is costly and does not take advantage of the heating value in the combustibles.

Another problem related to the hydrocarbon fumes is the formation of dioxins and furans. The hydrocarbons react with the halogens, typically contained in the salts that are used to produce the slag covering on the molten bath, to form dioxins and furans (see Chapter 2). These are carcinogens that are currently regulated in several countries. Methods to reduce dioxin and furan emissions include (1) to postcombust the hydrocarbons, (2) to quench the off-gases quickly within a certain temperature window, and (3) to use an activated carbon to adsorb the dioxins and furans. However, any of these methods would be at some additional cost to the aluminum producer.

The LEAM process, developed to minimize the pollutant emissions from rotary aluminum-melting furnaces, was first demonstrated in Germany where the environmental regulations are stringent. The first objective of the process was to at least maintain, if not improve, the existing yields in the conventional melting process, which is well known for its high yields. Other objectives included (1) to lower the emissions, especially hydrocarbons, dioxins, and furans, below the regulated limits and (2) to reduce the exhaust gas volumes to reduce the cost of the flue gas–cleaning system. This required reductions in the amount of filter dust collected in the cleaning system. All of these objectives had to be achieved at the lowest possible cost.

As shown in Figure 6.5, the first major change was relocating the burner from the charge door to the exhaust wall. This reversed-flame operation offered several

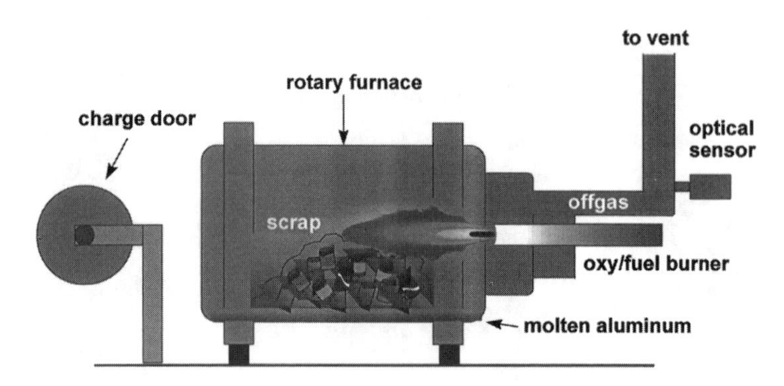

FIGURE 6.5 LEAM process.

benefits. The residence time of the combustion gases in the furnace increased, leading to increased thermal efficiency and higher destruction efficiencies for the volatiles. This configuration allowed the burner to be fired even during charging. Any volatiles coming from the newly charged scrap material had to pass through the flame and were postcombusted within the furnace. The heating value of the volatiles was now used, and the volatiles did not have to be removed in the off-gas-cleaning system. The overall melting cycle time was reduced because heat was now added during the charging cycle. This new configuration allowed the door to be properly sealed, which better controlled the excess air coming into the furnace.

Another major change was replacing the air/fuel burner in the conventional rotary aluminum melting furnace with an oxy/fuel burner. This provided several advantages, as described elsewhere in the book (Chapter 1). The reduced exhaust gas volume flow made it easier to incinerate the volatiles because they were in higher concentrations. Oxy/fuel burners also have a much wider operating range than air/fuel burners. They can easily operate with several times the stoichiometric amount of oxygen required for complete combustion. This extra oxygen was important for incinerating the volatiles which evolved from the dirty charge materials. Air/fuel burner operation became severely degraded when large quantities of excess air were supplied through the burner.

During furnace charging, the oxy/fuel burner was at low fire, whereas in the conventional process the burner must be off during charging since it is mounted on the door. A major advantage of the LEAM process is that the volatiles emitted during charging of dirty scrap are incinerated by the burner. In the conventional process, all the volatiles emitted during charging are exhausted and must be removed by the exhaust gas–cleaning system.

There are two levels of control in the LEAM process pertaining to coarse and fine adjustments of the operating conditions. The coarse adjustments are based on automatic set point conditions that are predetermined according to the phase of the melting cycle. The fuel and oxygen flow rates are predefined for low-fire and high-fire conditions.

The fine adjustments are made using a feedback control system. An important element of that system is an optical sensor which is used to control the hydrocarbon emissions. As more volatiles are released into the furnace, the sensor detects the

TABLE 6.2
Process Improvements Using the LEAM Process

	Air/Fuel	Oxy/Fuel	LEAM	Improvement, %
Tap-to-tap time (h)	10	7	7	30
Consumption (kWh/ton)	850	550	420	52
Thermal efficiency (%)	35	55	71	100
Relative metal yield (%)	0	0–0.5	0.5–2.0	
Slag (lb/ton)	770	770	620	20
Filter dust (lb/ton)	55	33	13	76

increased emissions and sends a signal to the controller, which increases the oxygen flow to the burner. As the quantity of volatiles being released from the dirty scrap begins to decline, the signal from the sensor automatically reduces the oxygen flow to the burner to the amount required to incinerate the reduced level of volatiles. Only the appropriate flow of oxygen is used. This is important because too much oxygen flow would oxidize aluminum and reduce the yield. Too little oxygen flow would not incinerate all the volatiles which would increase the load on the off-gas-cleaning system and waste the available energy in the volatiles. The optical sensor for the oxygen/fuel ratio correction is both very reliable and inexpensive. The sensor has a fast response time compared with typical gas analyzers, which require a sample to be extracted from the system and may have delay times of more than 30 seconds.

In the first commercial demonstration of the LEAM process in Germany, the diameter of the off-gas duct was reduced from about 31 to about 14 in. The time to clean the off-gas-piping system was reduced from 16 h every weekend to 2 h once a month. A new smaller scrubbing system was installed for the reduced off-gas volumes.

Many process improvements were achieved by converting to the LEAM process, as shown in Table 6.2. Three options were tested and compared:

1. Conventional air/fuel burner in the charge door,
2. Oxy/fuel burner in the charge door, and
3. The LEAM process.

Using oxy/fuel (options 2 and 3), the tap-to-tap time went from 10 to 7 h. This was primarily a consequence of the higher flame temperatures with oxy/fuel burners. The energy consumption using the LEAM process was less than half that of the conventional process which was again a consequence of using the more-efficient oxy/fuel burner.

Aluminum yields increased an average of 1% due to better control of the furnace atmosphere. Slag formation went down by 20%. There was a dramatic reduction in the filter dust generated during the melting process, which is especially significant where the dust is classified as a hazardous waste. In Germany in 1996, the disposal costs were about \$0.50/lb (\$0.23/kg) so this represented a large cost savings for the aluminum producer.

TABLE 6.3
Pollutant Emissions on a Volumetric Basis

	Air/Fuel	Oxy/Fuel	LEAM	German Limits
Hydrocarbons (mg/m^3)	0–1000	0–1000	0–30	50
NOx (mg/m^3)	50–200	200–3000	100–500	500
Dust (mg/m^3)	2–5	2–5	2–5	10
Dioxin/furan (ng-TE/m^3)	0.2–10	0.2–10	0.0–0.4	1

There were also significant reductions in the pollutant emissions. The regulated and measured values are shown in Table 6.3. For the air/fuel system, hydrocarbons were a big problem because they were difficult to control. NOx was not a problem because of the low melting temperature of the aluminum and the low flame temperatures. Dust emissions were effectively controlled by the baghouse. However, dioxins and furans were a big problem. Using oxy/fuel (option 2), the hydrocarbon, dioxin and furan emissions were similar to the base case air/fuel system. However, NOx was higher because of high levels of air infiltration due to the poor door seal in option 2. The infiltrated air passed through the very hot oxy/fuel flame which generated high NOx levels.

Using the LEAM process, hydrocarbons were well below the German limit and could even be further reduced by adjusting the sensitivity of the optical sensor. The NOx levels were reduced below the regulated limit by sealing the charge door and by putting a pressure control system onto the furnace to minimize air infiltration. Dust levels were again low. There was a significant reduction in the dioxin and furan emissions which are typically reported on a toxic-equivalent (TE) basis. There is currently some discussion in Germany about reducing the existing regulation for dioxins and furans down to 0.1 ng-TE/m^3. Some type of post treatment system would have to be added to the existing air/fuel system to meet that limit. The LEAM process may be able to meet that limit without any additional equipment.

The pollutant emissions in Table 6.3 are given on a volumetric basis. This does not reflect the dramatic reduction in the flue gas volume that occurs when replacing an air/fuel system with oxy/fuel. The pollutant emissions have been normalized in Table 6.4. This shows the dramatic reductions in hydrocarbons, dust, dioxins and

TABLE 6.4
Pollutant Emissions per Ton of Input

	Air/Fuel	Oxy/Fuel	LEAM	Improvement, %
Hydrocarbons (g)	87	60	6	93
NOx (g)	222	1360	214	4
Dust (g)	24	17	3	88
Dioxins and furans (mg)	36	25	0.2	99
CO_2 (kg)	238	154	114	52

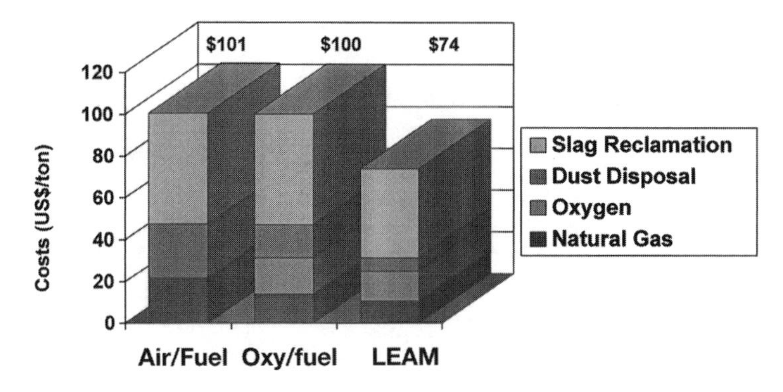

FIGURE 6.6 Production costs (Germany).

furans, and CO_2. CO_2 is considered to be a greenhouse gas that contributes to global warming and, as such, is regulated in many locations (see Chapter 2). There was also a slight reduction in NOx compared with the base case.

Figure 6.6 shows that the LEAM process can reduce the costs for the aluminum producer. In Germany, the costs to dispose of the filter dust and to reclaim the aluminum in the slag are high. By reducing those emissions, the cost of the oxygen is more than offset by the LEAM process improvements. Not included in the figure are the costs that could be incurred if a post treatment system were required to remove dioxins and furans. The actual savings for any given aluminum producer will depend on a variety of factors and are somewhat site specific. Other important factors to consider are the increase in metal yield and the ability to use cheaper grades of scrap.

6.2.1.2 Lead

Lead rotary furnaces are used either for melting battery scrap or smelting concentrates. Oxy/fuel technology was introduced about 20 years ago in lead scrap–melting furnaces. Most of the lead rotary furnaces have been converted to oxy/fuel in Europe. The benefits of oxy/fuel burners in secondary plants include up to 100% production increases; energy savings of up to 60%; lower CO, CO_2, and NOx; 85% lower combustion products; 50% lower furnace off-gas; and up to 60% lower particulate emissions.

When an air/oil burner of a 15 ton concentrate smelting rotary furnace is replaced with an oxy/oil burner, the following benefits have been realized: 27% reduction in smelting time, 30% reduction in tap-to-tap time, 59% oil savings per batch, 34% reductions in flue dust, higher tapping temperature, better working environment, less smoke exiting the furnace, lower noise, and overall furnace productivity improvements of 42%.

No major changes, other than installing the oxy/oil burner, which was much smaller in size, and an oxy/oil flow train, were needed to convert the furnace from air/oil to an oxy/oil system. This furnace was charged with oxide concentrates with coke and other fluxes. The increased furnace productivity allowed the plant to operate fewer rotaries and still exceeded the plant production obtained with the old air/fuel system.

TABLE 6.5
Performance of Rotary Copper Melting Furnace

	Air/Fuel	Oxy/Fuel	Difference (%)
Specific fuel consumption, Nm^3/t	102.1	40.2	–60.7
Specific oxygen consumption, Nm^3/t	—	72.9	—
Total energy consumption, kJ/kg	3633	2224	–38.8
Productivity, kg/h	1175	2595	+121.0

6.2.1.3 Copper

A rotary-type copper-alloy-melting furnace, rated for 1.2 ton/h, experimented with oxy/fuel combustion and productivity improvements and energy savings were realized.[4] The results of the test are tabulated in Table 6.5.

In a 20 t copper rotary furnace, where the dual fuel burner was retrofitted with a non-water-cooled oxy/fuel burner which provided the flexibility of firing air/fuel or oxy/fuel or any combination in between, furnace productivity improvements and fuel savings were realized. The oil leg of the dual-fuel burner was replaced with an oxy/natural gas burner. The firing rate was reduced from an equivalent of 600 m^3/h (21,000 ft^3/h) natural gas firing (50:50 gas:oil) to 400 m^3/h (14,000 ft^3/h) natural gas firing, The heat time was reduced from 7.5 to 6.5 h, with a fuel savings of 33%. In addition, fugitive emissions from the furnace during the early part of cycle were significantly reduced and refractory damage was controlled.

6.2.2 Reverberatory Furnace

6.2.2.1 Aluminum

The reverberatory furnace is used to remelt aluminum scrap, turnings, cans, etc., or it can also be used to smelt concentrates. Either way, oxygen is beneficial in the reverberatory furnace either in the general enrichment mode or air–oxy/fuel or oxy/fuel mode.

6.2.2.1.1 Oxygen enrichment

In the mid 1980s, oxygen enrichment tests were conducted in a nonrecuperative reverberatory aluminum remelt furnace. Pure oxygen was introduced into the combustion airstream. Oxygen analyzers were used to assure the proper enrichment level was maintained and the oxygen level in the flue was kept to approximately 2%. The results obtained are shown in Table 6.6.[5] During the test, no significant change in melt loss was found from the use of oxygen. However, productivity improvements of up to 27% were obtained with 7% oxygen enrichment.

6.2.2.1.2 Air–oxy/fuel

A technique of retrofitting conventional air/fuel burners with oxygen capability allows operators of reverberatory melting furnaces to easily obtain the advantage of oxygen-assisted melting. In most cases, the air inlet of the conventional burner is

TABLE 6.6
Oxygen Enrichment of an Aluminum Reverberatory Furnace

	Steady-State Melt Rate, tons/h (Δ%)	Corrected Specific Consumption, Btu/lb (Δ%)	Efficiency during Melt (Δ%)
Without recuperator	9.30 (base case)	1974 (base case)	25.1(base case)
3% Enrichment (Δ%)	10.99 (+18.2%)	1628 (–17.5%)	30.6 (+21.9%)
5% Enrichment (Δ%)	11.38 (+22.4%)	1569 (–20.5%)	31.7 (+26.3%)
7% Enrichment (Δ%)	11.76 (+26.5%)	1507 (–23.7%)	33.0 (+31.5%)

Δ% = percent change from base case of no recuperator and no enrichment.

Source: Boneberg, J. H., paper presented at Aluminum Industry Energy Conservation Workshop IX Papers, 1986. With permission.

retained and a new internal assembly including the oxygen connection is fabricated for installation on the hot furnace. The existing air/fuel controls are retained and the oxygen controls are interconnected. The ignition and flame supervision components are also retained although sometimes they must be relocated on the burner body. The resultant burner retains the capabilities of the air/fuel burner and can be operated as the original air/fuel burner or as an oxygen-assisted air/fuel burner.

The inclusion of some air in the combustion process helps to moderate the flame temperature to eliminate the possibility of localized overheating of the refractory or metal bath. Air surrounding the oxy/fuel flame cools the burner tile, prolonging the life of the burner tile. More importantly, the overall cost of operation is reduced compared with a 100% oxy/fuel burner for a desired production increase. Since the burner also can operate as an air/fuel burner without oxygen, the operating cost can be further reduced by switching to air/fuel when holding, alloying, or during production delays.

In many cases the conventional burner has an oil tube and an atomizing air section in the center of the burner. This can be removed and replaced with an independent oxy/fuel burner. In other cases, the gas tube is replaced with a combined oxygen–gas nozzle.

The economics of oxygen utilization depend upon the ability of the smelter to take advantage of an increased melt rate in the furnace. A major advantage of the retrofit approach for implementing oxygen-assisted combustion compared with full oxy/fuel conversion is that the amount of oxygen participation can be adjusted to the minimum necessary to provide the amount of productivity improvement desired. A typical example would be a plant that desires a 30% improvement in production because it has capacity upstream in feed preparation and downstream in casting to handle such an increase in furnace throughput. Typical assumptions are shown in Table 6.7.

By using the available heat calculated for the flue temperature and the heat contained in the metal at the pouring temperature (475 Btu/lb or 1100 kJ/kg), the heat lost to the environment from the furnace is calculated. Assuming the heat loss is constant, the amount of fuel required at various oxygen participation levels is

TABLE 6.7
Assumptions for a Typical Aluminum Melting Furnace

Parameter	Assumption
Flue temperature	2100°F (1150°C)
Natural gas cost	\$3.00/1000 ft³ (\$10.60/100 m³)
Oxygen cost	\$3.50/1000 ft³ (\$12.40/100 m³)
Desired productivity improvement	30%
Current firing rate for air/fuel melting	18×10^6 Btu/h (5.3 MW)
Current melt rate	9000 lb/h (4100 kg/h)

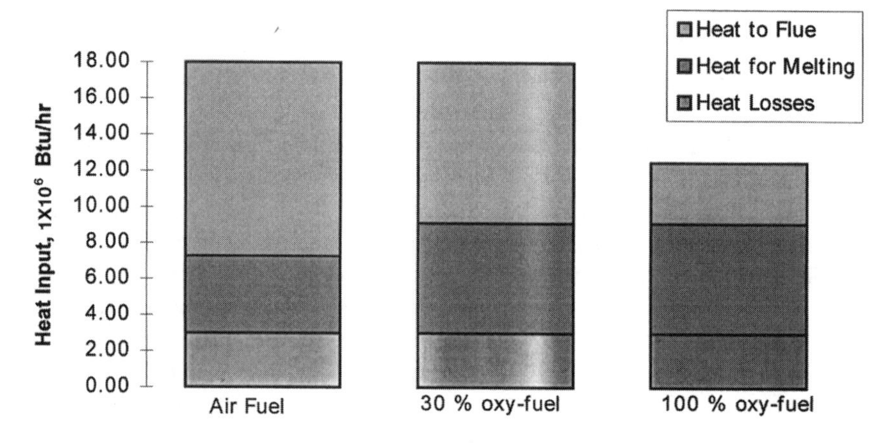

FIGURE 6.7 Heat balance comparison for various oxygen inputs.

calculated to give the desired 30% increase in production rate. The resulting heat balances are shown in Figure 6.7.

The cost per pound of metal melted is based on the rates of fuel and oxygen input. Results are shown in Figure 6.8. By keeping the fuel input at the current 18×10^6 Btu/h (5.3 MW) and replacing ~30% of the air with oxygen, the desired production goal was achieved. For full oxy/fuel, the fuel input was reduced to 12.5×10^6 Btu/h (3.7 MW). Note that although the fuel cost was reduced, the total cost increased with greater oxygen participation. Figure 6.9 shows the split in costs between the oxygen and fuel. Although thermal efficiency improved with increased oxygen participation, the overall cost increased because of the increased volumes of oxygen.

Some general conclusions can be drawn from this analysis: the lowest-cost option is always the lowest oxygen input that will give the desired production increase, and oxygen-enhanced combustion becomes more cost-effective as the processing temperature increases. For intermediate-temperature processes, such as aluminum melting, partial oxygen enrichment is generally most economical.

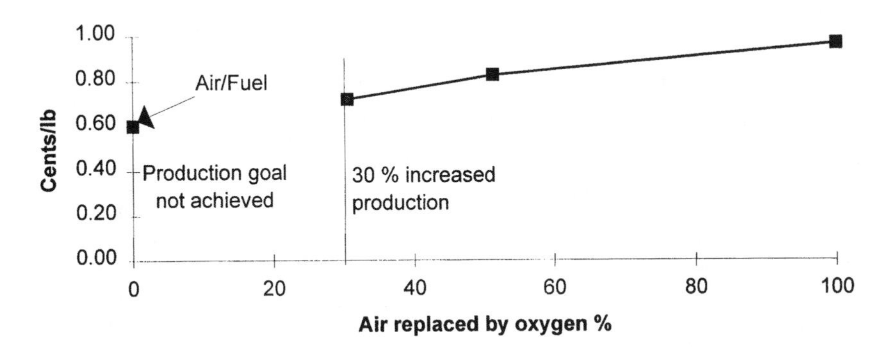

FIGURE 6.8 Fuel plus oxygen cost of melting as a function of oxygen participation.

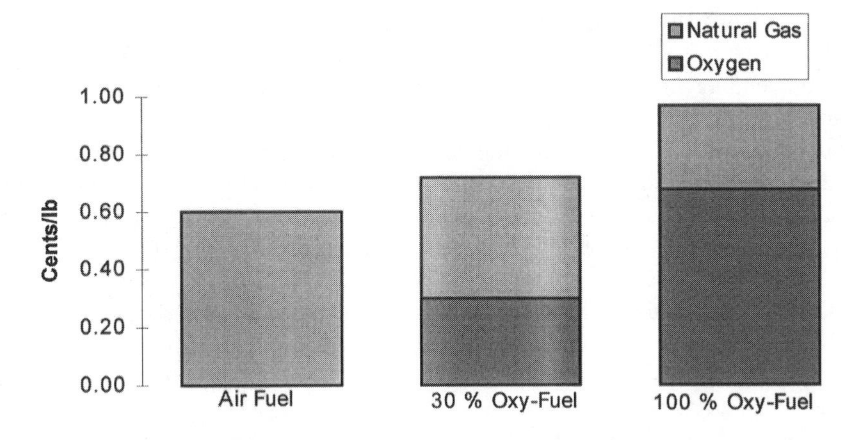

FIGURE 6.9 Cost split as a function of oxygen participation.

An aluminum smelter, with three open-well reverberatory furnaces melting scrap, which needed increased production, implemented oxygen-assisted combustion for increased furnace productivity, reduced cost per pound of material melted, and improved metal yield.[6] Two furnaces of 150,000 lb (68,000 kg) capacity each and one 70,000 lb (32,000 kg) capacity were sidewell-fed reverberatories. Four dual-fuel burners fired along one furnace wall in the larger furnaces, while the smaller furnace was equipped with two burners. For the first furnace, the EZ-Fire burners (see Figure 6.2) were installed within the existing air/fuel burners in a matter of hours without interrupting production. Subsequent installations were performed at regularly scheduled holiday shutdowns. Once the oxy/fuel burners were operating, the major change in furnace practice involved increasing the charge rate to keep up with the higher melt rate. By monitoring melt temperature, the furnace operators found that the normal charge rate could be increased by 25%.

During charging and melting, the EZ-Fire burners were fired 25% from the air/fuel burners and 75% from the oxy/fuel burners. During extended waiting periods, the oxy/fuel burners were shut off, but otherwise the burners were kept at high fire

TABLE 6.8
Performance of Typical Side-Well Reverberatory Aluminum Melting Furnace

Parameter	Units	Air/Fuel	EZ-Fire
Daily production	lb	160,000	195,000
Energy for melting	Btu/lb	2,800	1800
Tap-to-tap time	h	12	10
Fuel cost	$/lb	0.010	0.005
Oxygen cost	$/lb	—	0.004
Added production	lb/day	—	35,000
Cost savings (including labor and overhead)	$/lb	—	0.0075

throughout the charging/melting cycle and reduced to low fire during tapping. During start-up, each furnace was carefully monitored. At no time did the flue temperature, roof temperature, or curtain wall temperature increase beyond that experienced with air/fuel operation.

As a result of the faster melting rate, the total cycle time was reduced. This smelter, to accommodate work shifts and casting line utilization better, adopted new pouring schedules. The measured daily output of ingot and sow was therefore increased by 15 to 20% for each of the three furnaces.

Natural gas consumption was reduced from 2.8 ft^3 (0.08 m^3) to an average 1.8 ft^3 (0.05 m^3) per pound of aluminum melted. The oxygen requirement was 1.4 ft^3 (0.04 m^3) per pound of aluminum melted. Overall the average cost of processing a pound of aluminum was reduced by about 0.75¢ when labor and overhead costs were taken into account.

Of equal importance, there was no increase in refractory cost. The roof, flue, and curtain wall temperatures were closely monitored during the initial operation. No increase in the temperature of the monitoring thermocouples was noted. Inspection of the furnace refractory during a normally scheduled shutdown showed no signs of abnormal refractory wear.

Because the metal melted faster and spent less time in the furnace, dross losses decreased significantly resulting in a ¼ to ½% yield improvement valued at over $400/day. The detailed results of implementing oxygen-enhanced combustion as measured on one of the furnaces are summarized in Table 6.8.

Another aluminum producer of secondary ingots, billets, and slabs was interested in evaluating the EZ-Fire technology for both production enhancements and emissions reductions.[6] For the initial installation, a direct-charge reverberatory furnace was retrofitted. The furnace was charged with billet ends and primary and secondary sows.

The furnace combustion system contained two air/fuel-type burners firing at 14×10^6 Btu/h (4.1 MW), with air preheated to 500°F (260°C). The existing control system was designed to regulate firing rates to control the roof and bath temperature at desired set point. This system could meter the rate down below 5×10^6 Btu/h (1.5 MW) when the roof or bath temperature was attained. The EZ-Fire controls were incorporated into the existing furnace controls enabling the systems to work concurrently. Three oxy/fuel firing rates corresponding with three set point temperatures were utilized while the air/fuel firing rate was fixed to 5×10^6 Btu/h (1.5 MW). When the maximum set point temperature was reached, the oxygen system would turn off and the air/fuel system would take over.

As a result of the increased efficiency of the EZ-Fire system, the melt cycle was reduced from over 8 to 5½ h for the EZ-Fire system (a 30% melt rate increase). The reduction in melt time enabled the customer to optimize the workload for the entire billet-casting operation.

No negative effects were noted on the refractory. After an initial 3 month start-up period, the furnace was cooled down and the internals were inspected, with no refractory damage noted. It was even possible to reduce the furnace temperature set point after optimizing the charging and melting practice. Overall, the roof temperature was lowered by 100°F (56°C), while maintaining the 30% increase in melt rate. This reduction was expected to result in considerable savings in refractory expenses.

The optimized EZ-Fire system has been able to demonstrate emissions levels below some of the most stringent state and local emissions factors in the U.S. The system has demonstrated emissions levels under 0.3 lb NO_2/ton (0.15 g NO_2/kg) aluminum melted. Multiple emissions tests have seen levels consistently below 0.19 lb NO_2/ton (0.086 g NO_2/kg) aluminum melted.

6.2.2.2 Lead

In a lead reverberatory furnace, when 2% oxygen enrichment was practiced, the production increased by 6.8%. When the polypropylene-rich scrap content of the charge was increased from 30 to 60%, the burner firing rate was reduced from 18 to 14.4×10^6 Btu/h (5.3 to 4.2 MW), a 20% reduction in fuel. The increased use of polypropylene-rich scrap is an important benefit of the oxygen-enrichment practice. Although high temperatures were generated by oxidation of the plastic-rich scrap, the increased antimony content of the lead produced indicated the presence of reducing conditions in the furnace.

6.2.2.3 Copper

Several major copper smelters use oxygen in their reverberatory furnaces for increasing matte production in order to utilize excess downstream converter capacity. Alternatively, producing the same amount of matte from fewer reverberatory furnaces or fewer furnace hours can also be cost-effective.

Conventional pyrometallurgical smelting in reverberatory furnaces for recovery of copper from its sulfide concentrates uses large quantities of hydrocarbon fuels because the process makes little use of the energy available from oxidation of the sulfide charge. The use of oxygen in this process is a first step in retrofitting the furnaces to reduce overall energy costs and increase production.[7] In some cases, it may be desirable to enrich selectively some of the end-wall burners to higher levels than others, in order to concentrate the heat in colder or slower-moving portions of the melt. This selectivity is easily accomplished with either premixing oxygen and combustion air or undershooting oxygen enrichment. Both techniques are relatively

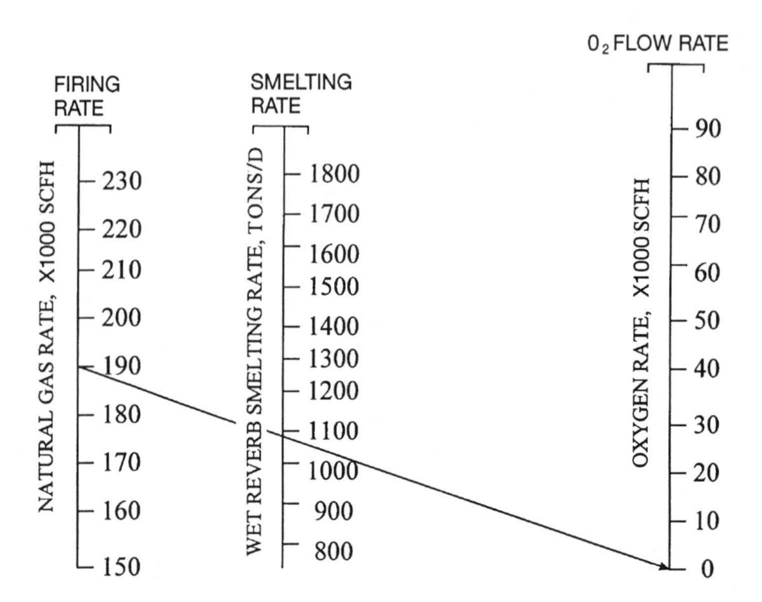

FIGURE 6.10 Nomogram of relationship between oxygen and fuel rate at constant smelting rate. (From Eacott, J. G., *Advances in Sulfite Smelting, Vol. 2: Technology and Practice,* Sohn, H. Y., George, D. B., and Zunkel, A. D., Eds., The Metallurgical Society of AIME, Warrendale, PA, 1983, 583–634. With permission.)

simple to install. Production increases of 20 to 30% may be attained on a routine basis using O_2 enrichment. Generally, 9% enrichment or 30% total O_2 in the combustion air is considered to be a practical maximum. In addition to a production increase and fuel savings, oxygen enrichment can provide additional cost savings in the gas-cleaning and sulfur fixation systems.

A mathematical model has been developed to predict the response of reverberatory furnaces to a variation in the fuel and oxygen rates[8]:

$$\text{Smelting Rate} = A\ (\text{Firing Rate}) + B\ (\text{Oxygen Flow Rate}) + C$$

where A, B, and C are constants, depending upon the charge composition, moisture content, heat losses, sulfur elimination, etc. of the particular furnace.

As the firing rate increases beyond a minimum rate to overcome heat losses, the smelting rate increases linearly with the heat input and the above relationship applies in the normal operating range. The nomogram in Figure 6.10[8] depicts the inverse relationship between oxygen and fuel rate at a constant smelting rate. This can be used to determine one of the three variables — smelting rate, oxygen rate, or firing rate — when two other variables are known or fixed. The concept of interchangeability between oxygen and fuel for a given smelting rate is illustrated in the nomogram.

After successful experience with oxygen enrichment of wall-mounted conventional air/fuel burners, several smelters have successfully tried roof-mounted

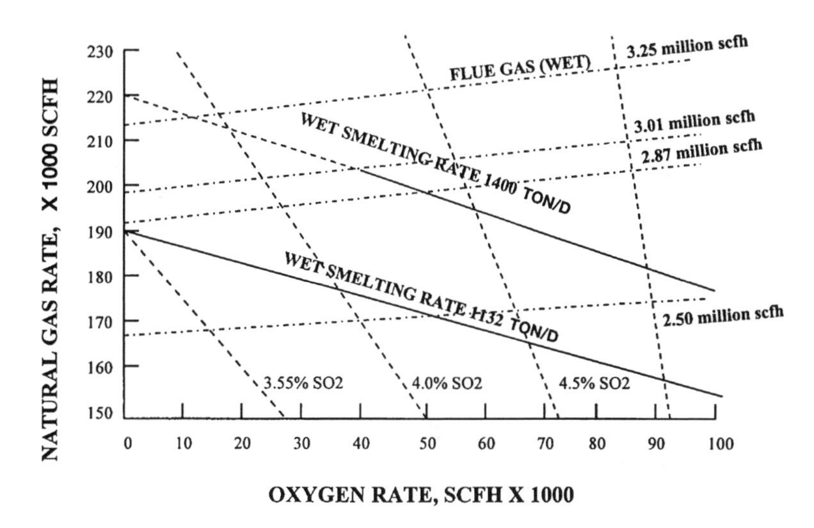

FIGURE 6.11 Reverberatory production and flue gas dependence on tonnage oxygen and fuel. (From Wrampe, P. and Nollman, E. C., Paper presented at TMS-AIME Workshop, No. A74-258. 1974.)

oxy/fuel burners, firing directly onto the charge. Substantial improvements in efficiency can be achieved with these burners because the heat is transferred to the unmelted charge by both radiation and convection. These burners can be used to supplement oxygen-enriched conventional burners to increase production further.

When using oxy/fuel burners, substantially greater benefits can be achieved. Data from INCO's Copper Cliff Smelter indicates a 45% increase in smelting rate with simultaneous reduction in fuel consumption of 55%.[9] The Noranda Horne Smelter reports a 60% increase in smelting rates and an 18% reduction in fuel consumption. The fuel rates are gradually being reduced further. Codelco's Caletones Smelter reports a smelting rate improvement over 100% and a 55% reduction in fuel. Their flue gas contains SO_2 concentrations of 6 to 8%.

6.2.2.3.1 Oxygen sprinkle smelting

In oxygen sprinkle smelting, a reverberatory furnace is retrofitted with roof-mounted burners that combust dry feed and oxygen. This concept, developed by Queneau and Schuhmann,[10] transforms an existing reverberatory furnace into an oxygen flash smelting unit with minimal capital investment. At a certain oxygen-to-concentrate ratio, autogeneous smelting conditions can be obtained. Based on heat balance calculations, the autogenous level can be attained when smelting to a matte grade of 64% Cu with 42% SO_2 in the off-gas. In order to reach autogenous conditions,[10] the feed rate needs to be almost doubled compared with conventional reverberatory smelting (matte grade 35% Cu). The oxygen-to-concentrate mass ratio should be 0.22:1. A small-scale oxygen sprinkle smelting test conducted at Phelps Dodge Corporation at its Morenci smelter[11] indicated that furnace throughput increases of 100% were possible with two thirds reduction in total smelting energy requirements

TABLE 6.9
Effect of Oxygen Enrichment of a Lead Blast Furnace

		O_2 Enrichment	
	Base	**3%**	**5%**
Production increase, %	Base	25.5	52.9
Coke savings, %	Base	9.9	16
Dust production, reduction (%)	Base	36.4	45.5
Pb in slag, %	1.50	1.40	—
	0.96	—	1.3

per unit of charge (including electricity and energy for oxygen production). SO_2 concentrations of 20 to 30% were expected.

6.2.3 Lead Cupola/Blast Furnace

In lead smelting, the oxygen-enriched blast has been used in lead cupola and slag fuming furnaces. Oxygen enrichment of the copper blast furnace is also common.

The cupola charge mix consists of reverberatory furnace slag, battery plates, coke, iron, and limestone. Injection of 2.5% pure oxygen into the cupola wind results in a 20% increase in lead production. Reductions in coke consumption with enrichment are normally in the range of 4 to 7% on a dry weight basis. At another location, 3% enrichment increased production 30% and reduced coke consumption by 33%. Good slag fluidity and a higher metal temperature resulting from oxygen enrichment facilitated tapping operations. Enrichment also allowed the tuyeres to remain cleaner. Care should be taken to maintain a CO-to-CO_2 ratio of 1:4 at the exit gas analysis to ensure smooth operation of a lead blast cupola. The results of the use of oxygen enrichment of the lead blast furnaces are shown in Table 6.9.[12]

6.3 ROASTING

The oxygen-enriched blast has been successfully tried in the zinc sulfide concentrates roasting in fluidized-bed, multiple-hearth, and suspension roasters. As early as 1956, Electrolytic Zinc Company of Australia improved calcine grade, increased oxidation rates, and facilitated use of cold blast to the hearth roasters for better temperature controls. Other nonferrous metals, such as molybdenum, copper, nickel, and lead, that use roasters to oxidize sulfide concentrates are technically and economically amenable to oxygen enrichment.

Fluidized-bed roasting accounts for most of the world zinc production today. Over the last 15 years, several zinc roasters have augmented their calcine production capacities with oxygen enrichment. These roasters typically encounter production increase limitations due to existing blower capacities, requirements for fluidization characteristics of the bed, and downstream gas-handling capacities. Oxygen enrichment provides roaster operators with a tool to increase roaster capacity in a cost-effective way. Several roaster plants utilize oxygen enrichment to improve calcine

TABLE 6.10
Oxygen Enrichment vs. Production Increase in Fluid-Bed Zinc Roasters

% O_2 Enrichment	Production Increase (%)
1.0	6–8
2.0	13–16
3.0	20–23

quality via reducing sulfide sulfur content, which results in higher zinc recovery in the downstream leaching operations.[13] Oxygen enrichment has also been used to process cheaper grades of fine concentrates that are difficult to process with air only.

To increase the current concentrate throughput rate, supplemental oxygen-enrichment practice is implemented (Table 6.10). This maintains the critical fluidization velocity essentially unchanged. Pure oxygen is injected into the air main through a diffuser, before the blower air reaches the gas plenum. The additional oxygen reacts with increased feed material to undergo the principal reaction, $ZnS + 1.5O_2 = ZnO + SO_2$. The higher oxygen content (p_{O_2}) of the process air increases the reaction kinetics of the above principal reaction, thereby improving the operating efficiency and reaction completion. This reduces the sulfide sulfur in the calcine (Table 6.11). Although the benefits are obvious, extreme care must be taken to control the heat regime in order to avoid overheating the roaster and causing incipient fusion of the feed materials. A thorough heat and mass balance is the first step to determine the additional cooling capacity required to handle the increased throughput rate. Adding bed coils, charging secondary materials, increasing moisture in the concentrate, introducing or increasing water spray, and humidifying the process air are some of the options that must be considered prior to injecting oxygen into the roaster process air.

The choice between an increase in calcine production rate and an improvement in product quality represents a classic operating trade-off in any production facility. Oxygen enrichment provides the capability to improve one or the other or a partial improvement of both. At a given enrichment level, greater productivity could be achieved if the calcine quality is maintained at the pre-enrichment level. Conversely, significant improvement in calcine quality can be achieved at the expense of a smaller increase in productivity. The choice must be made based upon the needs and relative economics of a given company.

6.4 OXIDE PRODUCTION

In lead oxide (litharge) production, molten lead is intermittently pumped into a reactor where a series of rakes, mounted on a shaft which turns at 200 rpm, violently agitate the metal. The agitation forces the molten metal to form particles and improves gas–metal reactions. The reactions are exothermic oxidation. Air pulled

TABLE 6.11
Oxygen Enrichment of Fluid-Bed Zinc Roasters — Plant Results

Throughput (t/day)	Air (Nm³/h)	Pure O_2 (Nm³/h)	O_2 Enrich (%)	Sulfide–Sulfur (%)
Plant A				
223	16,020	0	0.0	0.25
251	16,020	425	2.0	≤0.25
266	16,020	680	3.2	≤0.25
Plant B				
347.5	26,340	0	0.0	0.12
358	26,340	348	1.0	0.09
371	26,340	515	1.5	0.07
Plant C				
400	27,000	0	0.0	0.25
415	27,000	370	1.0	0.13
Plant D				
447	38,100	0	0.0	0.48
538	39,100	1300	2.5	0.37
Plant E				
750	47,660	0	0.0	0.52
762	49,580	645	1.0	0.30
780	46,950	600	1.0	0.64
850	52,000	1675	2.5	0.20

through the reactor by a fan is used to oxidize the lead as well as to carry the product out of the reactor. The production rate is controlled by the air flow through the reactor. On the one hand, higher air flow rates result in more production. On the other hand, higher air flows result in higher velocities which also could carry more free lead with the product.

With 3 to 5% oxygen enrichment, 25% production increases have been attained with simultaneous reductions in free metal from a typical 14% to 10–11%. Reactor temperatures are controlled by increasing cooling water flows. The reactor temperature was maintained at 1030°F (554°C). During the test, iron content was also reduced from 23 to 13 ppm.

Use of oxygen in the reactor calls for adequate consideration of airflow (air plus oxygen) volume, as this can imbalance the fluid dynamics, causing a certain distribution of particles sizes to become airborne and adversely affecting the residence time and product quality.

6.5 LEAD SOFTENING

The traditional method of softening lead with sodium nitrate can be improved by oxygen enrichment of the process air. When a 100,000 lb (45 t) heat was treated with a high level of oxygen-enriched air, the benefits shown in Table 6.12 were obtained.

TABLE 6.12
Lead Softening Results

	Na Nitrate	Oxy–Air
Softening time, h	24	8
Cycle time, h	48	30
Oxygen consumption, std. ft^3	—	4600
Na nitrate, lb	1300	100
External fuel	Yes	No
Sb % (start → finish)	Similar to air–oxygen technique	1.1 → <0.05
As % (start → finish)		0.05 → Nil
Sn % (start → finish)		0.25 → Nil

In addition to a productivity improvement, another important result obtained with the air/oxygen technique was the amount of sodium nitrate slag required to soften a kettle of lead completely. While the Na nitrate softening process required 1300 lb (590 kg) for a batch of 100,000 lb (45 t) to soften the kettle completely, the air–O_2 technique required only 100 lb (45 kg) of Na nitrate to lower the Sb level to an undetectable level. Minimizing the nitrate-containing dross, which is typically recycled through the blast furnace, will also minimize the nitrate-containing slag produced by the blast furnace. This nitrate-bearing slag is considered a hazardous waste and, therefore, creates very expensive disposal problems. In addition, since the kettles are to be heated to initiate the reactions in the standard nitrate process, while the air–O_2 technique requires very little external energy supply, a significant cost savings could also be realized with the air–O_2 system.

6.6 COPPER SMELTING

The copper-smelting process consists of two major operations: melting the charge materials and oxidation of the molten bath. Melting requires a supply of external heat energy, such as the burning of hydrocarbon fuels. For many years, melting has been done by supplying heat via combustion of fossil fuels with atmospheric air. With the availability of low-cost oxygen, air is now either totally or partially replaced by pure oxygen. The oxidation of a molten bath is a chemical phenomenon where an oxidant is required to remove sulfur from concentrates. Although oxidation is exothermic, external fuel is also required here unless the sulfur level is high enough to make the process autogenous. Oxygen is increasingly being utilized to replace combustion and process air in new smelting technologies, and to retrofit old smelters.

6.6.1 Oxygen in Process Air

The oxygen in the process air reacts with the sulfides of the concentrate. By enriching the process air with oxygen, nitrogen ballast is either reduced or eliminated, resulting in lower off-gas volumes with less sensible heat loss from the furnace in the waste gases. The higher p_{O_2} results in a greater driving force for the sulfide oxidation reactions. Consequently, both the reaction rate and temperature are increased. Increased matte grade, reduced dependence on external fuel, and increased SO_2 strength in the off-gas are some of the effects of oxygen enrichment of process air. Oxygen and fuel can be used interchangeably or complementarily. Oxygen enrichment can be used to reach the autogenous level where no additional fuel is required. Beyond this point, further O_2 enrichment will require that extra heat be dissipated in some way, e.g., increased water cooling, addition of reverts, or lowering of the matte grade.

6.6.2 Flash Smelting

In flash smelting, the concentrate is dispersed in an air or oxygen stream, and smelting and converting occur while the particles are in suspension. The major reasons that prompted the use of oxygen in flash smelting include the need for increased matte production from an existing smelter, the need for a more-efficient and lower-cost SO_2 recovery system because of increased SO_2 content of the furnace off-gas, and the need for an autogenous process to realize energy savings.[14] The level of oxygen use and associated economics depend on specific flash smelting process parameters. Some of the influencing factors are concentrate type, matte grade, fuel price, SO_2 recovery system, etc.

Since the oxidation of sulfides is the only source of heat in autogenous oxygen flash smelting, sufficient oxygen per ton of sulfide must be supplied to satisfy the heat balance for a given operation. When this is done, the matte grade is fixed. The oxygen rate cannot be changed without causing either an excess or deficiency of heat.[15]

Flash smelters achieve the combined pyrometallurgical operations of roasting, smelting, and partial converting in a single unit, as compared with separate units in traditional processes. The two basic types of flash smelting furnaces are discussed next.

6.6.2.1 Outokumpu Flash Smelting

During the earlier stages of flash smelting, fuel and preheated process and combustion air were used. However, modern practices use oxygen enrichment of the process air to produce higher grade matte (from 45–50% to 65–70% Cu) autogenously, resulting in substantial energy savings.[16] Similarly, SO_2 content of the off-gas is increased from 10 to 15% without oxygen, to as high as 30% with oxygen, due to the reduced gas volume. The high matte grade reduces the need for converter capacity, and the higher SO_2 content results in cost reductions in gas treatment and environmental control. An adjustment of oxygen in the process-air-to-concentrate ratio allows various types of concentrates to be processed to produce mattes of various grades.

As the specific capacity (ton/day/ft^3) is increased by increasing oxygen enrichment, the energy consumption per ton of concentrate decreases considerably, with oxygen enrichment in the range of 35 to 80% O_2 in air.[17] Concentrate burners are used to inject fluxed concentrate and oxygen-enriched air vertically downward from the top of a shaft in the Outokumpu design. Out of over 25 copper smelters using this technology, about 50% utilize oxygen enrichment of the process air.

6.6.2.2 INCO Oxygen Flash Smelting

The INCO process uses pure oxygen instead of air and is an autogenous operation. Concentrates are injected through two burners in each end wall of the furnace and combusted in a horizontal stream of oxygen. Oxygen flash smelting is very flexible and can treat feeds of various compositions to produce a wide range of matte grades. For producing higher grades, the oxygen-to-concentrate ratio is increased and a coolant is added, whereas for lower grades the ratio is decreased with the addition of a supplemental external fuel burner.

Metallurgical advantages from the use of oxygen include the absence of magnetite buildup as a result of a very low oxygen potential in the furnace, and the rapid establishment of slag–matte equilibrium.[18] Slag can be discarded directly as it contains about 0.8% Cu at a matte grade of 55% Cu. Other benefits related to oxygen use in the autogenous process are low off-gas volume (20% of other processes); low dust carryover due to low gas velocity in the furnace; high SO_2 content (80%) in the waste gas, offering increased flexibility in choosing an SO_2 recovery system; and relatively small size and cost for gas cleaning and treatment facilities. The use of oxygen makes the INCO process one of the lowest energy consumers.

It has been estimated that in both the INCO and Outokumpu processes, 1 tonne (1.1 ton) of oxygen replaces the combustion of about 0.20 to 0.25 tonne (0.22 to 0.28 ton) of oil.[19] This ratio can be used to assess the economics of replacing fuel by tonnage oxygen.

6.6.2.3 Kivcet Process

Developed in the former U.S.S.R., the Kivcet process is an autogenous oxygen flash-smelting process where high reaction rates and good heat and mass transfer can be achieved. Dry concentrate is introduced tangentially into a small water-cooled cyclone where it is flash smelted with pure oxygen. Similar to INCO oxygen flash smelting, the Kivcet process produces medium-grade matte (45 to 50% Cu) and high SO_2 content (80%) in the waste gas.[15] Since the oxygen potential of the cyclone can be controlled to a low level, the slag can be discarded for its low Cu content (0.35%). Oxygen consumption is reported to be 210 m^3/tonne (7100 ft^3/t) of copper concentrate.

6.6.3 Continuous Smelting

Of the various continuous copper-smelting processes for production of blister copper from dried copper concentrates, three processes are discussed here.

6.6.3.1 Mitsubishi Process

The Mitsubishi process is a multistep process where dried, fluxed concentrates and 9 to 14% oxygen-enriched air enter the smelting furnace through nonsubmerged vertical lances. Fuel is required in the smelting furnace to maintain the heat balance. In the converting step, 5 to 7% oxygen-enriched air is blown through vertical lances above the bath to oxidize the matte to blister copper continuously. Fuel is also required in the converting furnace to compensate for the heat deficit.[15]

The smelting unit can be made autogenous with 65% O_2 in air.[20] Trials at Naoshima[21] have demonstrated that throughput can be increased by 50% when the oxygen-enrichment level is increased from 7 to 18%. A concurrent fuel reduction of 70% of the normal requirement has also been achieved. SO_2 content of off-gas was increased from 14 to 24% with a volume reduction of 88%.

By employing separate furnaces for smelting and converting, the Mitsubishi process allows better control of oxygen potential, and, hence, of magnetite formation. Oxygen efficiency in the converting furnace is 85 to 90%.

6.6.3.2 Noranda Process

The Noranda Process can produce either blister copper or copper matte from sulfide concentrates. In matte production, when melting and converting are combined in a single reactor, fluxed concentrates are charged into a highly turbulent bath. The melt is oxidized to high-grade matte (70 to 75% Cu) with oxygen-enriched air injected through side-blown tuyeres. Overall oxygen efficiency is nearly 100%, and oxygen is consumed both within the bath in converting and above the bath in smelting.[15] The Noranda reactors in Utah (Kennecot Mineral Company) use about 34% O_2 in air, both for tuyeres and burners. With sufficient concentrate feed, the bath temperature is controlled to avoid refractory wear at the tuyere line.

The Noranda reactor can smelt concentrate autogenously with about 33 to 40% O_2 in air, depending on the fuel value of the concentrate. Above about 50% O_2 in the blast air, the fuel reduction is insignificant. Adjusting the oxygen-to-concentrate ratio, the process can be made very flexible. A significant production increase and fuel reduction can be achieved by enriching the tuyere air with tonnage oxygen. Figure 6.12 shows these parameters, as determined from Noranda operations.[22] SO_2 content of the effluent gases ranges between 10% with 2% oxygen enrichment at the Horne smelter and 17% with 13% oxygen enrichment at Kennecot's Utah smelter.

6.6.3.3 El Teniente Reactor

The El Teniente Reactor at the Caletones smelter in Chile has been developed for continuous smelting of concentrates using excess heat generated by oxidation of matte. Smelting of concentrates to produce high-grade matte (73 to 75% Cu) has become autogenous when 11% O_2-enriched air is used.[15] The oxygen efficiency in the reactor is 80 to 85%. The SO_2 content of the off-gases is 10 to 12%. Oxygen has been used at the Caletones smelter to increase production and reduce fuel consumption.

6.6.4 Economics of Tonnage Oxygen Usage in Copper Smelting

The energy and process efficiencies of any copper-smelting process are intimately tied with the use of oxygen. However, to determine the overall process economics, consideration must be given to the cost of oxygen (see Chapter 1). Because of the

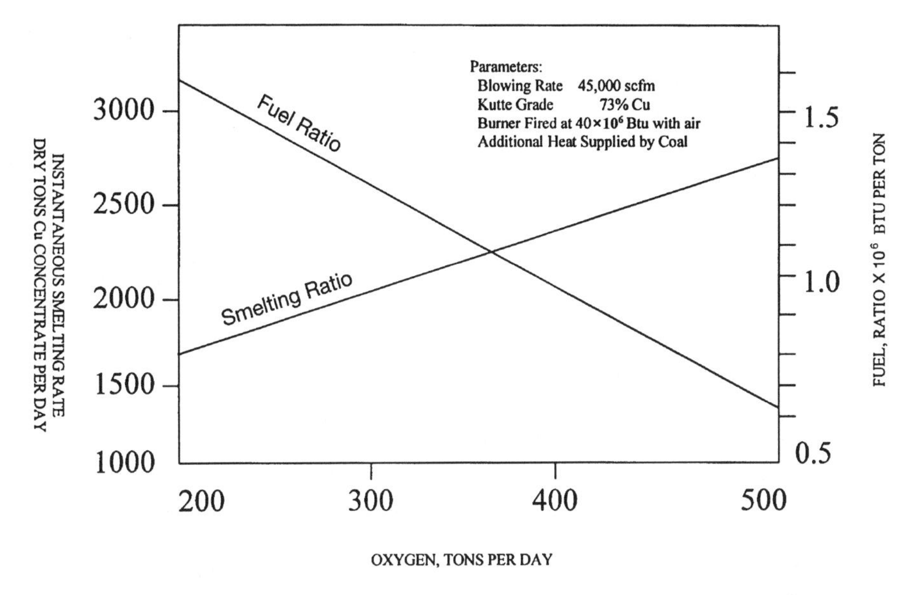

FIGURE 6.12 Noranda process — variation in smelting rate and fuel ratio with increasing tonnage oxygen. (From Bailey, J. B. W. and Storey, A. G., *CIM Bull.,* June, 142–148, 1980. With permission.)

interaction of various factors, the economic benefits of using oxygen will be discussed in general terms.

When retrofitting an existing smelter with oxygen technology, productivity improvement (and, hence, reduced unit cost of production) must be accounted for, in addition to fuel savings and other economic benefits resulting from process improvements. Capital, operating, and maintenance costs associated with alternative retrofitted systems or installation of a new reactor should be compared with the costs of oxygen-based technologies to determine the most-economic course of action. Compared with other methods of increasing production, such as adding furnaces, oxygen can result in both capital and operating savings. Methods of evaluating investment alternatives for a greenfield site will be different. For example, substituting oxygen (or oxygen-enriched air) for ambient air in a new smelter operation will result in smaller and/or fewer furnaces and gas cleaning and SO_2 recovery systems and result in lower costs for a given production capacity.

The favorable replacement ratio of oxygen to hydrocarbon fuels generally ensures a cost reduction to the smelter when the best possible oxygen supply mode is selected. One reference[23] shows typical North American unit costs which have been applied to published data for three types of smelting furnace. The reference shows a reduction in power and fuel costs, with increased oxygen enrichment.

The most significant energy savings in fuel-fired smelting comes from the use of oxygen enrichment of the process air. The autogenous flash smelting provides the maximum benefits.

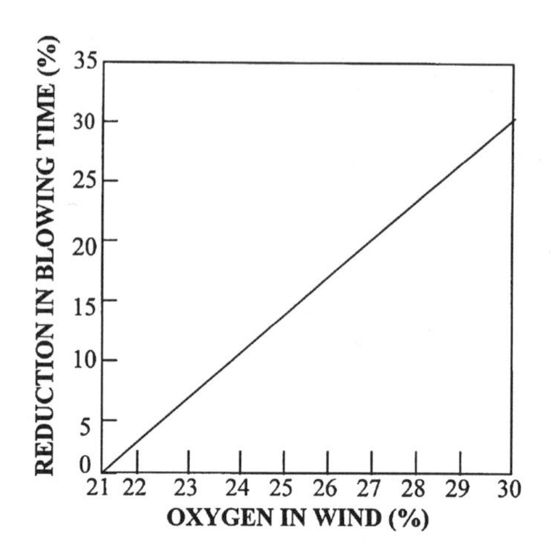

FIGURE 6.13 Effect of oxygen enrichment on blowing time in Pierce–Smith converter.[14]

6.7 COPPER CONVERTERS

6.7.1 Pierce–Smith Converter

Oxygen enrichment of copper converters has been practiced on a production scale since the late 1950s. Most major copper smelters currently use oxygen enrichment in their side-blown Pierce–Smith converters. Usually, oxygen is added to the tuyere air to a maximum enrichment limit of 9%. Above this level, refractory wear at the tuyere line becomes excessive unless sufficient cold dope is added to control bath temperature. In the side-blown Hoboken converter, oxygen is top blown through a lance directly into the copper bath to prevent refractory wear at the tuyere area. The incentives for oxygen enrichment of the converter blowing air or direct oxygen injection into the converter bath are[24]

1. To conserve heat in the charge to melt additional cold reverts, precipitates, or concentrates (during slag blow) and copper scrap (during copper blow),
2. To increase converter capacity by reducing the blowing time or increasing charge size, and
3. To increase the SO_2 content in the off-gas which can result in more-efficient and lower-cost SO_2 recovery systems. Higher SO_2 concentrations will not only allow smaller pollution control equipment with associated lower investment, but also reduce operating costs for the SO_2 recovery system.

These advantages can be attained simultaneously. The effect of oxygen-enrichment level on converter blowing time, assuming constant oxygen utilization efficiency, is shown in Figure 6.13.[14]

Nitrogen in the converter gas carries away about 45% of the heat provided. Based on converter heat balance calculations,[25] with 30% O_2 in the converter air (9% O_2 enrichment) about 1.8×10^6 Btu/ton (0.48 MW/t) oxygen can be saved at

80% O_2 efficiency (nearly 3×10^6 Btu/ton or 0.8 MW/t of oxygen at 50% efficiency). The energy saved can be used to melt cold dope.

At any given oxygen content of the converter blast, the ratio of heat released by the matte and the heat required for smelting concentrates determines the amount of concentrate that can be smelted per ton of matte without disrupting the converter heat balance. Smelters converting high-grade mattes at high oxygen efficiencies can particularly take advantage of using oxygen enrichment to process a significant amount of concentrate in the converter at a relatively low level of oxygen enrichment in the blast.

Depending on the type of dope, the amount of cold material that can be melted with oxygen-enriched air can range between 3 and 12 tons/ton of oxygen.[14] This can eliminate the need for installing additional primary smelting capacity because converters can be used for smelting concentrates. This can be a very cost- and energy-effective way of increasing production in a reverberatory smelter.

6.7.2 Hoboken Converter

Another major copper smelter has injected oxygen directly into the copper bath of a Hoboken–Siphon converter.[19] Consumable lances were used for submerged oxygen injection.

6.7.3 Top-Blown Rotary Converter

Another converting process, the top-blown rotary converter (TBRC), uses oxygen to attain high temperatures and control the atmosphere, resulting in greater operational flexibility. Top blowing oxygen through a water-cooled lance above the bath avoids any refractory wear, as can be experienced in Pierce–Smith converters with oxygen enrichment. Scrap copper may also be melted in the TBRC using oxy/fuel techniques. The Pierce–Smith converter by itself is not suitable for melting charge.

Average total oxygen consumption for the entire process is around 900 lb/ton (450 kg/t) of blister copper. In an INCO test, an oxygen-concentrate burner was inserted alongside the main oxy/fuel lance into a TBRC.[15] After initial heating of the vessel to 1250°C (2280°F), the oxy/fuel lance was withdrawn and the oxygen-concentrate burner was started. Oxygen flash smelting was autogenous when oxidizing to white metal. Oxygen-to-concentrate ratio was 0.38:1, on a weight basis, at a feed rate of 1.2 t/h (1.1 tonne/h). Oxygen efficiency was 92% and SO_2 concentration in the converter gas was >85%. By adjusting the O_2 to concentrate ratio and lowering the feed rate, a range of matte grades, from 61% Cu to white metal and through to blister copper, were produced in the pilot plant TRBC.

6.7.4 Operating Results of Enrichment

Typically, when oxygen is used to enrich the converter wind, or is injected directly into converter baths, enrichment levels of 25 to 30% oxygen are used during the first half or less of the slag blow. This part of the converter cycle produces the most heat for melting cold, copper-bearing materials. Also at this time, the efficiency of air and oxygen utilization is at a maximum. Conventional air is used for the remaining part of the blow when utilization efficiency is reduced because of a reduction in sulfur available for reaction. The amount of cold material, such as cold dope, which can be melted with oxygen, depends on the matte grade, desired liquid copper temperature, and degree of liquid copper temperature control.

Because of differences in these operating conditions, the amount of cold dope melted has varied over a wide range, from 3 to 8 tons of cold dope per ton of oxygen.

In addition to increasing converter melting abilities, oxygen also will decrease blow times. Enrichment levels of 25 to 30% can result in blow time reductions of 15 to 30%. Because oxygen blowing occupies only a portion of the total converter blowing and operating time, total converter time savings are well below this level.

Use of oxygen in a copper converter increases the SO_2 content of the waste gas. Results reported by Metallurgie Hoboken[19] indicate that, when using an equivalent oxygen enrichment level of 25.1%, the SO_2 content can be increased by nearly 2.0% from 8% to 10% SO_2. Use of higher oxygen enrichment levels results in a larger SO_2 concentration increase. It may be possible to mix this SO_2-rich gas with reverberatory gas low in SO_2 prior to feeding a sulfuric acid plant.

Converter Oxygen Results

Oxygen enrichment levels	25–30% oxygen
Blowing time reductions	15–30%
Increase in waste gas SO_2	At least 2%
Cold Material Melted	**Melting Ratio** (tons material/ton enrichment O_2)
Cold dope	3–8
Precipitate (cement) copper	4–8
Copper scrap	6–12

6.8 REHEATING

Today, the use of oxy/fuel burners is widespread for melting of various metals. Advances in burner design and manufacture are now extending the oxy/fuel burner application range further to include the heating or reheating of solid metal products during processing.

A novel concept of using a matrix burner in preheating copper strip at the continuous annealing furnace entrance, has been successfully tested.[26] The Rapidfire™ heating system, which is based on this novel burner design concept, has been introduced to the steel and copper industries. The burner contains a multitude of ports which produces a uniform, high-temperature oxy/natural gas flame (see Figure 6.14). Burners are positioned so that the flame impinges directly onto the solid surface of the part to be heated. This approach of applying high-intensity heat using a compact, variable-geometry burner permits the application of heat precisely where needed. This is of particular importance to rolling, forging, and heat treatment lines to increase line speed; maintain edge, corner, and surface temperatures of slabs; transfer bars between forming operations; reheat components after line stoppages; and heat bars and billets prior to hot working.

At a copper strip annealing line, if the strip temperature needs to be increased, Rapidfire burners can provide the extra heat that could not be supplied otherwise

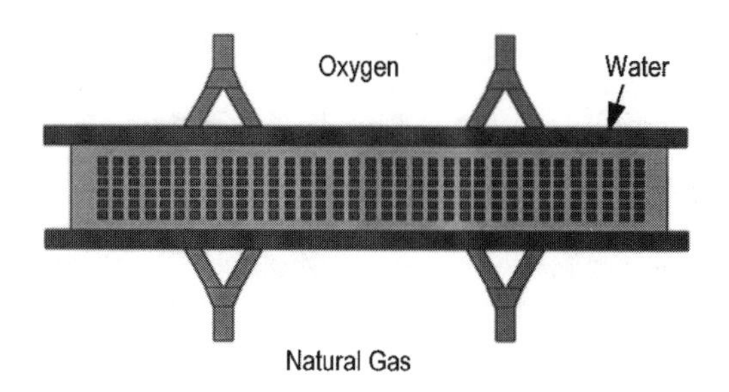

FIGURE 6.14 Rapidfire burner.

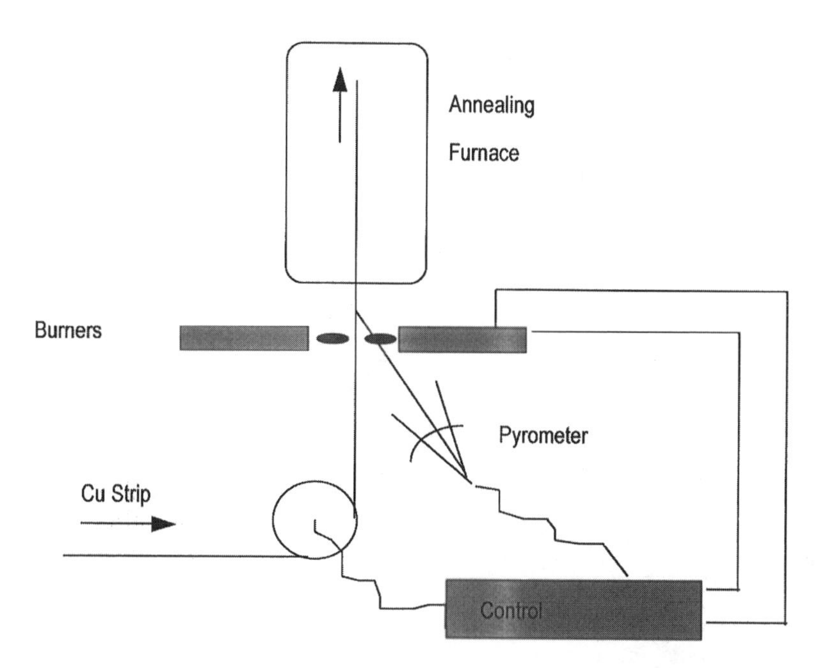

FIGURE 6.15 Rapidfire burner system control in copper strip heating.

by the existing annealing furnace. Typically, these burners are controlled by a signal from a pyrometer measuring the sheet temperature at the furnace entrance, the temperature set point decided by a control unit which monitors the line speed (see Figure 6.15). A line speed increase of 40% can be achieved, resulting in increased annealing line productivity.

The Rapidfire system uses water-cooled matrix burners of width similar to the sheet width and could be about 50 mm (2 in.) high. With the addition of automatic ignition and flame monitoring, the Rapidfire burners are fully automated — the burners only operate as and when dictated by the line speed. The matrix burner approach gives great flexibility to the shaping of the flame front which can be tailor-made to fit the component to be heated. This leads to a high degree of control over the heat input geometry which is important to prevent localized overheating or surface oxidation.

TABLE 6.13
Oxygen Applications in Nonferrous Metals Processing

O_2 Use	Smelting/Melting Process	Excess Energy Available	Use Pattern Continuity	Oxygen Pressure	Oxygen Purity
Concentrate combustion					
Burners	Flash Smelting — INCO, Outokunpu, FCR	Yes	Medium	Medium	70–100%
Submerged injection	Noranda continuous QSL, Mitsubushi	Yes	High	High Medium	70–100%
Oxy/fuel combustion	Conventional reverberatory (Al, Cu), TBRC, rotary (Cu, Al, Pb)	No	Low	Medium	90–100%
Enrichment	Conventional reverberatory (Al, Cu); rotary (Cu, Al, Pb); P-S converters (Cu); Roasters — Zn, Mo, Cu, Blast furnace; cupola — Pb, Cu	Yes/No	Medium/Low	Low	30–100%

6.9 CONCLUSIONS

Today, the nonferrous industry faces a double challenge to reduce costs while simultaneously meeting stringent pollution control standards. Cost savings can be achieved through improving productivity, reducing fuel consumption, and enhancing process efficiency. An improvement in process efficiency can also contribute toward cost savings in gas cleaning and SO_2 recovery systems. The use of oxygen helps to achieve these goals realistically and economically. In cases where a production increase is desired, use of oxygen is the simplest method of retrofitting an existing furnace. Typical examples are shown in Table 6.13.

Oxygen is playing a key role in the development of new and emerging processes where suspension of concentrates takes place in an oxygen-enriched or pure oxygen atmosphere. In the secondary industry, productivity, furnace efficiency, emissions controls, and recycling will continue to play a major role in global competitiveness. The availability of low-cost oxygen has not only prompted the commercialization of these newer processes, but also encourages the development of future oxygen-based autogenous smelting technologies and the upgrading of existing processes. Oxygen has been used primarily to increase productivity and to reduce fuel consumption, with a side benefit of increased SO_3 content in furnace off-gases which ease environmental control.

As new smelting processes are adopted, oxygen use is likely to accelerate since the investment for flash or continuous smelting and associated SO_2 control facilities is markedly reduced. Oxygen reduces nonproductive capital investment in pollution control facilities and increases the rate of return on new smelting investments.

REFERENCES

1. Bazarian, E. R., Heffron, J. F., and Baukal, C. E., Method for Reducing NOx Production during Air-Fuel Combustion Processes, U.S. Patent 5,308,239, May 3, 1994.
2. Paget, M. W., Heffron, J. F., Lefebre, M., and Bazinet, C., A novel burner retrofit used to increase productivity in an aluminum rotary furnace and reduce baghouse loading, paper presented at TMS Conference, 1994.
3. Baukal, C. E., Scharf, R., and Eleazer, P. B., Low emission aluminum melting process, paper presented at the 1996 Spring AFRC Meeting, May 6–7, Orlando, FL, 1996.
4. De Lucia, M., Oxygen enrichment in combustion processes: comparative experimental results from several application fields, *Trans. ASME,* 113, June, 122, 1991.
5. Boneberg, J. H., Oxygen enrichment test on a non-recuperative reverberatory remelt furnace, paper presented at Aluminum Industry Energy Conservation Workshop IX Papers, Aluminum Association, Washington, D.C., 1986.
6. Krichten, D. J., Baxter, W. J., and Baukal, C. E., Oxygen enhancement of burners for improved productivity, in *EPD Congress 1997, Proceedings of the 1997 TMS Annual Meeting,* B. Mishra, Ed., February 9–13, Orlando, FL, 1997, 665–672.
7. Saha, D., Heffron, J. F., Murphy, K. J., and Becker, J. S., The application of tonnage oxygen in copper smelting, paper presented at Fall Meeting of the Arizona Section of AIME, December 7, 1987.
8. Wrampe, P. and Nollman, E. C., Oxygen utilization in the copper reverberatory furnace: theory and practice, paper presented at TMS-AIME Workshop, No. A74-25B.
9. Antonioni, T. N., Blanco, J. A., Dayliw, G. J., and Landolt, C. A., Oxy/fuel smelting in reverberatory furnace at Inco's Copper Cliff smelter, paper presented at the 50th Congress of the Chilean Institute of Mining and Metallurgical Engineers, Nov. 1980.
10. Queneau, P. E. and Schuhmann, R., Metamorphosis of the copper reverberatory furnace: oxygen sprinkle smelting, *J. Metals*, Dec., 12–15, 1979.
11. Successful tests encourage Phelps Dodge Corporation to modify its copper smelter, *Chem. Eng.*, 89(3), 18–19, 1982.
12. Hase, E. A., Experiments with oxygen-enriched blast at the ASARCO East Helena lead smelter, paper presented at First Operating Metallurgy Conference and Symposium, Pittsburgh,
13. Saha, D., Becker, J. S., and Gluns, L., Oxygen-Enhanced Fluid Bed Roasting, Productivity and Technology in the Metals Industries, TMS, 1989.
14. Harbison, E. J. and Davidson, J. A., Oxygen in copper smelting, paper presented by Air Products & Chemicals, Inc., at the Fall Meeting of the Arizona section of the American Institute of Mining, Metallurgical and Petroleum Engineers, Dec. 1972.
15. Eacott, J. G., The role of oxygen potential and use of tonnage oxygen in copper smelting, in *Advances in Sulfide Smelting, Vol. 2: Technology and Practice,* Sohn, H. Y., George, D. B., and Zunkel, A. D., Eds., The Metallurgical Society of AIME, Warrendale, PA, 1983, 583-634.
16. Anderson, B., Hanniala, H., and Karlsson, P., A practical design paper on Outokumpu flash smelting technology, paper prepared for TMS-AIME Workshop, Feb. 1980.
17. Anderson, B., Hanniala, P., and Harkki, S., Use of oxygen in the Outokumpu flash smelting process, *CIM Bull.,* Sept., 172–177, 1982.
18. Antonioni, T. N. et al., Control of the INCO oxygen flash smelting process, copper smelting — an update, in *Conference Proceedings, TMS-AIME Meeting,* Dallas, Feb. 1982.
19. Biswas, A. K. and Davenport, W. G., *Extractive Metallurgy of Copper*, 2nd ed., Pergamon Press, New York, 1980.

20. Suzuki, T. et al., Recent operation of Mitsubishi continuous copper smelting and converting process at Naoshima, *CIM Bull.,* Sept., 51–76, 1982.
21. Suzuki, T. et al., Test operation for smelting more tonnages of copper concentrates at the Mitsubishi continuous copper smelting and converting process, paper presented at the 112th AIME Annual Meeting, March 1983.
22. Bailey, J. B. W. and Storey, A. G., The Noranda process after six years of operation, *CIM Bull.*, June, 142–148, 1980.
23. George, D. B., Taylor, J. C., and Traulsen, H. R., Copper Smelting — An Overview, *CIM Bull.*, pp. 1-15, June 1980.
24. Harbison, E. J. and Davidson, J. A., Oxygen in copper smelting, paper presented by Air Products & Chemicals, Inc., at the Fall Meeting of the Arizona Section of the American Institute of Mining, Metallurgical and Petroleum Engineers, Dec. 1972.
25. Hittl, A. E., Mann, C. A., and Szekely, A. G., Tonnage oxygen in the nonferrous industries, paper presented by Union Carbide Gas Products, Feb. 1969.
26. Smith, C. D. and Saha, D., Advances in oxy/fuel combustion, paper presented at International Wrought Copper Council Conference, Singapore, October 1997.

7 Glass

Prince B. Eleazer III and Bryan C. Hoke, Jr.

CONTENTS

0-8493-1695-2/98/$0.00+$.50

7.1 INTRODUCTION

Industrial oxygen has been used to enhance combustion in the glass industry for several decades. Most of these installations utilized supplemental oxygen/fuel (oxy/fuel) burners, premixed oxygen enrichment of the combustion air, or undershot lancing of oxygen to the port or burner. Supplemental oxy/fuel is the practice of installing one or more oxy/fuel burners into an air/fuel furnace. Premixed oxygen enrichment is the practice of introducing oxygen into the combustion air to a level of up to 27% total contained oxygen (i.e., 6% oxygen enrichment). The amount of oxygen enrichment is limited by materials compatibility issues in highly oxidizing environments. Undershot lancing is the practice of strategically injecting oxygen through a lance into the combustion region.

The frit industry, though, has not typically used any of these oxygen-enhancement techniques, but converted many plants to 100% oxy/fuel during the 1970s and 1980s. Frit furnaces generally produce less than 30 ton/day per furnace and are not continuous, with frequent startups and shutdowns of the furnace. Therefore, heat recovery/air preheat is uncommon. Without heat recovery, fuel savings alone justified the use of full oxy/fuel. In parallel with the conversions in frit, Corning converted a large number of its smaller specialty glass furnaces, mainly ones that had no heat recovery, to 100% oxy/fuel firing.[1]

The early oxy/fuel projects used high-momentum water-cooled burners modified from applications in the metals industries. The experience with frit conversions and with the partial oxy/fuel conversions in other glass segments revealed the need for burners to be developed specifically for glass.[2]

Coincident with this burner development, significant advancements in oxygen production technology and an increased focus on reduced air emissions occurred.[3] The most notable advancement in air separation is adsorption technology which lowered the cost of oxygen in the oxygen consumption range important for many furnaces. This improved the economics for converting to full oxy/fuel.

The glass industry has become one of the largest users of oxygen. Most of this oxygen is used in full oxy/fuel melters. However, the industry is still dominated by air/fuel furnaces. These air/fuel furnaces can also often benefit by judicious use of oxygen. This chapter will discuss some of the issues and benefits of partial oxy/fuel furnace conversions as well as full furnace conversions.

Virtually all segments of the glass industry have now implemented 100% oxy/fuel furnace technology. Table 7.1 summarizes the completed conversions in North America. The table includes "mixed melters," so-called because they get a large portion of their total melting energy from electricity.

The frit segment has the most furnaces converted to oxygen. The pressed and blown, fiberglass, and container industries have converted fewer furnaces but account for the most tonnage of glass produced by oxy/fuel combustion. As of the end of

TABLE 7.1
Summary of Oxy/Fuel Conversions in North America >30 ton/day

Segment	Number of Conversions	Most Common Driving Force	Second Most Common Driving Force	Factors Against
Fiberglass	21	Particulate	Capital reduction	Oxygen cost
Container	24	NOx	Particulate	Oxygen cost
Lighting/tableware/TV	25	Particulate	Capital reduction	Oxygen cost
Float	3	Production flexibility	Capital reduction	Risk, oxygen cost, glass quality
Frit	21	Fuel savings	Production increase	Oxygen cost
Sodium silicate	1	—	—	Need for large production swings
Specialty	15	Fuel savings	Production increase	Oxygen cost

1997, there have been no full oxy/fuel conversions of float tanks by the flat glass segment. However, several companies currently have active projects to investigate this option.

To date, the largest driving force for full oxy/fuel conversions has been to reduce particulate emissions. This may surprise many who consider NOx to be the primary reason for converting to oxy/fuel firing. Reduced NOx emissions and fuel savings are commonly cited as secondary benefits.[4] It is interesting to note that capital reduction was also an important factor in making a positive decision for oxy/fuel conversion.

Most of the 100% oxy/fuel conversions have occurred in the U.S. Nonetheless, many furnaces are now being converted in Europe and in Asia. In Europe, two primary factors have delayed implementation. The first is the way the regulations are written, and the second is related to timing of the regulations. European regulations are often very strict. However, many European countries have air regulations based on the concentration of components (i.e., ppm of particulates, NOx, etc.). Oxy/fuel combustion virtually eliminates the diluent nitrogen. Therefore, while the absolute quantity of emissions are reduced on a mass basis (i.e., pounds per ton of glass produced), the concentration may not significantly decrease. Exceptions to this rule are possible. For example, the TA Luft, Germany's air quality mandate, allows exceptions, but the local authorities must understand the difference and grant a variance.

Regarding timing of the regulations, many of the European Community countries have had strict regulations in place for several years. This led many glass producers to install posttreatment systems, especially for particulates. Many U.S. glass producers view oxy/fuel as a way to decrease capital expenditures by avoiding these posttreatment systems. Europeans, who already installed electrostatic precipitators or baghouses, have less incentive to convert to oxy/fuel.

TABLE 7.2
1997 Estimate of Oxy/Fuel Furnaces and the Cumulative Production Operating Globally

Glass Type	Number of Oxy/Fuel Furnaces	Approx. Volume of Glass Produced, t/day
Container	25	6,000
Television glass	19	3,600
Fiberglass	23	2,100
Lighting	10	1,200
Frit	33	500
Other/specialty	58	3,600
Total	148	17,000

Many Asian countries also are on a concentration basis rather than a mass basis for emission reporting. Another deterrent to conversions in many Asian countries is the high cost of electricity for producing oxygen relative to the cost of fuels.

A global estimate of oxy/fuel furnaces operating to date is in Table 7.2. This does not include furnaces that are currently under construction using oxy/fuel technology, which would increase the numbers to 157 furnaces producing over 19,000 t/day of glass.

7.2 TYPES OF TRADITIONAL GLASS-MELTING FURNACES

The type of furnace for melting glass typically depends on the type and quantity of glass being produced, and the local fuel and utility costs. While there are exceptions, the following discussion describes the primary furnace types and the glass segments that most commonly use each style.

7.2.1 UNIT MELTER

The term *unit melter* is generally given to any fuel-fired glass-melting furnace that has no heat recovery device. Generally, one is referring to an air/fuel-fired furnace when using this furnace term. However, most full oxy/fuel furnaces have no heat recovery system and are therefore, technically unit melters. Typically, the air/fuel unit melters are relatively small in size and are fired with 2 to 16 burners. Furnaces range in production from as large as 40 ton (36 t) of glass per day to as small as 500 lb (230 kg) of glass per day. Larger air/fuel unit melters are found in areas where fuel is extremely cheap. Frit, tableware, opthamalic glass, fiberglass, and specialty glasses with highly volatile and corrosive components are produced in unit melters.

Due to the very low energy efficiency and the use of individual burners, the air/fuel unit melters are very amenable to oxygen-enhanced combustion techniques, including supplemental oxy/fuel boosting, premixed oxygen enrichment, and full

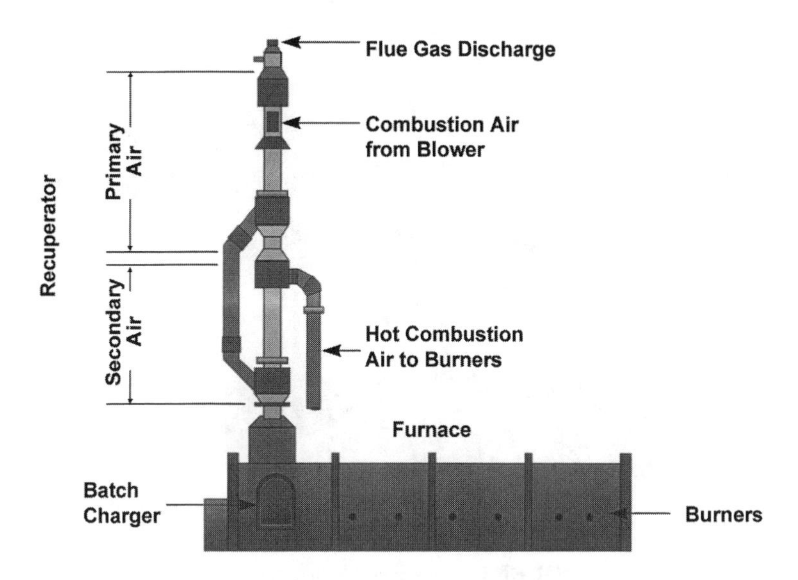

FIGURE 7.1 Typical recuperative melter (side view).

oxy/fuel combustion. Oxy/fuel unit melters have been built as large as 350 ton (320 t) per day of glass to as small as 500 lb (230 kg) of glass per day.

7.2.2 Recuperative Melter

A recuperative melter is a unit melter equipped with a recuperator. Typically, the recuperator is a metallic shell-and-tube-style heat exchanger that preheats the combustion air to 1000 to 1400°F (540 to 760°C). The furnace is fired with 4 to 20 individual burners. These furnaces range in size from as large as 280 ton (250 t) per day of glass to as small as 20 ton (18 t) per day of glass. These furnaces are common in fiberglass production but can also be used to produce frit. Some recuperative furnaces are used in the container industry, though this is not common. Furnace life is a function of glass type being produced. For example, a 6-year furnace life is typical for wool fiberglass. A typical recuperative melter is shown in Figure 7.1.

The recuperative melter is amenable to supplemental oxy/fuel technique or the premixed oxygen enrichment technique (see Chapter 1). Oxygen lancing is typically not used. In the supplemental oxy/fuel technique, an air/fuel burner is simply replaced by an oxy/fuel burner. When premix is applied, oxygen injection into the air main typically occurs downstream of the recuperator to avoid problems associated with air leaks in the recuperator. Care should be taken in locating the oxygen diffuser as discussed in Chapter 10.

These furnaces are good candidates for full oxy/fuel. Recuperative heat exchanger efficiencies are much lower than with regenerative furnaces, and therefore fuel savings can help to drive the conversion. Also, recuperative furnaces operate in a continuous and steady firing mode of operation similar to oxy/fuel furnaces.

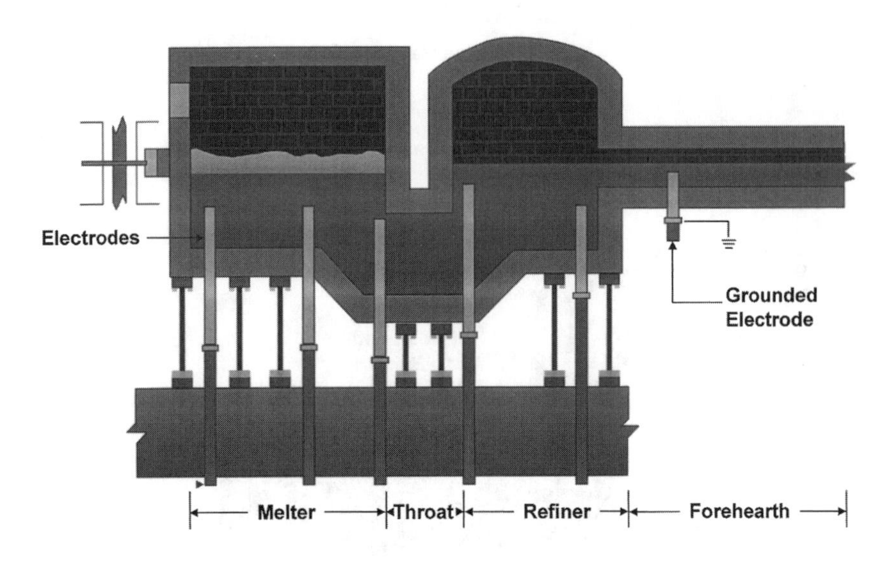

FIGURE 7.2 Typical electric melter.

7.2.3 All-Electric Melter

As the name implies, all-electric melters receive all of the energy for glass melting through electrical heating. Electric current is passed through the glass by means of electrodes. Because of the electrical resistance of the glass, the glass is heated by Joulean heating. Electrodes are typically made of molybdenum; however, tin oxide, platinum, graphite, and iron have also been used.[5] The electrodes are usually rod- or plate-type and can be located in the melter side walls or bottom.

The refractory tends to degrade much faster in these furnaces, resulting in very short furnace campaigns, typically less than 2 years. Most of these furnaces are less than 40 ton (36 t) of glass per day; however, furnaces as large as 200 ton (180 t) per day have been built.[6] A typical electric melter is shown in Figure 7.2.

Due to the design of these furnaces, there is typically no fit for oxygen-enriched combustion. One exception is "hot top" melters which provide some heat via burners located above the bath. In this latter case, supplemental oxy/fuel or premixed oxygen enrichment has been practiced.[7]

7.2.4 Regenerative or Siemans Furnace

The regenerative furnace was patented in the U.S. by Siemans Corporation in the late 19th century. While some design evolution has occurred, the basic concept has remained unchanged. In a regenerative furnace, air for combustion is preheated by being passed over hot regenerator bricks, typically called checkers. This heated air then enters an inlet port to the furnace. By using one or more burners, fuel is injected at the port opening, mixes with the air, and burns over the surface of the glass. Products of combustion exhaust out of the furnace through a nonfiring port and pass through a second set of checkers, thereby heating them. After a period of 15 to 30 min, a reversing valve changes the flow and the combustion air is passed over

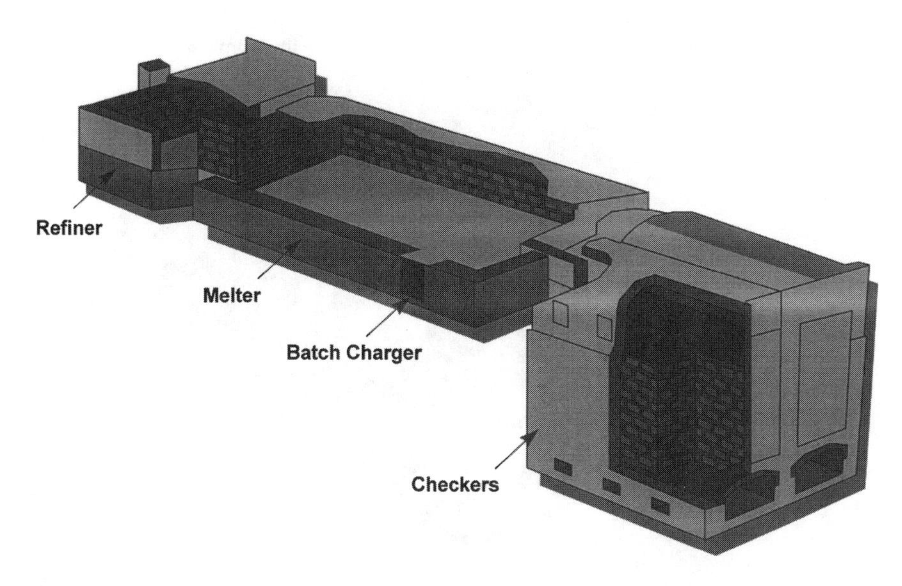

FIGURE 7.3 Typical end-port regenerative furnace.

the hot checkers that were previously on the exhaust side of the process. The fuel injection system also reverses. After reversing, the exhaust gases pass through and heat the checkers that had previously heated the combustion air.

The Siemans furnace is the workhorse of the glass industry. Most flat glass and container glass are produced in this furnace type. Regenerative furnaces are also used in the production of TV products, tableware, lighting products, and sodium silicates. There are two common variants of the Siemans furnace: the side-port regenerative melter, and the end-port regenerative melter.

7.2.4.1 End-Port Regenerative Furnace

End-port regenerative furnaces are typically used for producing less than 250 ton (230 t) of glass per day. In an end-port furnace, the ports are located on the furnace back wall. Batch is charged into the furnace near the back wall on one or both of the side walls. Figure 7.3 shows the layout of a typical end port furnace. These furnaces are commonly used for producing container glass, but are also used for producing tableware and sodium silicates. For container production, a furnace campaign typically lasts 8 years.

Undershot of oxygen through lances and supplemental oxy/fuel have been used successfully on this type of furnace.[7] Oxygen enrichment of the preheated combustion air has also been used on furnaces with damaged checkers (see Section 7.3.1.4.1).

7.2.4.2 Side-Port Regenerative Furnace

Side-port regenerative furnaces have ports located on the furnace side walls. Batch is charged into the furnace from the back wall. Figure 7.4 shows the layout of a

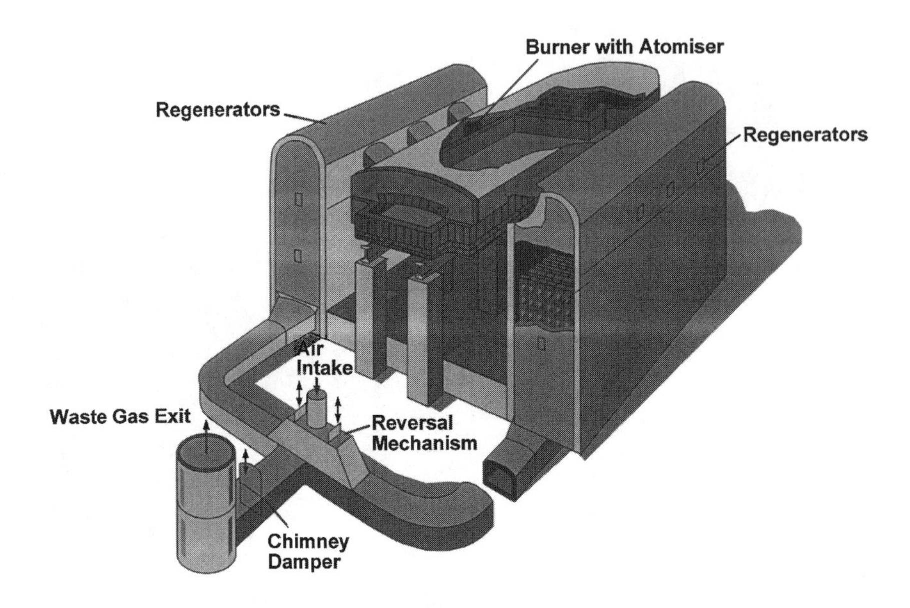

FIGURE 7.4 Typical side-port regenerative furnace.

typical side-port furnace. Side-port regenerative furnaces are typically used for producing greater than 250 ton (230 t) of glass per day. A side-port furnace for float glass commonly produces 500 to 700 ton (460 to 630 t) of glass per day. For container glass, side-port furnaces ordinarily produce between 250 to 350 ton (230 to 320 t) of glass per day. These furnaces are commonly used in container and float glass production, but are also used for the production of tableware and sodium silicates. For container production, a furnace campaign typically lasts 8 years and for float glass production can last as long as 12 years.

Undershot and supplemental oxy/fuel oxygen enrichment have been successfully used on this type of furnace.[8] Premix enrichment has also been used on furnaces with damaged checkers. These furnaces have also been converted to full oxy/fuel.

7.3 OXYGEN-ENHANCED COMBUSTION FOR GLASS PRODUCTION

This section first discusses some of the unique benefits of oxygen-enhanced combustion in glass production. It also presents some potential problems of which operators should be aware as they consider oxygen injection into the furnace.

7.3.1 Benefits

The general benefits of oxygen-based combustion have previously been discussed. This section will focus on some of the specific benefits to the glass producer. Oxygen is typically used in glass making for one or more of the following benefits:

- Fuel savings due to improved furnace efficiency;
- Production increase resulting from improved heat transfer;
- Reduced electricity costs by substituting combustion energy for electric boost energy;
- Extended furnace life by overcoming combustion air throughput limitations caused by plugged checkers or failing recuperator;
- Extended furnace life by substituting combustion energy for electric boost energy, thereby reducing refractory wear caused by electric boost;
- Reduced pollutant emissions such as NOx, particulates, and CO_2;
- Improved glass quality resulting from improved furnace temperature profile, lower volatilization, and better batch line control; and
- Decreased capital cost by reducing or eliminating posttreatment systems and/or heat recovery systems.

More-detailed descriptions of these benefits are given in the sections that follow.

7.3.1.1 Fuel Savings

Fuel savings are directly related to furnace efficiency. Less fuel is used if energy losses are reduced. By substituting oxygen for air, energy losses from gases leaving the flue are significantly decreased. For the air/fuel case, hot nitrogen leaving the system carries significant energy with it. This is discussed in detail in Chapter 1.

Other factors also contribute to improved furnace efficiency. Higher oxy/fuel flame temperatures improve heat transfer to the glass melt by radiation. Tighter control of stoichiometry in the flame region allows the furnace to run with a lower excess oxygen ratio. For full oxy/fuel, radiation heat losses through ports are reduced since fewer ports are needed.

Typical savings range from 5 to 45%/ton of glass produced.[9] The wide range is a function of furnace design and the design of the heat recovery system. Another key factor is whether the furnace is fully converted to oxy/fuel melting or only partially converted. For recuperative furnaces, converting to full oxy/fuel improves fuel efficiency by 30 to 50%. For regenerative furnaces, fuel efficiency improvements typically range from 10 to 30%.

7.3.1.2 Production Increase

Production increase is the most common motivation for adding oxygen-based combustion technologies to an air/fuel furnace. These technologies are a low-capital-cost method for achieving production increases. Any of the methods described in the introduction, including full oxy/fuel, could be used, depending on the goals and furnace design. It should be noted that even though oxygen-enhanced combustion allows for increased energy to the glass, other limitations in the furnace design may prevent production increase. Computational fluid dynamics (CFD) simulations of the combustion space coupled to the glass melt can be used to determine whether or not the flow patterns in the glass would change in a way that would prevent increased production.[14]

Premixed oxygen enrichment is the simplest method to apply. To use this method, only a basic flow control skid to control pressure and flow and a diffuser for proper distribution of oxygen into the air main are required. This method is not foolproof, however. Problems such as shortening of the burner flame and increased temperature of the burner nozzle can occur if not properly implemented. There is also a limit on the level of oxygen enrichment that can be used because of the increased oxidizing nature of the combustion air. Typically, oxygen enrichment is limited to 6% on nozzle mix burners and 3% on regenerative-style burners because of operational concerns. Experience has shown that higher levels of enrichment cause burner nozzles to wear at an accelerated rate.

Undershot oxygen enrichment and supplemental oxy/fuel are more common methods of achieving production increase in an existing furnace. From an equipment standpoint, undershot enrichment is less expensive since no burners or additional flow controls are required. It is not trivial, however, to adjust the excess air and balance fuel. Typically, undershot enrichment is focused on the burners near the furnace hot spot.

Supplemental oxy/fuel has also been used for production increase and with great success, especially in flat glass furnaces. The advantage of this technique is that the additional energy can be focused directly where it is desired without concern about balancing other burners.

Conversion to full oxy/fuel also provides an opportunity for production increase. The change in pull rate achieved with an oxy/fuel furnace, in comparison with an air/fuel furnace, varies depending on furnace type. Pull rate increases of up to 60% have been observed for unit melters. Cross-fired regenerative furnaces have seen increases as little as 10%. End-fired regenerative furnaces converted to oxy/fuel increase pull capacity by 20%. Recuperative melters typically achieve a 30% pull rate increase.

These improvements in pull capacity are due to the improved heat flux density to the glass, which results in a faster melt rate. In addition, the ability of an oxy/fuel furnace to distribute heat input to the most ideal locations within the furnace helps to improve pull capacity. Some claim that the water content in the glass increases because of the different combustion atmosphere for oxy/fuel.[1] More water or hydroxyl ions in the glass decrease the viscosity of the melt allowing improved circulation of the melt.

7.3.1.3 Reduced Electricity Costs

Glass producers sometimes use electric boost to increase the energy to a furnace beyond what the air/fuel-based combustion system is capable of delivering. Electric boosting also affects the temperature and flow patterns in the melt. This method of melting can be very efficient since the energy is released directly in the melt. However, electric melting is a very expensive method of melting.

Proper use of oxygen enrichment can deliver many of the benefits of electric boost while decreasing melting cost. Typically, electric boost reduction is accomplished by either undershot enrichment or supplemental oxy/fuel burners. Undershot is more common on side-port regenerative furnaces, where installation of burners between ports is very difficult. For other furnaces, either technique has been used depending on other furnace issues.

Since substitution of top firing for electric boost also affects the temperature and flow patterns in the melt, CFD modeling can be used to evaluate changes in the flow and temperature patterns in the melt to assure no negative impact to product quality.

7.3.1.4 Extended Furnace Life

Depending on the type of glass produced, a typical furnace campaign is between 3 and 10 years. The furnace efficiency, and therefore the production capacity, typically degrades as the furnace ages. Deterioration of the heat recovery system is often the major cause for reduced furnace efficiency. For electrically boosted furnaces, erosion of the refractory may also affect furnace life. Oxygen-based technologies can play an important role in maintaining furnace capacity during the last years of furnace life.

7.3.1.4.1 Plugged checkers

As a regenerative furnace ages, the thermal efficiency and airflow capacity of the regenerators decrease. This is due to blockages that occur within the checker package. These blockages are typically caused either by collapse of part of the brick structure or by plugging of the flow passages. Checker collapsing is caused by bricks deteriorating in a section due to temperature, corrosion, or thermal cycling. Plugging is caused by condensation of volatile species contained in the glass, such as sodium and sulfates which combine in the furnace atmosphere to form sodium sulfate. It is also caused by entrainment of batch particles and the carryover of these particles into the checker system.

7.3.1.4.2 Failing recuperator

Like regenerators, metallic recuperators may plug due to condensation of volatile species. However, deterioration due to corrosion of the metallic tubes within the recuperator by chemical reaction with the condensables from the furnace is more common. Corrosion causes the formation of holes in the tubes resulting in air leakage. When air leaks, more air must be supplied to the cold air intake of the recuperator to provide the same amount of oxygen to the burners. Sometimes it is not possible to increase the air intake enough to supply sufficient oxygen to the burner because of air loss and pressure drop. When this is the case, high-purity oxygen can be injected into the airflow to meet the oxygen demand.

Usually, the oxygen is supplied via a premix technique using a diffuser in the air main downstream of the recuperator. However, an alternative to premix is supplemental oxy/fuel, where some of the air/fuel burners are replaced with oxy/fuel burners. Burners are typically exchanged at either the hot spot or the charge end of the furnace.

7.3.1.4.3 Refractory erosion by electric boost

Although electric boost can be beneficial for heat input to the glass and establishing good convective currents in the glass, it can also promote refractory erosion. The electric boost elevates the temperature of the glass in the region of the electrode resulting in lower glass viscosity and lower glass density. The glass flow in the region of electrodes is relatively high because of buoyancy-driven flow of lower-viscosity glass. The high flow rate of hot glass increases the rate of local erosion of the refractories near the electrodes, thereby decreasing furnace life. In fact, there have been instances where refractory has completely eroded and the melt drained. By reducing or eliminating electric boost through the use of oxygen-based technologies, refractory erosion can be reduced and furnace life extended.

7.3.1.5 Reduced Pollutant Emissions

Glass furnaces emit some pollutants into the atmosphere. This is inevitable because of the high temperatures at which the furnaces operate, the materials used to produce glass, and the large amount of energy required to melt the batch constituents. Modifications to air emissions regulations are an important driving force for change in the glass industry. While many segments of the glass industry can easily meet regulations for particulate and NOx emissions, those with the highest volatile contents are having to address emissions now.

It is widely known that a few states, like California and New Jersey, have stringent air quality regulations. But many states have laws that make it difficult to produce glass with conventional air/fuel melting practices while meeting the regulations. Therefore, 100% oxy/fuel conversions have not been limited to the few states known traditionally for stringent air quality regulations. In 1997, 16 states were known to have had at least one oxy/fuel conversion, and several states have had multiple sites with conversions. In addition to this list, conversions are at present being planned in four additional states.

The Federal Clean Air Act Amendment of 1990 (CAAA) will significantly change the air quality regulations in most states since the CAAA gives even stricter guidelines for each state than previous regulations. Glass melting is primarily affected by Titles I, III, and V. Title I seeks to control urban air quality problems in nonattainment areas. Three air pollution problems are covered: (1) smog/ozone caused by nitrogen oxides (NOx) and volatile organic compounds (VOCs), (2) carbon monoxide (CO), and (3) particulate matter. Title III seeks to control hazardous air pollutants (HAPs, also called air toxins) that are hazardous to humans and the environment. HAPs are typically carcinogens, mutagens, and reproductive toxins. Relevant pollutants for glass manufacturing include lead, selenium, and chromium. Title V introduces a national operating permit program for any major sources subject to Title I or Title III. Major sources must obtain operating permits to ensure plant compliance with applicable requirements. The permits will require more monitoring and reporting than existing plans in most states.[10]

Full oxy/fuel combustion greatly reduces the emissions of NOx and particulates. NOx emissions are reduced by over 80% for an oxy/fuel furnace compared with its air/fuel counterpart. This is due mainly to the fact that there are simply fewer nitrogen molecules around to react and form NOx.

7.3.1.5.1 NOx

NOx has been found to be a major contributor to acid rain and ozone formation. Ozone is a major problem in many world cities. As noted above, the Clean Air Act

TABLE 7.3
NOx Emissions

Furnace	Air/Fuel lb NO/ton	Oxy/fuel lb NO/ton	Change, %	Excess Oxygen in the Stack, %
A	4.5	1.5	–67	1.5
B	16.3	8.8	–46	12.0
C	18.0	0.5	–97	0.5
D	13.1	0.7	–95	1.0

has dictated that several states in the U.S. must implement very tough NOx standards. For example, the California South Coast Air Quality Management District requires NOx emission to be less than 429 g/t (0.86 lb/ton) of glass by 2003. This level is achievable with oxy/fuel technology, but is very difficult to achieve with traditional preheated air systems for glass.

Table 7.3 shows NOx emission data from several furnaces converted to full oxy/fuel. The data in the table show significant changes in NOx emissions before and after the conversion. For furnaces C and D with minimum air leakage, the measured NO emissions were around 0.5 lb NO/ton (0.25 kg/t) of glass. Furnace A uses air bubbling to enhance glass convection in the tank which provides more nitrogen for NO formation. Furnace B reduced NO emissions by about 50%. The number with oxygen is still very high (8.8 lb NO/ton or 4.4 kg/t of glass). This can be explained by the way the furnace operates. As shown in Table 7.3, the furnace operates with high excess oxygen. This is done to maintain glass quality and reduce CO emissions. A lead glass furnace needs to maintain highly oxidizing conditions in the furnace. This is achieved by controlled airflow to purge windows on the peepholes as well as a furnace camera. Finally, the batch chemistry for this furnace has significant quantities of nitrate compounds, which is an additional source of nitrogen for NO formation.

7.3.1.5.2 Particulates

Particulates in glass furnaces come from two sources: (1) physical entrainment of the batch into the flue system and (2) volatilization of glass and batch constituents that recombine and condense upon cooling. Physical entrainment of particles is reduced when an air/fuel furnace is enhanced with oxygen. The flue gas volume decreases, thereby lowering the gas velocity across the furnace and entering the flue system.

Although volatile species concentrations are typically higher for oxy/fuel relative to air/fuel, the mass flow rate of the volatile species exiting the furnace is reduced because of the significantly reduced total mass flow. If you assume a 30% decrease in fuel usage for oxy/fuel relative to air/fuel, the mass flow rate through an oxy/fuel furnace will be roughly five times less than the air/fuel furnace. In this case, the mass fraction of particulate-forming species in an oxy/fuel furnace must be five times greater than in the air/fuel counterpart for particulate pollution to increase.

TABLE 7.4
Particulate Emissions

Furnace	Air/Fuel (lb particulates/ton)	Oxy/Fuel (lb particulate/ton)	Change (%)
A	3.3	1.3	–60
B	5.2	2.6	–50
C	3.7	0.8	–78
D	1.0	0.3	–70

As an added benefit, posttreatment systems, such as electrostatic precipitators and baghouses, are more efficient at the lower oxy/fuel flow rates, further reducing particulate pollution. This is due to the higher residence time through the emission control device and the more-consistent temperature of the exhaust stream. Particulate data for several fully converted furnaces are given in Table 7.4. For the furnaces with particulate emission control, the particulate measurements were made before the electrostatic precipitator or baghouse. Note from Table 7.4 that the reductions in particulate emission were significant. In most cases, particulate emissions for the oxy/fuel case were reduced by greater than 50% compared with the air/fuel case.

This is important especially for processes using expensive volatile components such as lead oxide since it can significantly improve the economics in favor of oxy/fuel firing. The impact of conversion will vary with furnace and glass type. However, oxy/fuel combustion nearly eliminates particulates attributable to carry-over or entrainment of batch particles.

7.3.1.5.3 Carbon dioxide

Many scientists have attributed global warming and the greenhouse effect to increases in carbon dioxide in the atmosphere. Because all combustion-fired glass melters are fired with carbon-based fuels such as natural gas or oil, one of the products of combustion is CO_2. Obviously, the more one saves fuel, the more one reduces CO_2 emissions. As discussed previously, oxygen-based combustion can reduce fuel consumption by 5 to 45%. This directly relates to a corresponding CO_2 reduction of 5 to 45%.

7.3.1.6 Improved Glass Quality

There are numerous factors in oxy/fuel glass melting that can lead to higher-quality glass. It would be difficult to attribute the results to any single item, as many are interdependent.

7.3.1.6.1 Improved placement of energy input

Oxygen-based combustion delivers several control benefits that allow the operator to improve glass quality. Operators of regenerative furnaces are limited in tailoring the furnace profile by the flow patterns within the checkers. As noted previously, these flow patterns can change as the furnace ages. Oxy/fuel burners allow the operator to place heat exactly where the glass maker desires the energy along the furnace length. Undershot enrichment allows air/fuel furnaces to have additional flexibility on balancing fuel and excess oxygen. Both of these techniques result in a better furnace profile and much better control of the batch and scum line within the furnace.

7.3.1.6.2 Less loss of volatiles

As discussed in the section on particulate pollution, the rate of volatilization of the more volatile glass species is reduced for oxy/fuel. Without this preferential loss of some components from the surface of the melt, the concentration in the melt is more likely to be homogeneous. Concentration nonhomogeneity can result in cord defects or other visual distortions in the glass product.

7.3.1.6.3 Improved operational consistency

Oxy/fuel furnaces are capable of having very consistent oxygen and fuel flow, allowing for very consistent and constant conditions within the furnace. Air/fuel furnaces have fan systems where volume flow varies with temperature changes. Further, regenerative furnaces are subject to uneven flow conditions through the left and right checker packages, leading to nonsymmetric firing of the furnace. With all regenerative-type air/fuel furnaces, this reversal of heat input every 20 to 30 min creates pressure swings in the melter which affect glass level. This can result in gob weight variations and glass bottle wall thickness variations. Eliminating these variations is particularly critical in high-quality glass products like TV or lightweight, narrow neck-press-and-blow objects.

7.3.1.6.4 Higher water content and improved workability

Glass makers refer to workability as improved glass wall thickness distribution in the final product and reduced splits and checks. Higher workability has been reported by many glass producers using oxy/fuel. It is the general belief by glassmakers that the higher water content in glass produced by oxy/fuel is a result of this significantly higher water content in the combustion space. An oxy/fuel furnace provides a higher partial pressure of water vapor over the glass which increases the amount of water, or hydroxyl ions, dissolved in the glass. This increased water content in the glass lowers its viscosity and enhances glass convection currents. This, in turn, leads to more dwell time in the furnace which directly reduces seed and stone counts. Lower viscosity also permits bubbles in the glass to rise to the glass surface faster, in accordance with Stokes law.

7.3.1.7 Decreased Capital Cost

Many of the methods of oxygen enrichment can be implemented with low capital cost for the benefit gained. For example, many furnaces have achieved a 15 to 20% production increase with the simple addition of two oxy/fuel burners. In this case, liquid oxygen is commonly used to supply the oxygen. Typically, the oxygen supplier supplies the storage tank and proper vaporization system for a monthly service charge. Therefore, the furnace operator only needs to supply a small concrete pad, some house line piping which is typically less than 2 in. (5 cm) in diameter, oxygen and natural gas flow controls, and the burners. The associated costs would be significantly less than any major furnace modification that could provide similar results.

Capital savings are one of the largest contributors to the positive economics of a full oxy/fuel furnace. The regenerator checker packs and ports are not required in an oxy/fuel furnace. This is a substantial savings in capital with reduced refractory and labor to build the regenerators and ports.

An air reversal system is required in a regenerative air/fuel furnace. Oxy/fuel furnaces do not reverse, reducing capital requirements and eliminating a maintenance concern. With an 80% reduction in furnace exhaust, the flue opening and exhaust system can be greatly downsized, saving additional capital.

7.3.2 Potential Problems

7.3.2.1 Overheating of Refractories

If high-momentum oxy/fuel burners are used, the wall opposing the burner may overheat and damage may occur to the refractory.[2] To minimize this problem, low-momentum burners are recommended. Overheating can still occur with low-momentum burners if the flame impinges on the opposite wall. In this case, the firing rate must be reduced.

If the wall on which the burner is mounted is overheating, the flame is simply too hot and/or too short. This can occur when applying premix, supplemental oxy/fuel, or full oxy/fuel. If the enrichment technique is premix, a lower level of enrichment is required. If an oxy/fuel burner is being used, a lower-momentum, slower-mixing burner is recommended.

Crown overheating is commonly caused either by burner lofting or by two opposing burner flames colliding and deflecting toward the roof. If the flame appears to be lifting toward the crown due to buoyancy, then the burner is likely to be operating below the lower firing rate of the burner. A different burner size or higher firing rate is required. If the problem is due to opposed burners, the burner firing rate must be decreased. Generally, opposed burners are not recommended for furnaces less than 24 ft (7.3 m) in width.[3]

7.3.2.2 Refractory Corrosion

Advanced refractory corrosion has been noted in some furnaces with full oxy/fuel combustion.[11] This phenomenon is due to the condition discussed in Section 7.3.1.5.2. The problem can be minimized by selecting the appropriate refractory, keeping joint sizes small, and minimizing air infiltration.[12]

7.3.2.3 Overheating of Glass

Fast-mixing oxy/fuel burners can overheat the glass locally, leading to numerous glass defects such as cord and seeds. On the melt surface, foam formation or reboil is sometimes observed. To reduce the possibility of glass overheating, the burners should not be angled toward the glass.

7.3.2.4 Burner Deterioration

There are two common types of burner deterioration that occur when using oxygen-enhanced combustion: corrosion and oxidation. Corrosion is limited to oxy/fuel burners, and oxidation can be a problem for air/fuel burners operating under premix conditions.

Corrosion is caused by condensation of glass volatiles that have flowed into the burner region. If the burner is colder than the condensation point of the volatiles, the volatiles will condense on the burner and chemically attack the burner. This phenomenon is common for water-cooled burners since they are relatively cold and typically high momentum. High-momentum burners generally induce a flow pattern that promotes movement of the furnace gases back to the burner tip. To minimize this problem, non-water-cooled, low-velocity burners are recommended.[11]

Oxidation of the metallic burner tips is sometimes accelerated for air/fuel burners when subjected to higher oxygen levels, as in premixed oxygen enrichment. The flame is often shorter, resulting in an increased burner tip temperature for some burner designs. Higher temperature in conjunction with higher oxygen concentration accelerates oxidation of the metal (steel). If this occurs, then the level of oxygen enrichment must be decreased, the burner type must be changed, or the method of oxygen enrichment must be changed.

7.3.2.5 Higher NOx Emissions

As mentioned earlier, higher flame temperatures are coincident with higher oxygen concentrations. As discussed in Chapter 2, the formation of thermal NOx is increased by higher temperatures. Therefore, in air/fuel furnaces, thermal NOx generation can increase with the addition of oxygen. To reduce NOx production, reduce the excess oxygen to the oxygen-enhanced burners or apply methods to delay mixing of the fuel and oxidant.

7.3.2.6 Permitting

A potential problem of using oxygen-enriched combustion is the need for modified or new operating permits. Depending on the specific application and region of the country, the use of oxygen may be considered a significant-enough departure from the existing operating conditions that the air emission permits need to be reevaluated. This may be a lengthy and expensive process. The switch from a highly polluting air/fuel operation to a cleaner oxy/fuel operation can actually be stalled by the permitting process.

7.4 PROCESS ECONOMIC REVIEW

The overall cost of oxy/fuel varies with each case. The following is an economic review for converting an air/fuel furnace to 100% oxy/fuel. Generally, it is very difficult to predict accurately the economics of using oxygen before the conversion because most rebuild projects include many factors that may affect the analysis. These include comparison of different furnace sizes and pull rates, different insulation packages installed, etc. Even after the conversion, there are economic issues that are hard to quantify, such as easier furnace operation, better turndown, and minimized downtime. Those benefits do not show up on a spreadsheet calculation, but they indirectly affect the process economics, for example, through lower maintenance costs and longer furnace life, and they are greatly appreciated by the furnace operators.

Due to the confidential nature of the subject we are unable to present here specific economic data about actual furnace conversions. Instead, a generalized analysis of oxy/fuel process economics is given based on the following premises:

- Furnace size: 250 ton/day regenerative
- Air/fuel furnace efficiency: 4.2×10^6 Btu/ton glass
- Emission control equipment:
  - Ammonia DeNOx and electronic precipitator for the air case
  - A lower-capacity electronic precipitator for the oxygen case
- 8-year furnace life/capital depreciation
- Discount rate = 0% (*Note:* This simplifies the analysis, but the oxygen case is penalized)
- Cost of electricity varies from $0.03 to 0.07/kWh
- Oxygen cost varies with electricity (electricity is roughly ⅓ of oxygen cost)
- Cost of natural gas varies from $2 to 5/1000 scf (standard cubic feet)
- Fuel savings with oxy/fuel varies from 10 to 40%

For this example, we will assume a 10% greater pull capacity for the oxy/fuel furnace. This benefit can be enumerated in three different ways. First, the oxy/fuel furnace can be the same size as the air/fuel furnace and operated at a lower crown temperature for the same glass pull rate. This will increase the life of the furnace by up to 10%, or roughly 1 year. Second, the oxy/fuel furnace could be built for the same pull rate as the air/fuel furnace, hence a 10% smaller footprint than the air/fuel furnace. This would reduce refractory costs and use less space. Third, the oxy/fuel furnace can be built the same size as the air/fuel furnace. This would provide the glassmaker with 10% added production, or, in this analysis, 25 ton/day (23 t/day), if required. This added capacity could be provided with an air/fuel furnace in later years, but would require oxygen boosting or electric boosting. This analysis assumes oxygen boosting is required starting in year 4 to obtain the increased production on the air/fuel furnace.

An example calculation is shown in Table 7.5. It was assumed that the electricity is $0.05/kWh, natural gas is $3.5/1000 scf, and fuel savings of 26% has been achieved after converting to oxygen firing. This number is at the high end for a regenerative furnace and relatively low for a recuperative furnace. It may be seen from the table that with this fuel savings, air and oxygen economics are about the

TABLE 7.5
Comparison of Process Economics with Air and Oxygen for a 250 ton/day Glass Melter

Costs	Unit	Air/Fuel Furnace	Oxy/Fuel Furnace
Capital Costs			
1. Heat recovery	$\$ \times 10^6$	1.4	—
2. Melter rebuild	$\$ \times 10^6$	1.0	1.0
3. Emission control equipment (particulate and NOx)	$\$ \times 10^6$	2.1	0.6
4. Total Capital Cost (1 + 2 + 3)	$\$ \times 10^6$	**4.5**	**1.6**
Operating Costs			
5. Natural gas cost/year	$\$ \times 10^6$/year	1.34	0.98
6. Emission control equipment operating cost/year	$\$ \times 10^6$/year	0.13	0.03
7. Air or oxygen cost	$\$ \times 10^6$/year	0.07	0.84
8. Total Operating Cost/Year (5 + 6 + 7)	$\$ \times 10^6$/year	**1.54**	**1.86**
Other Costs			
9. Capital depreciation/year	$\$ \times 10^6$/year	0.56	0.20
10. Total Production Cost/Year (8 + 9)	$\$ \times 10^6$/year	**2.10**	**2.06**
$/ton glass		**$23/ton**	**$23/ton**

same for the given assumptions. With any fuel saving beyond 26%, the oxy/fuel case becomes more economical than the conventional air furnace. Fuel savings below 26% make the air/fuel case more economical.

The spreadsheet exercise does not reflect the economics of any particular case. Instead, it was included in trying to answer one of the key questions: Can fuel savings and emission reduction offset the cost of using oxygen? The resulting break-even point of 26% for the given assumptions seems to be relatively high for a highly efficient regenerative-type furnace. A similar analysis when applied on a 100 ton/day (90 t/day) recuperative furnace yields just a few percent more in fuel savings as the break-even point. In this case, oxy/fuel is clearly more economical since most conversions of recuperative furnaces resulted in fuel savings beyond 30%.

The results of the above analysis will vary with the ratio of natural gas and electricity costs. This is illustrated in Figure 7.5. It may be seen that the maximum benefit from using oxygen is when the cost of natural gas is high and the cost of electricity is low. In that case the break-even point is at about 17% fuel savings for the 250 ton/day (230 t/day) regenerative furnace and the assumptions given above. The economics are much more severe for low-cost natural gas and expensive electricity, for which the furnace would need to make almost 40% in fuel savings to justify using oxygen.

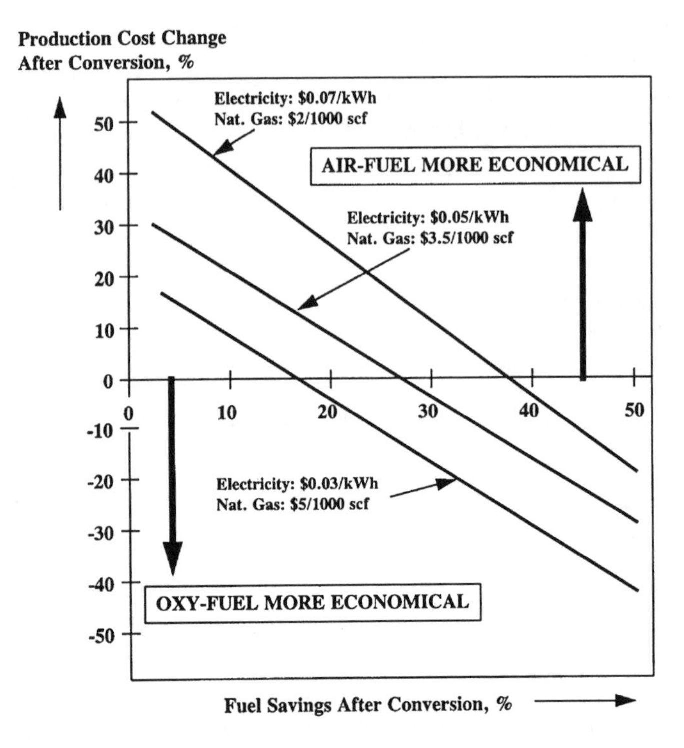

FIGURE 7.5 Effect of energy cost on process economics.

A similar chart, shown in Figure 7.6, presents the effect of installing flue gas treatment equipment on the total production cost. For constant energy costs ($0.05/kWh electricity and $3.5/1000 scf natural gas) a conversion of 250 ton/day (230 t/day) regenerative furnace with emission control equipment would need to realize about 10% less fuel savings than a furnace without such flue gas treatment to achieve break-even economics.

From this analysis, one may conclude that the oxy/fuel technology will be easiest to justify for air/fuel furnaces with inefficient or no heat recovery, with high-cost electricity, or with a need for posttreatment of the flue gases. And this is indeed the trend seen in the glass industry with the first conversions occurring in the frit and specialty glass segments, followed by those having particulate emission problems. As more oxy/fuel conversions are coming onstream, glass manufacturers are starting to realize the full potential of this technology making it even easier to justify on the combined technology and economics basis.[13]

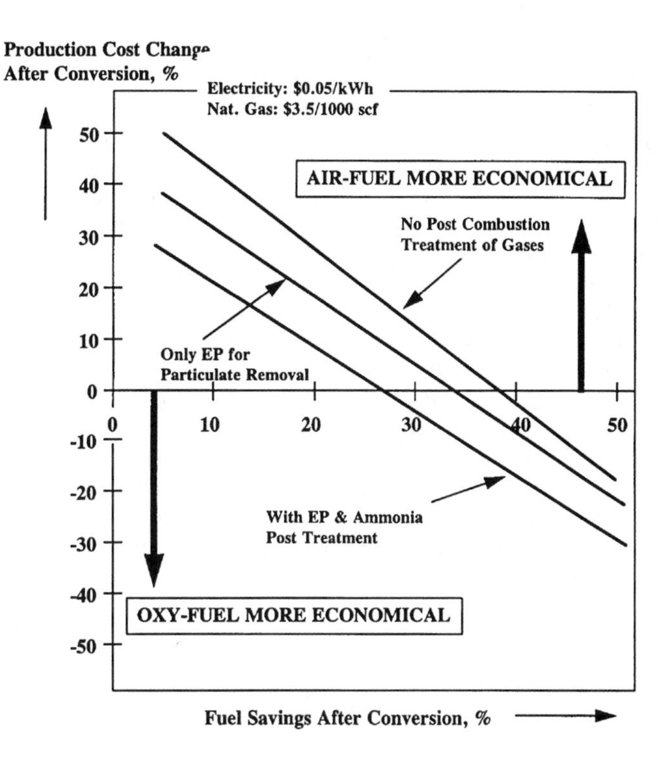

FIGURE 7.6 Effect of NOx and particulate control cost on process economics.

REFERENCES

1. Brown, J. T., 100% oxygen-fuel combustion for glass furnaces, in *Collected Papers from the 51st Conference on Glass Problems*, The American Ceramic Society, Westerville, OH, 1990, 202–217.
2. Shamp, D. E. and Davis, D. H., Application of 100% oxygen firing at Parkersburg, West Virginia, in *Collected Papers from the 51st Conference on Glass Problems*, The American Ceramic Society, Westerville, OH, 1990, 218–239.
3. Slavejkov, A. G., Baukal, C. E., Joshi, M. L., and Nabors, J. K., Oxy/fuel glass melting with a high performance burner, *Am. Ceram. Soc. Bull.*, 71(3), 340, 1992.
4. Eleazer, P. G., Brown, G., and De Sanctes, S., New oxy/fuel burner provides benefits to glass melters, *Ceram. Ind.,* Feb. 24–29, 1994.
5. Tooley, F. V., Ed., *The Handbook of Glass Manufacture*, 3rd ed., Vol. I, Ashlee, 1984, 400-9–400-15.
6. Moore, R. D. and Davis, R. E., Electric furnace application for container glass, paper presented at 47th Conference on Glass Problems, Nov. 1986.
7. Ertl, D. and McMahon, A., Conversion of a fiberglass furnace from 100% electric to oxy/fuel combustion, in *Proceedings from 54th Conference on Glass Problems*, Oct. The American Ceramic Society, Westerville, OH, 1993, 186–190.

8. Hope, S. and Schemberg, S., Oxygen-fuel boosting on float furnaces, *Int. Glass Rev.*, Spring/Summer, 1997.
9. Eleazer, P. B. and Slavejkov, A. G., Clean firing of glass furnaces through the use of oxygen, in *Proceedings from 54th Conference on Glass Problems*, The American Ceramic Society, Westerville, OH, Oct. 1993, 159–174.
10. Destefano, J. T., The Clean Air Act: past, present, and future, in *Collected Papers from the 52nd Conference on Glass Problems,* The American Ceramic Society, Westerville, OH, Nov. 1991, 146–152.
11. Shamp, D. E. and Davis, D. H., Application of 100% oxygen firing at Parkersburg, West Virginia, *Ceram. Eng. Sci. Proc.*, 12(3–4), 610–631, 1991.
12. Horan, W. J., Slavejkov, A. G., and Chang, L., Heat transfer optimization in TV glass furnaces, in *Conference Proceedings from 56th Conference on Glass Problems,* The American Ceramic Society, Westerville, OH, Oct. 1995, 141–151.
13. Bazarian, E. R., Eleazer, P. B., and Slavejkov, A. G., Economics drive the conversion of glass tanks to oxy/fuel, *Am. Ceram. Soc. Bull.*, 73(2), 1994.
14. Hoke, B. C. and Marchiando, R. D., Using computational fluid dynamics models to assess melter capacity changes when converting oxy-fuel, in *Proceedings of the 18th International Congress on Glass,* The American Ceramic Society, San Francisco, July, 1998.

8 Waste Incineration

Charles E. Baukal, Jr.

CONTENTS

0-8493-1695-2/98/$0.00+$.50

8.1 INTRODUCTION

Incinerators are designed to burn and, in many cases, destroy waste materials which may sometimes be contaminated with hazardous substances. The waste materials usually have some heating value. However, nearly all incineration processes require a substantial amount of auxiliary heat which is commonly generated by the combustion of hydrocarbon fuels such as natural gas or oil. Most combustion processes use air as the oxidant. In many cases, these processes can be enhanced by using an oxidant that contains a higher proportion of O_2 than that in air (see Chapter 1).

Incineration is a common method for treating waste materials. Historically, air/fuel combustion has been used in waste processing to provide heat for thermal destruction of solid, liquid, and gaseous waste streams. Examples include medical and municipal waste (solid), spent solvents (liquid), and off-gas or vent streams (gaseous). Oxygen-based combustion systems are becoming more common in waste-processing applications.[1] As will be shown later, when traditional air/fuel combustion systems have been modified for oxygen-enhanced combustion (OEC), many benefits can be demonstrated. Typical improvements include higher destruction and removal efficiencies of the waste, increased thermal efficiency, increased processing rates, lower NOx and particulates emissions, and less downtime for maintenance.

8.2 OXYGEN IN WASTE INCINERATION

This section first discusses some of the unique benefits of using OEC in waste incineration. It also presents some potential problems. Specific examples of OEC in incineration are then given.

8.2.1 Specific Benefits

The general benefits of OEC have been previously discussed (Chapter 1). This section discusses the benefits that are specific and possibly unique to incineration applications (see Figure 8.1). The benefits are generally discussed here. Specific examples of actual improvements are discussed later.

8.2.1.1 Overcoming Thermal Limitations

OEC has been used in several different types of incineration applications to overcome thermal limitations in the process.[2] One example is a fixed rotary incinerator where kiln instabilities were produced by variations in the incoming waste, which was sometimes cold and wet. The existing combustion system in the primary combustion chamber (PCC) was unable to handle these transient variations. O_2 was injected into the kiln through a lance. The O_2 flow was automatically controlled, based on feedback from the temperature at the exit. This improved the kiln stability. It also improved the refractory life in the afterburner due to the more uniform temperature profile in the overall system. In general, OEC can be used to increase incinerator throughput of low-heating-value waste materials.

Another application of OEC to overcome thermal limitations is in thermal pyrolysis of municipal solid waste or refuse-derived fuel (RDF).[3] The N_2 in air

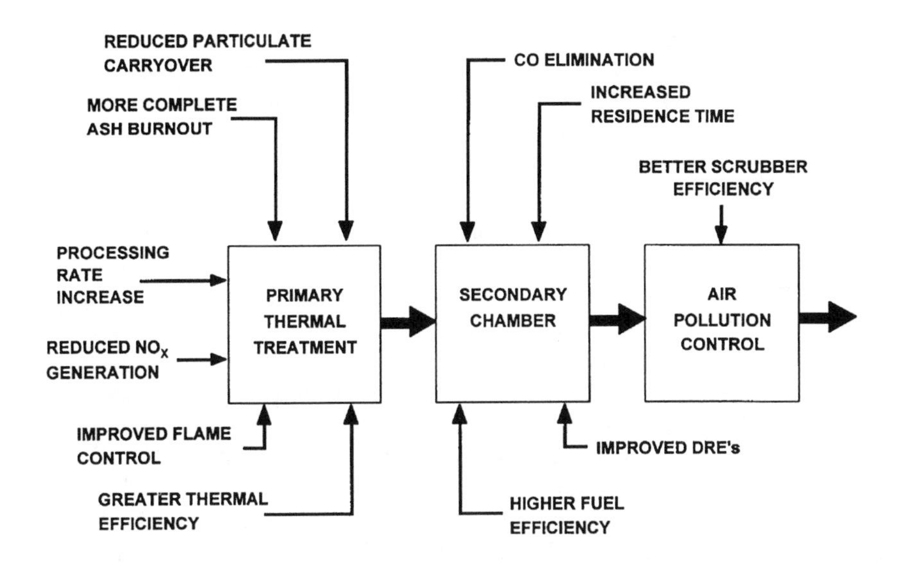

FIGURE 8.1 Benefits of using oxygen in waste incineration applications.

impedes the pyrolysis process which is commonly used to recover chemicals and energy. By using oxygen enrichment, high-quality char or gas can be produced from the high-ash and high-moisture-content fuels. OEC can increase the heating value of the gas produced in gasification of municipal solid waste (MSW) by enhancing the devolatilization and evolution of the gaseous products.[3]

In many cases, the volatilization characteristics of waste materials are unpredictable so that incinerators are generally run with large amounts of excess air in order to ensure complete destruction of the volatiles.[4] This excess air puts a large heat load on the incinerator. When some or all of the air is replaced with oxygen, some or all of the ballast nitrogen is eliminated which reduces the heat load on the furnace and increases the thermal efficiency of the process. Eliminating ballast nitrogen also improves the destruction efficiency of the volatiles so that less excess oxygen is usually required which further reduces the unproductive heat load on the incinerator. For an existing incinerator, this reduction in heat load may either mean more waste can be processed or the auxiliary fuel consumption can be reduced. For a new incinerator, the equipment size and cost can be reduced.

8.2.1.2 Lower Pollutant Emissions

Many of the benefits associated with OEC relate to increasing the partial pressure of O_2 in the incinerator. The combustion process becomes more reactive (see Figure 2.15) which tends to increase the destruction efficiency of any hydrocarbons in the system. This lowers the pollutant emissions.

8.2.1.2.1 Higher destruction-and-removal efficiency

The destruction-and-removal efficiency (DRE) is typically regulated for hazardous pollutants as defined by the Resource Conservation and Recovery Act (RCRA).

These pollutants include, for example, polychlorinated biphenyls (PCBs) and principal organic hazardous constituents (POHCs) that may be found, for example, in contaminated soil. OEC can increase the DRE in an incineration process. This is accomplished by a combination of increased residence time within the incinerator and higher gas temperatures. These are two of the "three T's of incineration." The higher residence time is a result of the dramatically reduced flue gas volume that occurs when some or all of the combustion air, in an existing system, is replaced with oxygen. For a given PCC, the gas velocities through the system will be lower with OEC compared with air/fuel combustion. Therefore, the residence time will increase, which improves the mass transfer within the system. This increases the destruction efficiency of the hazardous pollutants in the process. The higher flame temperatures associated with OEC ensure that all hydrocarbons in the incinerator will be well above their ignition temperatures. Therefore, if there is sufficient oxygen for combustion, the hazardous pollutants should combust. As previously discussed (see Section 1.3.3.3), the turbulence in the flame may be higher for OEC compared with conventional air/fuel combustion. However, the overall turbulence level in the combustion chamber will be reduced because of lower gas velocities. This must be offset by the increased residence time and higher gas temperatures in order to ensure adequate DRE. As will be shown, the DRE generally increases using OEC.

8.2.1.2.2 Lower combustibles

Combustible gases, such as carbon monoxide and hydrogen, may be produced when fossil fuels are incompletely burned. These are sometimes referred to as products of incomplete combustion (PICs) or as unburned hydrocarbons (UHCs). There are regulatory limits on the amount of PICs that may be emitted into the atmosphere. High levels of PICs may occur when there is insufficient oxygen for complete combustion, which may be caused by fuel-rich combustion. High PIC levels may also occur when there is insufficient mixing between the fuel and the oxidizer, which may happen in a poorly mixed diffusion flame. In addition, high PICs may occur when there are transient increases in either the fuel flow rate or in the amount of hydrocarbons in the material being processed, or if the oxidizer flow is not properly adjusted and distributed. For example, in an incinerator the heating value of the material being incinerated may vary considerably with time.

Transient feeds of high-heating-value waste may lead to spikes or "puffs" of PICs in the exhaust gases. There are two common ways of handling this problem. One is to set the oxidizer flow rate high enough to handle the worst case. However, that means there will be excess oxygen for the vast majority of the time. This lowers fuel efficiency and also produces higher NOx emissions. Therefore, running with excess oxygen is not generally desirable. A better method for handling puffs is to have either a manual or automatic control system that can compensate for the changes. An example of a manual system would be one in which the material to be processed is analyzed before it enters the combustion system. Then the combustion system is adjusted according to the incoming material. Another way is to mix the incoming materials in such a way as to keep the heat content nearly constant. This method is expensive and labor intensive. In an automatic control system (see Figure 8.2), high levels of PICs, usually in the form of CO, are usually sensed in

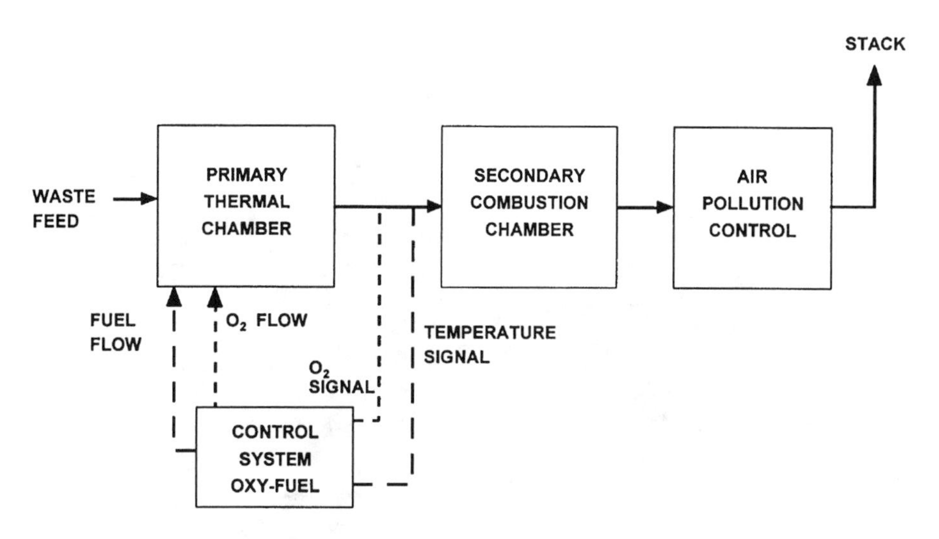

FIGURE 8.2 Feedback control system for puff suppression.

the PCC. The PIC measurement system must respond quickly in order for this to be used in a real-time, feed-forward control system. Adjustments are then made in the secondary combustion chamber (SCC) to ensure complete combustion before the gases exit the exhaust system. These adjustments usually include increasing the oxidizer flow, somewhere downstream of the PCC. Oxygen has successfully been used to control these puffs in incinerators.[5] Oxygen is much more effective at controlling puffs than air because it has much higher reactivity since it does not have the additional diluent N_2 contained in air. The reduction in nitrogen in the combustion products also allows the system to react more quickly because there is less dilution so that adding oxygen to handle puffs produces a quicker response than adding air, which contains nearly 80% nitrogen. A coincidental benefit is that the gas flow through the combustion system is generally slower for an air/fuel incinerator that has been retrofitted with OEC compared with the original air/fuel system which means that there is more time for the control system to react to the transient puffs.

Two examples of using O_2 to control CO spikes both concern fixed incinerators in which the incoming waste feed would be automatically stopped, because of regulated limits, if high levels of CO were detected.[2] O_2 was added to the combustion air in the afterburner of a fixed hearth incinerator to bring the total O_2 in the oxidizer up to a minimum of 27%. This nearly doubled the throughput by virtually eliminating the CO spikes. In a fixed rotary kiln incinerator, pure O_2 was injected at the firing end of the kiln. This virtually eliminated the CO spikes, which increased the processing rates by 25 to 40%, depending on the waste composition.

8.2.1.2.3 Higher ash burnout

Because of higher temperatures and more chemical reactivity, OEC can increase the completeness of combustion of the solids being processed in an incinerator. An example is shown in Figure 8.3. This may have a number of important benefits. In certain geographic locations, regulations limit the amount of hydrocarbons that may

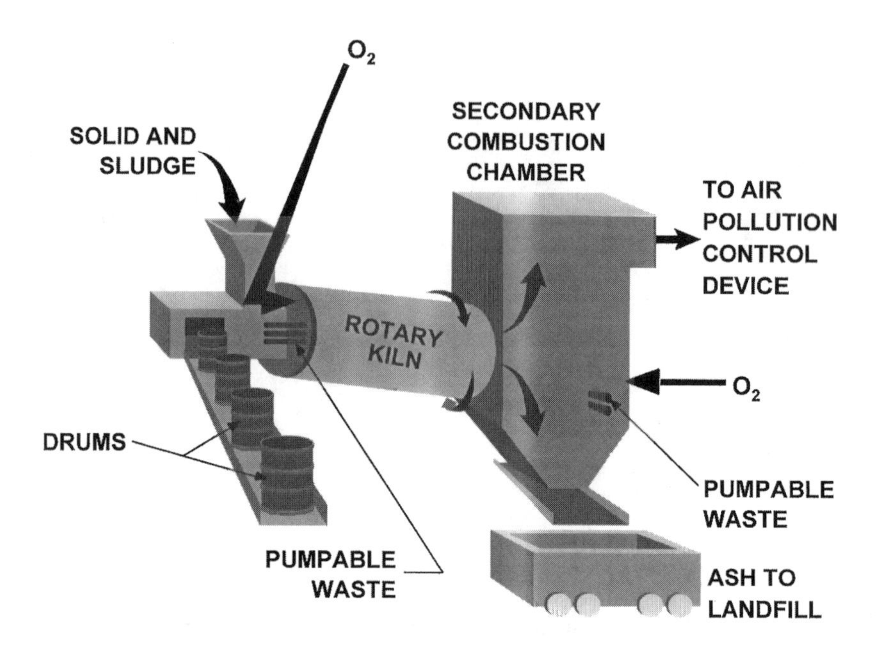

FIGURE 8.3 OEC in a rotary kiln incinerator.

be present in both the bottom ash and the fly ash. If the level is too high, then the ash may need to be either reprocessed or be classified as hazardous. In either case, this is usually costly. Oxygen has been used to solve this problem. Less ash is produced and the ash that is produced contains fewer hydrocarbons.

A recently patented technology describes the use of oxygen to remove unburned carbon in fly ash.[6] The ash is fed through a heated stainless steel chamber. Oxygen is used to accelerate the combustion of the carbon. The carbon content of the ash can be brought down below 0.7% so that the fly ash can be sold for use in making cement, instead of being sent to a landfill at a substantial cost.

8.2.1.2.4 Lower particulate emissions

Lower particulate emissions can result when an existing air/fuel incineration process is enhanced with oxygen. Because of the large reduction in flue gas volume, the gas velocity through the incinerator is reduced. This tends to reduce the amount of solid particles that are entrained by the gas flow through the incinerator. Therefore, there will usually be a substantial reduction in the particulate emissions from the incinerator. This benefit is particularly important when a high-ash waste is being processed. The quantity of fly ash that is produced is generally reduced. This decreases the load on the particulate removal system which might consist of an electrostatic precipitator or a baghouse.

8.2.1.2.5 Increased pollutant removal efficiency

OEC greatly reduces the flue gas volume in a combustion system by the removal of diluent N_2. This results in the concentration of the off-gases, so that any pollutants are then contained in a smaller off-gas volume flow. This makes it easier to remove them with existing off-gas treatment equipment, such as scrubbers, electrostatic precipitators, or baghouses. Also, the treatment equipment generally becomes more efficient at removing the pollutants because of the higher pollutant concentrations.

8.2.1.2.6 Reduction of stack opacity

OEC has been used to reduce the opacity in the exhaust gases of a waste chemical boiler which had no scrubbing system to remove PICs that were caused by inadequate residence time in the boiler.[2] O_2 was used to atomize the waste liquids. This not only reduced the opacity, but it also increased the overall thermal efficiency of the boiler.

8.2.1.3 Waste-Processing Flexibility

OEC can give the incinerator operator more flexibility in the types of wastes that can be processed. One important example of the variation in waste composition is the moisture content. For example, in a municipal waste incinerator, the incoming moisture may vary dramatically depending on the weather. If it has been warm and dry for several days, the trash will contain less moisture than if it has been raining for several days. Wet trash presents a much larger heat demand for the incinerator, compared with dry trash. Oxygen can be used to supplement an incinerator when more-intense energy release is needed to vaporize large quantities of water contained in the waste. OEC is also effective for incinerating high-moisture (>20%), low-heating-value (<500 Btu/lb or <1 MJ/kg) wastes, such as the contaminated soil at a Superfund site.[7] In a process that will be discussed in more detail later, OEC has been used to add the capability of sludge disposal at a MSW incinerator facility.[8]

8.2.1.4 Improved System Stability

OEC has been used in a commercial hazardous waste incinerator to improve the system stability.[9] The primary advantage of the OEC system was the reduction in excursions and system upsets that required operator intervention. With the old air/fuel system, the excursions were caused by the non-steady-state nature of the containerized waste feed. That system was unable to compensate for the nonuniform release of hydrocarbons as the containers opened inside the incinerator. The waste feed would be cut off whenever these excursions led to low oxygen concentrations in the exhaust. These excursions were minimized by injecting O_2 into both the PCC and the SCC, through lances cooled by a water/glycol mixture (see Figure 8.4). O_2 flow in the afterburner was controlled by an automatic feedback control system, set to a desired level of O_2 in the exhaust. Reducing the excursions caused the average solid waste feed to increase by nearly 11%.

8.2.1.5 Reduced Maintenance

Reduced maintenance generally results in a higher on-line factor.[7] Maintenance costs may be reduced when using OEC in waste incineration. One cost reduction that has been demonstrated is a reduction in dust buildup in an SCC, which is discussed in more detail in Section 8.2.3.3. Another reduction in maintenance costs may result

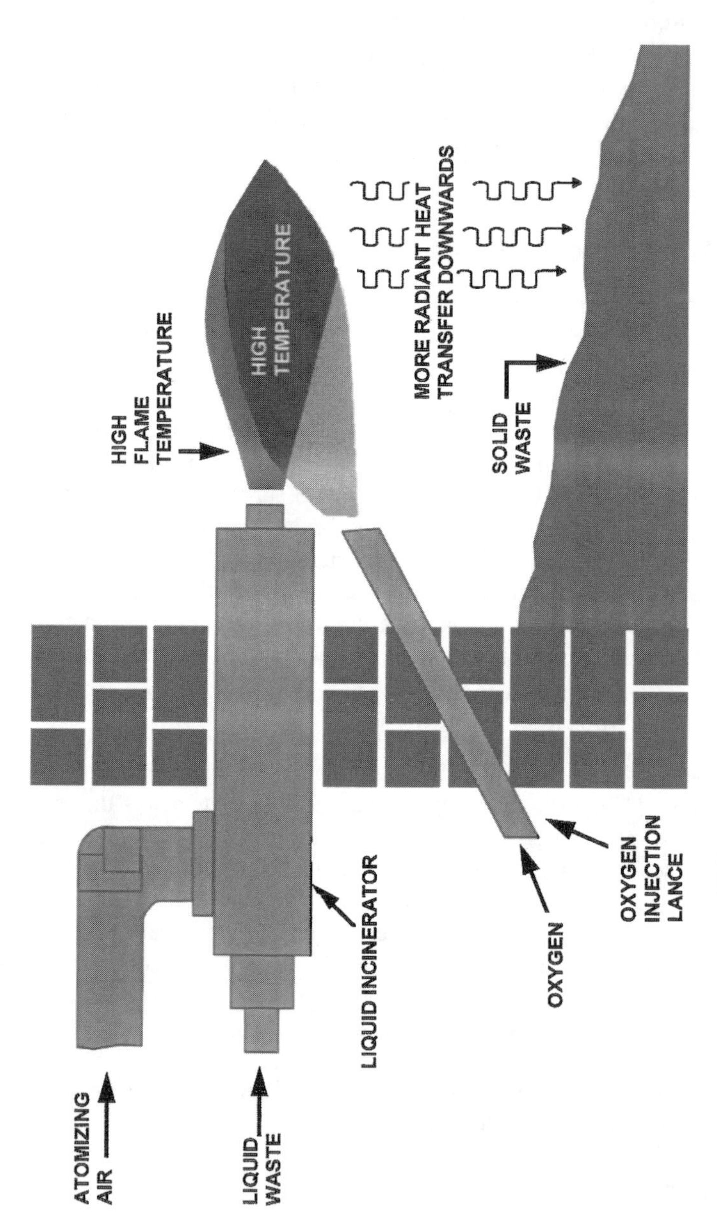

FIGURE 8.4 Oxygen lancing in waste incineration.

from reduced particulate emissions. The filter bags in the baghouse may require less frequent cleaning or replacing, compared with conventional air/fuel systems.

An oxy/fuel burner has been used in one case to remove slag buildup in the afterburner of a slagging rotary kiln incinerator.[2] High gas velocities through the kiln carried some slag through to the afterburner. This slag had to be periodically removed with a jackhammer by shutting down the system for several days. An oxy/fuel burner was positioned so that its flame could melt any accumulated slag, without shutting down the incinerator.

8.2.1.6 Reduced Capital Costs

There are two cases that need to be considered for OEC in an incinerator: retrofitting an existing system or designing a new system. In the case of a retrofit, there are additional capital costs associated with OEC for the oxygen piping and control systems. However, these may be offset by reductions in the capital costs of other equipment.[8] For example, a gain in additional throughput may not require upgrading other equipment, like the exhaust gas system. The exhaust gas equipment includes the exhaust fan, the ductwork, and any posttreatment equipment, like scrubbers, baghouses, or electrostatic precipitators. In order to process more material using the original air/fuel technology, the exhaust gas system would have to be expanded. These high capital costs may be avoided with OEC.

The second case to be considered is designing a new incineration process to include oxygen enhancement. The savings in capital equipment may be very substantial. The incinerator itself can be smaller, for a fixed processing rate, compared with an air/fuel system. The associated downstream equipment may be smaller as well.

8.2.1.7 Increased Boiler Efficiency

In a waste-to-energy MSW plant, OEC can increase the efficiency of the boiler. This may occur for several reasons. As previously shown, the burner efficiency increases due to the removal of diluent N_2. The hydrocarbons contained in the waste material are more completely burned. This not only reduces the pollutant emissions, but it also increases the useful heat released from the waste that may be utilized by the boiler. The heat transfer efficiency may also increase, as previously discussed, which improves the boiler efficiency. OEC can produce a cleaner burn, leading to cleaner boiler tubes and higher boiler efficiency.

8.2.2 Potential Problems

As with most technologies, there are some potential risks. The potential problems of using OEC in incineration are discussed here. These may all be overcome with the proper choice of equipment and operating conditions.

8.2.2.1 Overheating

There are several potential problems that may result from OEC. Depending on the burner design, oxygen-enhanced flames may be much hotter than air/fuel flames.

This could lead to potential equipment damage due to overheating. Refractory damage may occur as a result of localized heating. A related problem is slagging. This may occur if the waste material is overheated and begins to vitrify. Both of these overheating problems may be avoided by the choice of a properly designed burner, adequate waste flow through the incinerator to absorb the heat, and the proper choice of auxiliary heat input into the incinerator.

8.2.2.2 Reduced Turbulence

As previously discussed, oxygen-enhanced flames are generally more turbulent than air/fuel flames, due to the increased flame velocities. However, the exhaust gas volume is reduced with OEC. For a system designed for air/fuel combustion, the gas velocities through the incinerator would then be lower with OEC. This reduces the overall turbulence of the exhaust gases which could reduce the DRE. However, this is generally more than offset by the increased time and temperature in the incinerator, so that the DRE still increases compared with the original air/fuel system. Specific examples are given later.

8.2.2.3 Reduced Convection

Another issue to consider when using OEC in an incinerator is the potentially detrimental effects of reducing the amount of convection in the combustion chambers that results from the reduction in the nitrogen in the combustion products. Convection is important to both the heat and mass transfer within the incinerator. The gas circulation helps to homogenize the temperature in the combustor which minimizes hot spots that can damage refractories and cause nonuniform heating. A reduction in the gas volume could also reduce the mass transfer from the waste to the gas stream as volatiles are carried away from the surface of the waste. The gas flow over the waste material in the incinerator carries away the volatiles so they can be incinerated and so that more volatiles can diffuse to the surface of the waste. Therefore, a reduction in the gas circulation within the gas space of the combustor could potentially reduce the heat and mass transfer in the incinerator.

This problem is partially offset by the greater volume expansion of the gases because of the higher gas temperatures associated with OEC. Higher gas expansion increases the gas velocity in the chamber which increases convection. Proper selection of equipment can also overcome this potential problem so that the gas circulation in the incinerator is maintained using OEC.[10] New incinerator designs for OEC could have a smaller combustion chamber for a given waste throughput, as compared with air/fuel systems, in order to achieve the desired average gas velocity.

8.2.2.4 Higher NOx Emissions

Thermal NOx is exponentially dependent on the gas temperature (see Chapter 2). If there are significant sources of N_2 in an incinerator using OEC, high NOx emissions may be produced (see, for example, Reference 5). There are many possible sources of N_2. Probably the most common is air infiltration into the combustion system. In most cases, attempts are made to minimize infiltration. However, infiltration is difficult to eliminate completely. Another source of N_2 is from the fuel. In the U.S., natural gas normally contains up to 5% N_2. In Europe, it may contain up to 15%. Other fuels, like oil and coal, may have organically bound N_2. This may lead to so-called fuel NOx. Another source of N_2 may be the oxidizer. In some applications, O_2 is blended with air, which means that there will be large quantities of N_2 in the combustion system. In other cases, a lower-purity O_2 may be used for economic reasons. Still another source of N_2 may be the waste feed material. It may contain organically bound N_2. All of these sources of N_2 may contribute to possible increases in NOx emissions, depending on the actual incinerator and operating conditions.

8.2.2.5 Permitting

A potential problem of using OEC in waste incineration applications is the need for modified or new operating permits. Depending on the specific application, OEC may be considered a significant-enough departure from the existing operating conditions that the existing permits need to be re-evaluated. This may be a lengthy and expensive process. Also, since OEC is still relatively new in incineration, this may further complicate the permitting process.

8.2.3 Examples of OEC in Incineration

This section discusses some actual examples of using OEC in MSW incinerators, mobile incinerators, transportable incinerators, and fixed incinerators.

8.2.3.1 Municipal Waste Incinerators

A number of benefits of using OEC in MSW incinerators have been cited.[8] The economic incentives include increased waste-processing capacity, greater thermal efficiency, increased production in a waste-to-energy facility, reduced demand on the exhaust system, and a smaller air pollution control system. Increased capacity may be particularly important for many waste processors which are at their maximum capacity, since it is usually difficult to obtain permits to build new facilities. The environmental incentives include improved ash burnout, lower hydrocarbon emissions, lower CO, greater flexibility and control, and the ability to burn low-heating-value wastes such as dewatered sludge.

One of the earliest tests of oxygen enrichment in an MSW incinerator occurred at the Harrisburg, PA Waste-to-Energy Facility in 1987.[11] The combustion air was enriched with 2% O_2 ($\Omega = 0.23$). The average test results are given in Table 8.1. The waste throughput, steam production, boiler efficiency, and sludge throughput all increased. The NO_2 and SO_2 emissions increased while the volatile organic compounds (VOC) and CO emissions decreased. The flame stability improved with OEC. The combustibles in the ash decreased by 48%. The cost of the O_2 was easily offset by the improved system performance. However, the increase in NO_2 and SO_2 may be a problem for most existing facilities.

OEC has been used in an MSW incinerator to overcome thermal limitations.[2] O_2 was injected, through a diffuser, into the air plenums beneath the waste bed. It was also injected, through a lance, directly onto the bed (see Figure 8.5). This resulted

TABLE 8.1
OEC Performance at the Harrisburg Waste-to-Energy Facility

Parameter	Units	Air/Fuel	OEC
Waste throughput	buckets/h	8.5	8.9
Steam production	lb/h	68,000	75,700
Boiler efficiency	%	53.8	57.2
NO_2	ppm @ 7% O_2	43	61
VOC	ppm @ 7% O_2	14	0.13
SO_2	ppm @ 7% O_2	155	176
CO	ppm @ 7% O_2	12	10
Sewage sludge throughput	ton/h	0.5	2.25

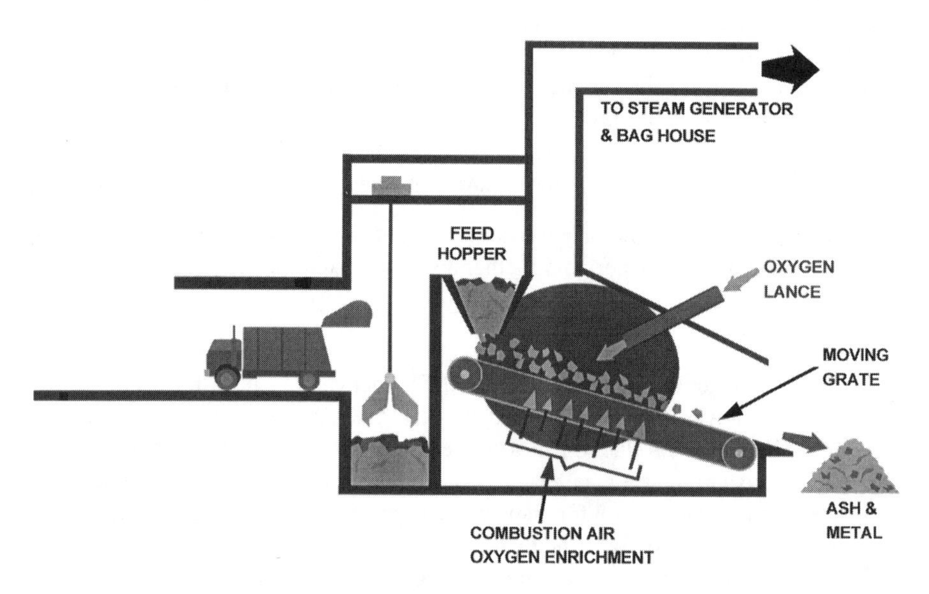

FIGURE 8.5 Oxygen enrichment in an MSW incinerator.

in a 10% increase in the waste-processing capacity, an increase in steam production, better overall boiler efficiency, and more-complete burnout of the ash. Significant cost savings were realized in ash disposal due to its lower volume and increased density.

The U.S. Environmental Protection Agency (EPA) sponsored a demonstration program to investigate OEC in a pilot-scale incinerator.[12] With only 3% O_2 enrichment ($\Omega = 0.24$), the waste-processing rate increased by 24%. OEC did not seem to have any effect on the metal content of the ash. There were some concerns about higher hydrocarbon emissions at the higher throughputs. Further research was recommended. One important commercial consideration was the impact this technology might have on permitting. New or amended permits normally require a lengthy and usually costly review process.

TABLE 8.2
OEC Performance at the Denney Farm Superfund Site

Parameter	Units	Air/Fuel	Oxy/Fuel
Contaminated soil throughput	lb/h	1478	4000
Specific fuel use	MMBtu/ton soil	11.9	4.7
Kiln superficial velocity	ft/s	8.1	3.3
SCC residence time	s	2.6	3.2
Quenched gas volume	dscfm	3250	2250

8.2.3.2 Sludge Incinerators

The biggest challenge in sludge incineration is the large amount of energy required to evaporate the large quantity of water contained in the sludge. The heating value of the sludge is minimal. Therefore, large quantities of auxiliary fuel are required. Oxygen has been used to increase the capacity of a multiple-hearth sludge incinerator by 35 to 55% at a sewage treatment plant in Rochester, NY.[13] Oxygen was injected into the sludge-drying zone at a rate of 1 ton/h through a series of lances. The amount of auxiliary natural gas fuel used to dry and burn the sludge was reduced by 57%. Emissions (per mass of dry sludge) of total hydrocarbons, NOx, and CO were reduced by 58, 62, and 39%, respectively.

8.2.3.3 Mobile Incinerators

Mobile incinerators are commonly used to clean up contaminated soil and water at Superfund sites. The entire incineration system, including the PCC, the SCC, and the pollution control equipment, is small enough to be transported over the road. It can be quickly set up and is usually preferred for smaller-size cleanups. One of the first applications to use OEC in incineration was at the Superfund cleanup site at the Denney Farm in McDowell, MO starting in 1987.[14] The EPA mobile incinerator was used, along with an OEC incineration technology that was later awarded the prestigious Kirkpatrick Chemical Engineering Achievement Award for the results at this cleanup.[15] Dioxin-contaminated liquids and solids were successfully treated. The OEC system showed impressive performance compared with the original air/fuel system, as shown in Table 8.2. The throughput was increased 171%, the specific fuel consumption decreased 61%, the residence time in the SCC increased by 21%, CO spikes were reduced, while NOx levels were unaffected.

A more recent example is a trial burn to destroy PCB-containing electrical transformers and related contaminated materials.[16] The waste material was fed into one end of a rotary kiln. A single oxy/fuel burner, located at the kiln entrance, fired cocurrently with the feed material. The ash was collected at the kiln exit. The combustion and process off-gases from the kiln were fed into the secondary chamber, which operated at a higher temperature than the primary chamber, to maximize destruction of any remaining combustible gases. A block diagram of the process is shown in Figure 8.6. The input parameters for the trial burn are listed in Table 8.3.

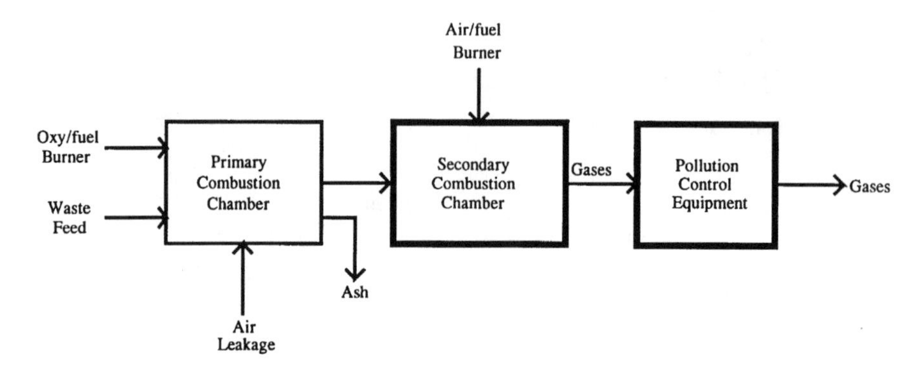

FIGURE 8.6 Mobile incineration system block diagram.

TABLE 8.3
Trial Burn Input Parameters

Parameter	Units	Series A	Series B	Series C
Primary Combustion Chamber				
Fuel	—	C_3H_8	C_3H_8	None
Oxidizer	—	O_2	O_2	None
Firing rate	MMBtu/h	2.7	2.7	0
Excess O_2	%	15	15	N/A
Air infiltration	lb/h	3500	3500	3500
Soil feed rate	lb/h	3030	3030	0
PCB in soil	wt%	1	1	N/A
Incoming soil temperature	°F	59	59	N/A
Incoming soil moisture	wt%	8.9	8.9	N/A
Secondary Combustion Chamber				
Fuel	—	C_3H_8	Oil	Oil
Oxidizer	—	Air	Air	Air
Firing rate	MMBtu/h	7.8	8.5	10
Excess air	%	10	10	10
PCB in fuel oil	wt%	N/A	1	42

Before the contaminated soil was processed, the system was tested using surrogate wastes to ensure the emission requirements could be met. Three different series of tests were conducted using various combinations of fuels and wastes. An oxy/propane burner in the PCC was used to incinerate PCB-contaminated soil for the series A and B tests. The only difference between A and B was that propane and oil with 1% PCB, respectively, were used as fuels in the SCC. In series C, the PCC was not operated, while the fuel for the SCC was oil with 42% PCB.

TABLE 8.4
Trial Burn Results

Parameter	Units	Series A	Series B	Series C	Regulatory Limit
PCC gas temperature	°F	1500	1500	N/A	N/A
SCC gas temperature	°F	2250	2250	2250	N/A
Ash temperature	°F	150	150	N/A	N/A
Ash moisture	wt%	30	30	N/A	N/A
PCB DRE	%	99.999972	99.999993	99.999996	99.9999
Combustion efficiency	%	99.9929	99.9926	99.9976	99.9
Stack particulate	mg/Nm^3	0.8377	3.320	2.556	50
Stack HCl	mg/Nm^3	N.D.	N.D.	N.D.	75
Stack CO	mg/Nm^3	9.95	10.6	7.4	114
Blowdown PCB	mg/l	N.D.	N.D.	N.D.	0.2
Blowdown 2,3,7,8 TCDD TE	ng/l	N.D.	N.D.	N.D.	0.12
Ash PCB	mg/kg	N.D.	N.D.	N.D.	0.5
Ash 2,3,7,8 TCDD	mg/kg	N.D.	N.D.	N.D.	1

N/A = not applicable; N.D. = not detectable by instruments; TCDD TE = tetrachlorodibenzodioxin toxic equivalent.

The trial burn results are listed in Table 8.4. The values for each series are the average of three sets of data taken on different days. The ash or cleaned soil exited the kiln and went into a water bath. This provided a seal for the kiln to minimize air infiltration and also quenched the outgoing soil. The blowdown refers to the scrubber effluent.

All of the regulatory requirements were satisfied. In many cases the emissions were below the detectable limit of the instruments. Fewer particulates were generated when propane was the fuel in the SCC, compared with oil. For series A and B, the PCC used about one quarter and the SCC three quarters of the total fuel input. Oxygen was successfully used in the commissioning of a new mobile waste incinerator. PCB contaminants were efficiently destroyed using several configurations for waste feeds and fuels, in both the PCC and SCC. All measured pollutants were well within regulatory limits.

Vesta Technology, LTD (Fort Lauderdale, FL) is a hazardous waste incineration company that provides on-site services throughout North America. Vesta's mobile systems use high-temperature rotary kiln incinerator technologies, coupled with innovative, proprietary designs for flue gas scrubbing. This proven technology destroys PCBs, dioxins, oil sludges, and other hazardous wastes with DREs that exceed RCRA and Toxic Substances Control Act (TSCA) standards. Vesta has used this technology with both air/fuel and oxy/fuel systems.

The decision to try oxygen was mainly influenced by the requirement to reduce the flue gas volume within the system. The result was twofold. First, the lower flue

TABLE 8.5
Operating Data from Two Superfund Sites Using a Mobile Incinerator

Parameter	Units	Air/Fuel	Oxy/Fuel
Waste feed	lb/h	2280	3400
Kiln temperature	°F	770	925
SCC	°F	1990	2010
NOx	ppmvd	255	245
NOx as NO_2	lb/h	4.4	2.1
NOx emissions	lb/ton soil	3.8	1.3
CO	ppmvd	10	25
Oxygen	% dry	10	10

gas velocity in the kiln resulted in less particulate carryover to the SCC and the air pollution control system. Actual site operations showed that SCC clean-outs were reduced with the lower flue gas flow, while actually increasing soil-processing rates. Second, the flue gas residence time in the SCC increased by 50% which was expected to give higher DREs.

Representative data from two similar remediation projects using the same equipment are shown in Table 8.5. The instantaneous soil throughput rates increased by 50% using oxygen, compared with the air/fuel base case. Of greater importance, however, was the overall average hourly throughput which increased by 150% due to the elimination of downtime for SCC particulate clean-out when using oxy/fuel. This rate enabled Vesta to complete the remediation project 28 days ahead of schedule.[17] The hourly NOx emissions were reduced by 52%, while the pounds of NOx per ton of soil processed were reduced by 66%.

8.2.3.4 Transportable Incinerators

Transportable incinerators are also commonly used at Superfund cleanup sites. They are larger in size than mobile incinerators and take longer to set up. Therefore, they are used at larger sites because of their increased processing capacity. The trend is to use these, instead of mobile incinerators.[18] OEC has been used to maximize the transportable incinerator throughput by reducing the gas volume and improving the heat transfer performance. At the Bayou Bonfouca Superfund site in Saint Tammany Parish, LA, cost savings were estimated to be nearly $3 million using OEC instead of air/fuel.

The example given here shows how oxygen reduced particulates in a transportable incinerator. Williams Environmental Services (WES), located in Auburn, AL, used an incineration system consisting of a cocurrent rotary kiln, a hot cyclone, a secondary combustion chamber, a quench tower, baghouses, an induced draft fan, an acid gas absorber, and an exhaust stack. Air/fuel and oxy/fuel were used in this incineration system at two Superfund sites: a bankrupt wood-treating operation in

TABLE 8.6
Transportable Incinerator Data

Parameter	Units	Prentiss Woodtreating Prentiss, MO (Air/Fuel)	Bog Creek Farm Howell, NJ (Oxy/Fuel)
Waste feed	ton/h	15.2	18.5
Kiln temperature	°F	1570	1400
SCC	°F	1710	1900
NOx	ppmvd	58	120
NOx as NO_2	lb/h	8.55	10.0
	lb/ton soil	0.56	0.54
CO	ppmvd	40	<5
SO_2	ppmvd	<10	<1
Oxygen	% dry	11.5	13.9
Particulates[a]	@ 7% O_2	0.0125	0.0051

[a] Federal particulate standard is 0.08 gr/dscf corrected to 7% oxygen in the flue gas.

Prentiss, MO and at the Bog Creek Farm site in Howell, NJ. At the Prentiss site, the kiln was equipped with two air/fuel burners while at the Bog Creek Farm site, a single oxy/fuel burner was installed. The Prentiss site contained 9200 ton (8300 t) of creosote-containing soil. The Bog Creek Farm site contained 25,000 ton (23,000 t) of soil that contained VOCs including benzene derivatives, chlorinated hydrocarbons, and semivolatile organics such as naphthalene and phthalates.

OEC was selected for the Bog Creek site, due to the particulate emissions criteria set by the New Jersey Department of Environmental Protection (NJDEP). The primary problem encountered during the start-up at the Prentiss site was higher-than-expected fines carryover. About 50% of the ash output was from the air pollution control system. The Bog Creek site was located near the New Jersey coastline where the soil is naturally sandy. The existing particulate emission limit of 0.03 grains per dry standard cubic foot (gr/dscf) or 0.07 grams per dry standard cubic meter (g/dscm) was reduced to a more rigorous standard of 0.015 gr/dscf (0.034 g/dscm) for this site, to prevent contaminated sand from entering the atmosphere.

At a soil feed rate of 20 ton/h (18 t/h), the equivalent Prentiss data indicated the estimated combined emissions (soil and metals) would have to be reduced by 65% to fall below the newly prescribed limit. OEC technology was selected to meet the tougher particulate standard, without causing a delay in the schedule.[19] The Bog Creek site was the first Superfund site to use OEC incineration in the northeast region of the U.S. and the first North American site to use OEC in a commercial, transportable incinerator for the entire project.

Table 8.6 shows a comparison of the data for the air/fuel and oxy/fuel sites. The processing rate increased using OEC. All emission and ash requirements were satisfied. The site was cleaned up 60 days ahead of schedule. In this case, there was

TABLE 8.7
Fixed-Based Resource Recovery Process

Parameter	Units	Air/Fuel	Oxy/Fuel
Flue gas temperature	°F	580–600	720–770
NOx	ppmvd	120–165	185–480
NOx as NO_2	lb/h	2.1–4.1	1.7–3.2
NOx emissions	lb/ton soil	1.1–1.3	0.8–1.2
Oxygen	% dry	15.0–15.4	16.7–19.6

no reduction in NOx by using oxy/fuel which was probably due to high air leakage into the kiln as noted by the increased oxygen in the flue gas.

8.2.3.5 Fixed Hazardous Waste Incinerators

OEC has been used to reduce NOx in a fixed-based resource recovery process.[20] Giant Resource Recovery (GRR) is a subsidiary of the Giant Group, Ltd., which is involved with cement manufacture and the use of waste materials as fuel and raw materials supplements. GRR processes creosote-contaminated soil through countercurrent rotary kilns. By a patented process, the decontaminated soil is then used as a raw material for cement production, thus replacing a certain portion of the traditional feed material stream. The combustion products are ducted into the cement kilns.

The processing rates of the kilns were limited by two factors. Since the contaminated soil was high in moisture content and low in heating value, more heat transfer was required to increase the throughput while ensuring the creosote concentration of the soil did not exceed permitted levels at the kiln discharge. If the creosote concentration is too high, the soil has to be reprocessed before being sent to the cement kiln. Second, the flue gas volume needed to be minimized to prevent upsetting the cement kiln operation. These criteria were met by using an oxy/fuel burner.

Table 8.7 shows a comparison of representative data taken on side-by-side rotary kilns processing the same soil. The kiln back end temperature increased over 100°F (56 K) using an oxy/fuel burner. Over a range of several tests, NOx emissions per ton of material processed were from 5 to 35% less using oxy/fuel. In this case, large amounts of air infiltration, evidenced by high O_2 concentrations in the flue gas, limited the NOx reduction using oxygen.

Oxygen has been injected into rotary kiln incinerators and the secondary combustion chamber to reduce CO emissions by more than 60% while increasing waste throughput by more than 15% at a German merchant incinerator.[21] In the process, 530 ft^3 (15 m^3) of oxygen is injected for each 84 lb (38 kg) drum of waste material.

Oxygen injection is planned to be used on two incinerators being designed for Duisburg, Germany to be commissioned in 1997.[21] Between 7000 to 14,000 ft^3/h (200 to 400 m^3/h) of oxygen will be injected into the primary combustion chambers of the incinerators to reduce the fuel consumption in the secondary combustion chambers. An added benefit will be a reduction in the size of the gas-handling system.

8.3 ECONOMICS

Cost is a key consideration in determining whether or not to use O_2 in any given combustion application. The cost to use air for combustion is very low. It consists basically of the cost of the blower and the electricity to operate it. Some filtration or cleaning may also be required. Therefore, the benefits of using O_2, instead of air, must offset the added cost of the O_2. This section has been broken into operating and capital costs, although the actual categorization of a particular cost may vary for a given application. The purpose of this section is neither to justify the use of OEC nor to put specific costs on any particular item. The purpose is to help a potential user consider all the relevant costs and to give some general trends. Both general and specific benefits of OEC have been previously discussed in Chapter 1. No attempt will be made to assess the value of those benefits. It will be up to the potential user to calculate the actual costs and benefits, in order to determine whether or not it is economically advantageous to use OEC.

8.3.1 Operating Costs and Savings

8.3.1.1 Costs

There are several potential operating costs in any given OEC system. By far the largest is the cost of the O_2. It is generally very difficult to give detailed information about this cost. One reason is that little is published on these costs. Another reason is that there are many variable factors that greatly affect the final price. Certainly one important factor is supply and demand. At the time this chapter was written, the demand for oxygen in the U.S. exceeded the supply. Therefore, oxygen prices were increasing. However, provisions were being made by the industrial gas suppliers to increase their capacities to meet the demand. This is expected to improve pricing for the end user. Another important factor that affects the pricing is the method of supply. On-site O_2 is typically less expensive than merchant O_2. Cyclical usage of O_2 generally costs more than a fairly steady requirement. If O_2 is delivered, then the transportation costs are important. There are also some general competitive conditions that may affect the price. These include things like the term of the supply contract, any performance guarantees, the local O_2 availability, and the total potential size of the customer's requirements at all sites. Therefore, it is not possible to give exact pricing. However, generalizations about relative costs are given here. Ultimately, the best sources for actual cost information are the industrial gas suppliers.

Besides the cost of the O_2, there are other potential operating costs that may need to be considered. There may be minor maintenance requirements for the oxygen equipment. However, this should be comparable with an equivalent air/fuel system. If the burner(s) or injection lance(s) needs water cooling, there may be some costs associated with the water.

8.3.1.2 Savings

There are many potential savings that may be realized with an OEC system. One typically large saving is a reduction in the fuel consumption per unit of waste processed. Another may be reduced incinerator maintenance, as has been shown in many incinerator applications. If one of the objectives of using OEC is to increase the throughput rates, then there are usually substantial labor savings per unit of waste processed. Because of the potential for dramatically lower pollutant emissions, an OEC system may save the ongoing expense of a posttreatment system. Also, the existing posttreatment system may become more efficient because of the lower exhaust volumes and higher pollutant concentrations, since a large amount of N_2 has been removed.

In many cases, OEC may also increase the revenue for waste processors. They may be able to treat waste that they were previously unable to process. This might include things like sewage sludge and other types of low-heating-value, high-moisture-content wastes. At Superfund cleanup sites, the waste processor can usually increase profits by cleaning up the contaminated waste faster than scheduled. Examples of this have already been discussed.

8.3.2 Capital Costs

There may be two major types of capital costs for an OEC system. This first concerns the O_2 supply system. There will be some expense for the piping from the supply system to the delivery location. This will vary depending on the length and size of the pipe. There may be some costs for a concrete foundation, fencing, and some wiring for the O_2 supply or storage system, depending on which supply method is used. In a limited number of cases, the waste processor may purchase the O_2 generation system. In most cases, this system is owned and operated by the industrial gas supplier. If waste throughput is increased, there may be some added costs for other components in the incineration system. For example, the waste-handling system may need to be upgraded to handle the added throughput.

There may be some potential capital cost savings with an OEC system. These depend on whether the incineration system is being redesigned or retrofitted for OEC. A newly designed system may be much smaller than a comparable air/fuel system, for a given waste throughput. The OEC system is usually much more efficient and produces much less exhaust gas. This means that the incinerator, as well as all the associated downstream equipment like the afterburner, posttreatment system, and exhaust ductwork, may be smaller. For an OEC retrofit, the waste processor may save the cost of enlarging the existing posttreatment system. Normally, if the throughput is increased, the posttreatment system must be enlarged. However, OEC may actually reduce the required exhaust gas volume requiring treatment, even when the waste throughput is increased.

8.4 FUTURE OF OXYGEN-ENHANCED INCINERATION

There are a number of factors that make the future look bright for OEC in incineration. In separate reports on the industrial uses of OEC sponsored by the U.S. Department of Energy[22] and the Gas Research Institute,[23] incineration was not even mentioned as a potential market. Since those reports were written, OEC has been applied to a wide variety of incineration applications. This trend is expected to continue because of several factors. The economic benefits of using OEC in incineration have been well proven. This should drive more waste processors to consider its use in their applications. As the technology becomes more well known, it should be easier to get the necessary permitting. New advances are continually being made in existing and new (see, e.g., Reference 24) oxygen generation technologies. This should reduce the cost to the end user, which should expand the number of amenable applications.

In addition to the growing number of commercial applications, the popularity of OEC in incineration is evidenced by the level of research activity. Research continues into new applications, such as the TAZAS process in Germany,[25] the Pyretron burner system,[5] a novel flue gas elimination system,[26] a fuzzy logic control system in the SCC for transient puff suppression,[27] and a gasification process for solid waste decomposition.[28] The research activity is evidenced by the increase in patent activity for using OEC in incineration. Typical examples include using oxygen to combust wet solid waste,[29] to coprocess sludge in a municipal waste incinerator,[30] and to incinerate very high moisture content liquid wastes.[31]

However, there are some factors that may adversely affect the future of OEC in incineration. One is the general uncertainty of incineration itself. It continues to be difficult to obtain permits for new facilities and modifications for existing permits. Some of the environmental activist groups continue to question the use of incineration. Public perception will continue to be an important factor in incineration. The current shortage of O_2 in the U.S. may limit the expansion of OEC in incineration until new capacity can be constructed. In some incineration applications, O_2 requirements may be transient. If O_2 is in short supply, O_2 costs may be higher for that type of application because of the preference for applications with a constant ongoing O_2 requirement.

8.5 CONCLUSIONS

There are numerous examples of using OEC in waste incineration applications. In some cases, low-level O_2 enrichment of an existing air/fuel combustion system can have dramatic results. In other cases, oxy/fuel burners may be used to provide needed solutions to specific problems. OEC has been applied to both the PCC and the SCC. It has been used in both portable and fixed incinerators in a wide range of configurations.

Many of the benefits of using OEC are related to improvements in the three T's of incineration: time, temperature, and turbulence. The residence time in an existing air/fuel incinerator may be significantly increased by OEC because of the reduction in the flue gas volume. The flame temperature dramatically increases with OEC. The flame turbulence may increase as a result of the higher flame speeds of OEC. However, the average turbulence level in the combustion chamber usually decreases because of the reduced gas velocities. This reduction in average turbulence is usually offset by the increased residence time and the increased gas temperatures since DREs generally increase with OEC.

The future of OEC in incineration appears bright. Lower-cost O_2, coupled with significant environmental and operating benefits of OEC, make OEC an attractive technology. Research continues to assess its effectiveness in a wide range of incineration applications. The major hurdle that may limit its implementation is the potential problem of getting new or revised operating permits. At this time, OEC may not be viable for all incineration applications. However, it can be used economically to solve some specific problems and enhance many processes.

REFERENCES

1. Fusaro, D., Incineration technology: still hot, getting hotter, *Chem. Proc.*, 54(6), 26, 1991.
2. Reese, S. D., Diverse experience using oxygen systems in waste incineration, paper presented at the Fourth Annual National Symposium on Incineration of Industrial Wastes, February 28 to March 2, Houston, TX, 1990.
3. Gupta, A. K., Thermal destruction of solid wastes, *J. Energ. Resour. Technol.*, 118, 187, 1996.
4. Gitman, G., Zwecker, M., Kontz, F., and Wechsler, T., Oxygen enhancement of hazardous waste incineration with the PYRETRON thermal destruction system, in *Thermal Processes*, Vol. 1, H. M. Freeman, Ed., Technomic Publishing, Lancaster, PA, 1990, 207–225.
5. U.S. Environmental Protection Agency, American Combustion Pyretron Destruction System — Applications Analysis Report, U.S. EPA report no. EPA/540/A5-89/008, Office of Research and Development, Cincinnati, OH, June 1989.
6. Martinez, M. P., Apparatus and Process for Removing Unburned Carbon in Fly Ash, U.S. Patent 5,555,821, Sept. 17, 1996.
7. Acharya, P. and Schafer, L. L., Consider oxygen-based combustion for waste incineration, *Chem. Eng. Prog.*, 91(3), 55, 1995.
8. Shahani, G. H., Bucci, D., DeVincentis, D., Goff, S., and Mucher, M. B., Intensify waste combustion with oxygen enrichment, *Chem. Eng.*, Special Suppl. to 101(2), 18, 1994.
9. Davidson, S. L., Fryer, S. R., and Ho, M.-D., Optimization of process performance of a commercial hazardous waste incinerator using oxygen enrichment, in *Proceedings of the 1995 International Incineration Conference,* Bellevue, WA, May, 1995, 631.
10. Ding, M. G., The use of oxygen for hazardous waste incineration, in *Thermal Processes*, Vol. 1, H. M. Freeman, Ed., Technomic Publishing, Lancaster, PA, 1990, 181–190.
11. Strauss, W. S., Lukens, J. A., Young, F. K., and Bingham, F. B., Oxygen enrichment of combustion air in a 360 TPD mass burn refuse-fired waterwall furnace, in *Proceedings of the 1988 National Waste Processing Conference, 13th Bi-Annual Conference,* Philadelphia, PA, May 1–4, 1988, 315.
12. CSI Resource Systems and Solid Waste Association of North America, Evaluation of Oxygen-Enriched MSW/Sewage Sludge Co-Incineration Demonstration Program, U.S. Environmental Protection Agency report EPA/600/R-94/145, Office of Research and Development, Cincinnati, OH, Sept. 1994.
13. Parkinson, G., Oxygen enrichment enhances sludge incineration, *Chem. Eng.*, 103(12), 25, 1996.
14. Ho, M.-D. and Ding, M. G., Field testing and computer modeling of an oxygen combustion system, *J. Air Pollut. Waste Manage.*, 38(9), 1185, 1988.
15. Chopey, N. P., The tops in chemical engineering achievement, *Chem. Eng.*, 96(12), 79, 1989.

16. Baukal, C. E., Schafer, L. L., and Papadelis, E. P., PCB cleanup using an oxygen/fuel-fired mobile incinerator, *Environ. Prog.*, 13(3), 188, 1994.
17. Griffith, C. R., PCB and PCP destruction using oxygen in mobile incinerators, in *Proceedings of the 1990 Incineration Conference,* San Diego, May 14–18, 1990.
18. Acharya, P., Fogo, D., and McBride, C., Process challenges in rotary kiln-based incinerators in soil remediation projects, in *Proceedings of the 1995 International Incineration Conference,* Bellevue, WA, May, 1995, 637.
19. Romano, F. J. and McLeod, B. M., The use of oxygen to reduce particulate emissions without reducing throughput, in *Proceedings of 1990 Incineration Conference,* Paper 3.3, San Diego, May 14–18, 1990.
20. Baukal, C. E. and Romano, F. J., Reducing NOx and particulate, *Pollut. Eng.*, 24(15), 76, 1992.
21. Fouhy, K. and Ondrey, G., Incineration: turning up the heat on hazardous waste, *Chem. Eng.*, 101(5), 39, 1994.
22. Chace, A. S., Hazard, H. R., Levy, A., Thekdi, A. C., and Ungar, E. W., Combustion Research Opportunities for Industrial Applications — Phase II, U.S. Department of Energy report DOE/ID-10204-2, Washington, D.C., 1989.
23. Williams, S. J., Cuervo, L. A., and Chapman, M. A., High-Temperature Industrial Process Heating: Oxygen-Gas Combustion and Plasma Heating Systems, Gas Research Institute report GRI-89/0256, Chicago, July 1989.
24. U.S. Department of Energy, Assessment of Thermal Swing Absorption Alternatives for Producing Oxygen Enriched Combustion Air, Report no. DOE/CE-040762T-H1, Washington, D.C., April 1990.
25. Ringel, H. and Herbermann, M., Experiments on the incineration of hazardous waste with pure oxygen, paper presented at the 1991 Incineration Conference, Knoxville, TN, May, 1991.
26. Kephart, W., Angelo, F. and Clemens, M. K., Thermal oxidation vitrification flue gas elimination system for hazardous, mixed, and transuranic waste processing, in *Proceedings of the 1995 International Incineration Conference,* Bellevue, WA, May, 1995, 211.
27. Lemieux, P. M., Miller, C. A., Fritsky, K. J., and Chappell, P. J., Development of an Artificial-Intelligence-Based System to Control Transient Emissions from Secondary Combustion Chambers of Hazardous Waste Incinerators, in *Proceedings of the 1995 International Incineration Conference,* Bellevue, WA, May, 1995, 527.
28. Stahlberg, R., Feuerriegel, U., and Runyon, D. J., Thermoselect-energy and raw materials recovery process foundation for the continuous conversion of wastes, in *Proceedings of the 1995 International Incineration Conference,* Bellevue, WA, May, 1995, 535.
29. Ho, M., Method for Combusting Wet Waste, U.S. Patent 5,000,102, 1991.
30. Goff, S. P., DeVincentis, D. M., Wang, S.-I., Bucci, D. P., Romano, F. J., Shahani, G. H., and Foder, M., Process for Combusting Dewatered Sludge Waste in a Municipal Solid Waste Incinerator, U.S. Patent 5,405,537, 1995.
31. Chapman, R.D., Oxygenated Incinerator, U.K. Patent Appl. 9326000.8, 1995.

9 Safety Overview

Mark A. Niemkiewicz and J. Scott Becker

CONTENTS

9.1 INTRODUCTION

There is a substantial amount of literature concerning the safety of combustion processes. The majority of this literature is dedicated to air-based combustion applications, and it includes information for the supply and safe handling of many fuels, such as natural gas, propane, and fuel oils. Less information exists concerning combustion with oxygen-enriched air or with pure oxygen. The purpose of this chapter is to provide a safety overview of the issues that are important for the practical use of oxygen in oxygen-enriched combustion applications. However, those who are planning to implement such a process should contact their industrial gas supplier for detailed guidelines and recommendations concerning the safe use of oxygen.

9.2 THE COMBUSTION TRIANGLE

The most basic aspects of combustion safety relate directly back to the combustion triangle (Figure 9.1). The triangle illustrates that the simultaneous presence of a fuel, an oxidant, and an ignition source is required to create and sustain a fire. Removal of any one of the three requirements will prevent a fire from occurring or will cause an existing fire to be extinguished. Note that the flame itself can be considered a source of ignition, once a fire has been established.

0-8493-1695-2/98/$0.00+$.50

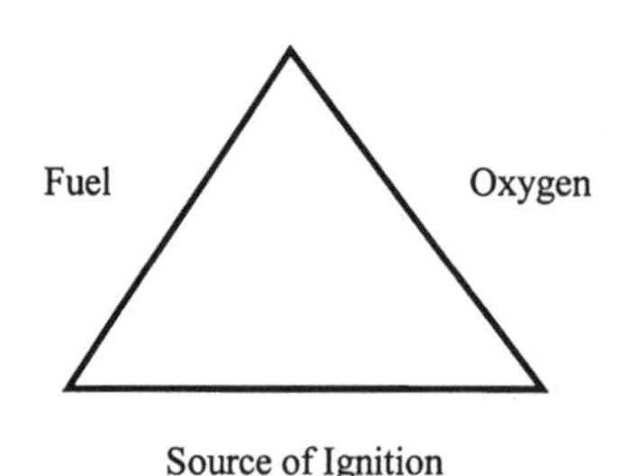

FIGURE 9.1 The combustion triangle.

9.3 DESIGN ENGINEERING

In most practical combustion installations there are two separate parts of the equipment system: (1) the burner itself and (2) all of the peripheral equipment necessary to control the burner operation efficiently and safely. The control equipment includes fuel and oxidant pressure and flow controls; automatic shutoff controls; flame supervision equipment; furnace purge equipment; and other related devices. The safety issues of the burner are significantly different in character as compared with the safety issues of the control equipment. In many ways, these two safety issues are diametrically opposed. Within the burner, fire is a desired condition, whereas within the control system and surrounding environment, fire is to be avoided entirely.

The burner design engineer accepts the presence of fire and attempts to safely contain and manage it. The engineer intentionally provides an ignition source and strives to control carefully the deliberate combination of the fuel and oxidant so that the desired results of the combustion process are achieved. The critical safety issues are maintaining flame stability (e.g., loss of flame, oscillating flame, etc.); controlling the flame shape and location; avoiding excessively high temperatures and corrosion (i.e., of the burner itself and the mounting block/tile); minimizing undesirable combustion products (e.g., NOx, CO, etc.); and avoiding undesirable melting of the rest of the furnace, refractory, and equipment. Temperature concerns become important with oxygen or oxygen-enriched combustion due to the increase in flame temperature and velocity as compared with air-based combustion (see Chapter 1). An example is the comparison of a typical air/methane flame temperature of 3200°F (2000 K) with that of a typical oxygen/methane flame temperature of 4500°F (2800 K).

Conversely, the combustion controls system engineer seeks to isolate, contain, and control the flow of the fuel and the oxidant. Isolation and containment involves the separation of the fuel and the oxidant by designing the pipeline in a manner that prevents leakage and premature mixing of the two gases. Included in this effort is the design and selection of the oxygen piping components and materials so that they are compatible with the contained oxygen. A material's compatibility with oxygen is a function of its ability to combust or not combust in an oxygen-enriched environment. Keep in mind, oxygen by itself is not dangerous; it will not cause combustion unless a fuel is present. On the other hand, the combination of oxygen with a fuel has the potential to be very dangerous (i.e., causing rapid combustion and/or an explosion). Many materials which are typically not combustible in air will combust rapidly in an oxygen-enriched atmosphere (e.g., greases, rubber compounds, carbon steel pipe, etc.). In addition, the safety hazards of the fuel can be significantly different in character from those of the oxidant. The critical safety issues faced by the combustion controls system engineer include isolation and containment of the fuel and the oxidant (e.g., positive shutoff valves); ignition avoidance; ratio control of the fuel flow rate to the oxidant flow rate; flame supervision; removal of combustible gases from enclosed spaces prior to ignition; ignition control; and the design of safety shutdown logic that is interlocked with undesirable events (e.g., high temperature, high flow rates, flameout, etc.).

9.4 FIRE HAZARDS

A fire hazard can be defined as any fluid, product, piece of equipment, process, etc. that has the potential to cause or contribute to a fire. The fuel delivery system is a primary fire hazard because of the potential for a leak of the fuel into the surrounding environment or atmosphere which is almost always air. This is an immediate hazard because of the possibility of an uncontrolled fire or an explosion depending upon the size and location of the leak and the availability of an ignition source.

Leakage of air from the supply or control system for an air-based combustion process is obviously not a fire hazard. However, an environmental release of oxygen from an oxygen-based combustion process is a significant hazard. There is the possibility of a fire due to the combustion of the oxygen supply piping and control equipment that is located within the vicinity of the oxygen leak. The possibility also exists for extending the fire hazard by consuming anything in the vicinity that can become a fuel for the fire and/or consuming the surrounding facility. The probability of a fire and the extent of a fire are dependent upon the size of the leak, the speed with which the oxygen becomes diluted into the surrounding atmosphere, and the availability of ignition sources. Fires that involve a pure oxygen source can be abrupt, rapid, and very destructive. Figure 9.2 illustrates the results of such a fire (i.e., the complete destruction of an automobile) due to a localized oxygen-enriched environment created by a leak from a nearby oxygen storage facility.

Instances where there might be the simultaneous release of a fuel and an oxidant can be very hazardous. An example of a simultaneous release is when the failure to contain a fuel causes a fire, and the fire leads to the rupture of a nearby oxygen line, thereby providing pure oxygen to the already established fire. Such a scenario can greatly magnify the devastating results of the fire.

Traditional gaseous-fuel fire safety techniques such as flammability-limit control (avoiding an accumulation of flammable mixtures of a fuel and an oxidant), ignition controls (avoiding the ignition of a flammable mixture of gases), fire extinguishment (methods to cause combustion to cease), and damage controls (methods to contain a fire or explosion and minimize the potential damage) are used by engineers to minimize the fire hazards associated with oxygen-enhanced combustion. An example of the application of a damage control technique would be the installation of a concrete enclosure that surrounds an oxygen compressor. The purpose of the enclosure would be to contain an oxygen compressor fire and limit the extent of the damage. Note that damage control techniques are beyond the scope of this book.

FIGURE 9.2 Photographs of a car damaged by an oxygen-enriched fire.

9.5 FLAMMABILITY-LIMIT CONTROL

In most combustion applications, the primary approach to ensure a safe process is to avoid the accumulation of a fuel and an oxidant that is within the "flammability limits" of the mixture. There are three gas-phase combustion flammability limits that are typically cited in the literature. They are the lower flammability limit (LFL), the upper flammability limit (UFL), and the minimum oxygen for combustion ($MinO_2$). The LFL is defined as the lowest percentage of a fuel mixed with an oxidant that will support the propagation of a flame. The UFL is defined as the highest percentage of a fuel mixed with an oxidant that will support the propagation of a flame. Inversely, the $MinO_2$ is defined as the lowest possible percentage of oxygen, in any oxygen/fuel/diluent mixture, that will support the propagation of a flame.[1] LEL (lower explosive limit) and UEL (upper explosive limit) are sometimes used interchangeably with LFL and UFL, respectively. Also, the term $MinO_2$ is sometimes referred to as the "critical oxygen content."

The flammability limits of a gas-phase mixture are the threshold compositions at which sustained combustion just occurs.[2] The energy level at which the combustion reaction begins to occur is known as the minimum ignition energy of the particular fuel/oxidant mixture. Propagation of a flame occurs when the combustion reaction of a few molecules of the fuel and oxidant provide enough energy (in the form of velocity and temperature) to continue to self-ignite neighboring fuel and oxidant molecules. The energy necessary to initiate the combustion reaction must be provided by an ignition source (e.g., a match, spark plug, pilot flame, static electricity, etc.) unless the temperature of the mixture is raised above the autoignition temperature of the mixture. The autoignition temperature (AIT) can be defined as the temperature at which a flammable mixture will automatically combust without a specific ignition source.

Figure 9.3 illustrates the flammability limits for methane in air (at room temperature and atmospheric pressure) as a function of the percent of oxygen enrichment.[3] At 21% oxygen (typical air), the LFL is approximately 5%, and the UFL is approximately 14%. At 100% oxygen, the LFL has not changed, but the UFL is approximately 59%. This increase in the UFL represents a flammability range for methane in pure oxygen that is six times larger than the range for methane in air. This is a clear illustration of the increase in the potential for a fire associated with an oxygen-enriched combustion process.

Figure 9.4 illustrates the effect of pressure on minimum oxygen requirements for propagation of natural gas, ethane, and propane in air–nitrogen mixtures.[4] The $MinO_2$ for natural gas at room temperature and atmospheric pressure is 12%. The $MinO_2$ for propane at room temperature and atmospheric pressure is 11.5%.

There has been a strong research effort spanning nearly 200 years that has attempted to understand the theoretical conditions at which gas-phase flammability can occur. However, for most practical applications, flammability limits have been assigned based on experimental testing and laboratory work. Flammability limit data are dependent upon the chemical behavior of the combustion reaction; the pressure and temperature of the mixture; the geometry and surface-area-to-volume ratio of the apparatus; the presence of a diluent, inhibitor, or other materials; and many other variables. Figure 9.5 shows the effects of temperature on the limits of flammability for methane in air.[5] An increase in temperature has the effect of lowering the LFL and raising the UFL, therefore increasing the flammability range. Figure 9.6 illustrates the effects of pressure on the flammability limits for natural gas in air.[1] An

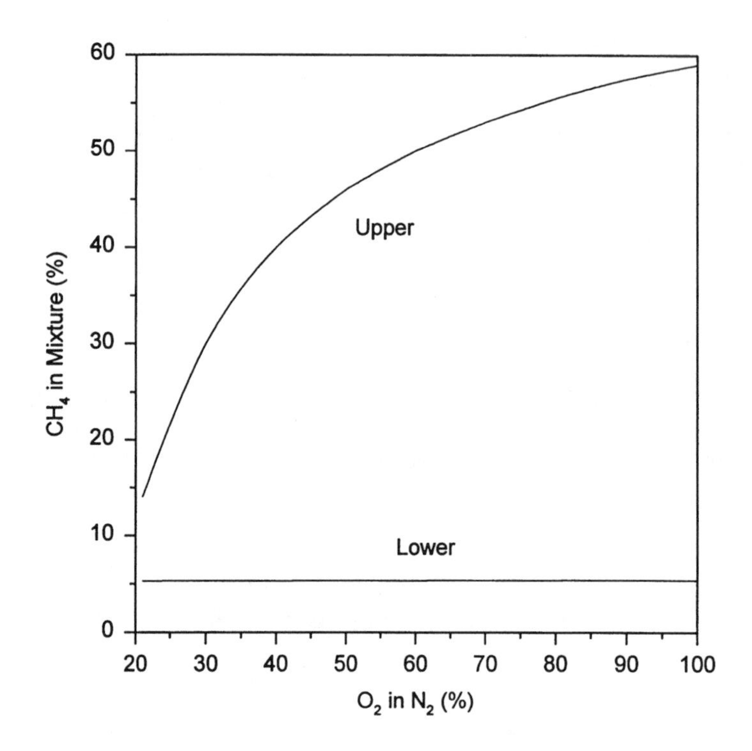

FIGURE 9.3 Flammability limits of methane in air/oxygen as a function of oxygen enrichment (at STP). (Adapted from Turin, J. J. and Huebler, J., Report no. I.G.R.-61, American Gas Association, 1951.)

increase in pressure can also have the effect of increasing the flammability range for methane in air. These trends are typical of most flammable mixtures. Consequently, appropriate safety factors should be applied when flammability data is used for fire prevention analyses and procedures.[6]

Combustion may also occur in mixed-phase systems (solids with liquids, solids with gases, liquids with gases). The theory of mixed-phase combustion is a much less studied subject, and consequently flammability data are almost exclusively determined by experimental testing. The primary hazard with most oxidant delivery systems is that of mixed-phase combustion; the oxidant is a gaseous material which may support combustion of the system which is either a liquid (oils, greases, coolants) or a solid (piping, valves, instruments).

Practical flammability limit techniques for oxygen-enriched combustion processes are based upon the prevention of any mixtures of fuel and oxygen (or oxygen-enriched air) throughout the entire system, with the only exception being at the burner during a controlled operating sequence. The oxygen and fuel supply systems, pipelines, and control equipment must always be installed in a well-ventilated space to prevent the buildup of flammable mixtures caused by small leaks. Note that every attempt should be made to continually stop any small leaks that may have developed. Newly installed fuel and oxygen pipelines are usually purged with an inert gas, such as nitrogen, in an effort to avoid the short-term possibility of a flammable mixture

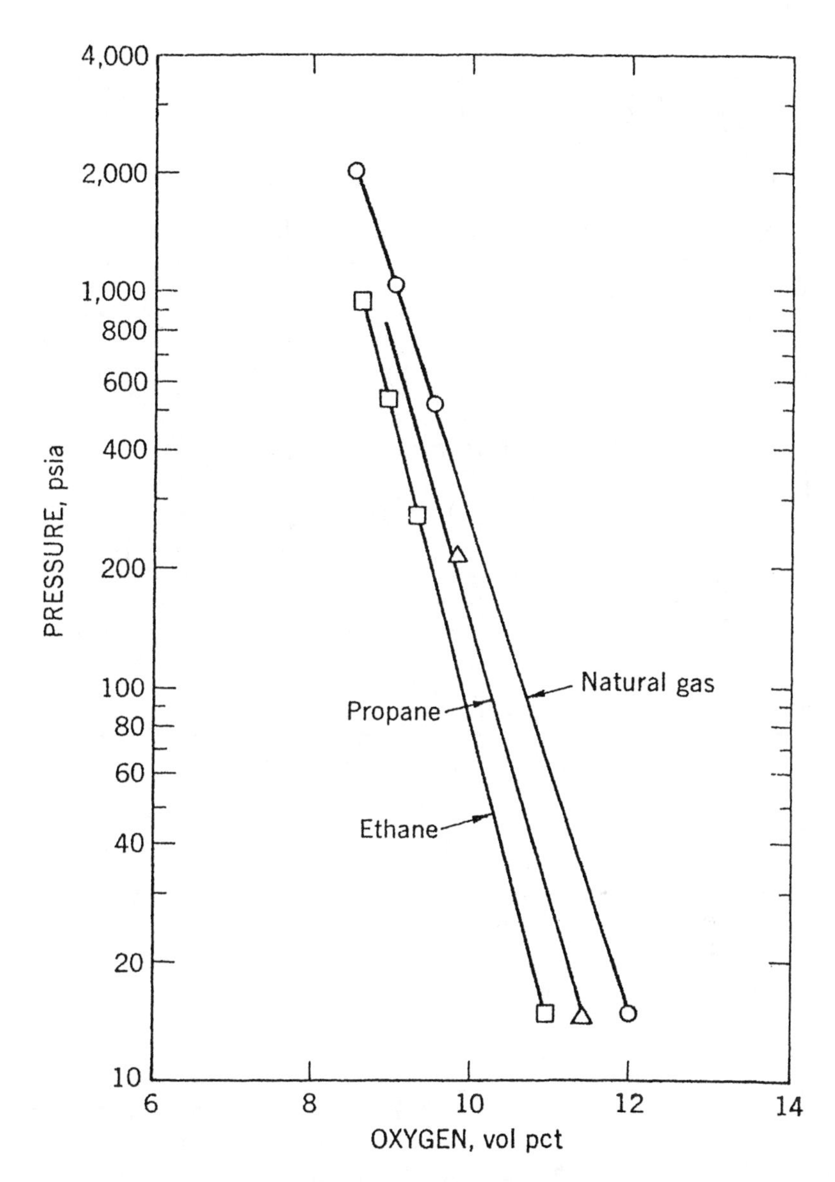

FIGURE 9.4 Effect of pressure on minimum oxygen requirements for propagation of natural gas, ethane, and propane in air–nitrogen mixtures at 25°C (77°F). (From Kuchta, J. M., Bulletin 680, U.S. Department of the Interior, Bureau of Mines, 1952.)

from occurring during commissioning of the equipment. This practice is also used for maintenance purposes or for preparing the system for a long-term shutdown. Manual isolation valves with lockout/tag-out capability are also required to be installed in the pipelines to prevent the inadvertent flow of gases during maintenance operations. Additionally, check valves must be installed in the fuel and oxygen

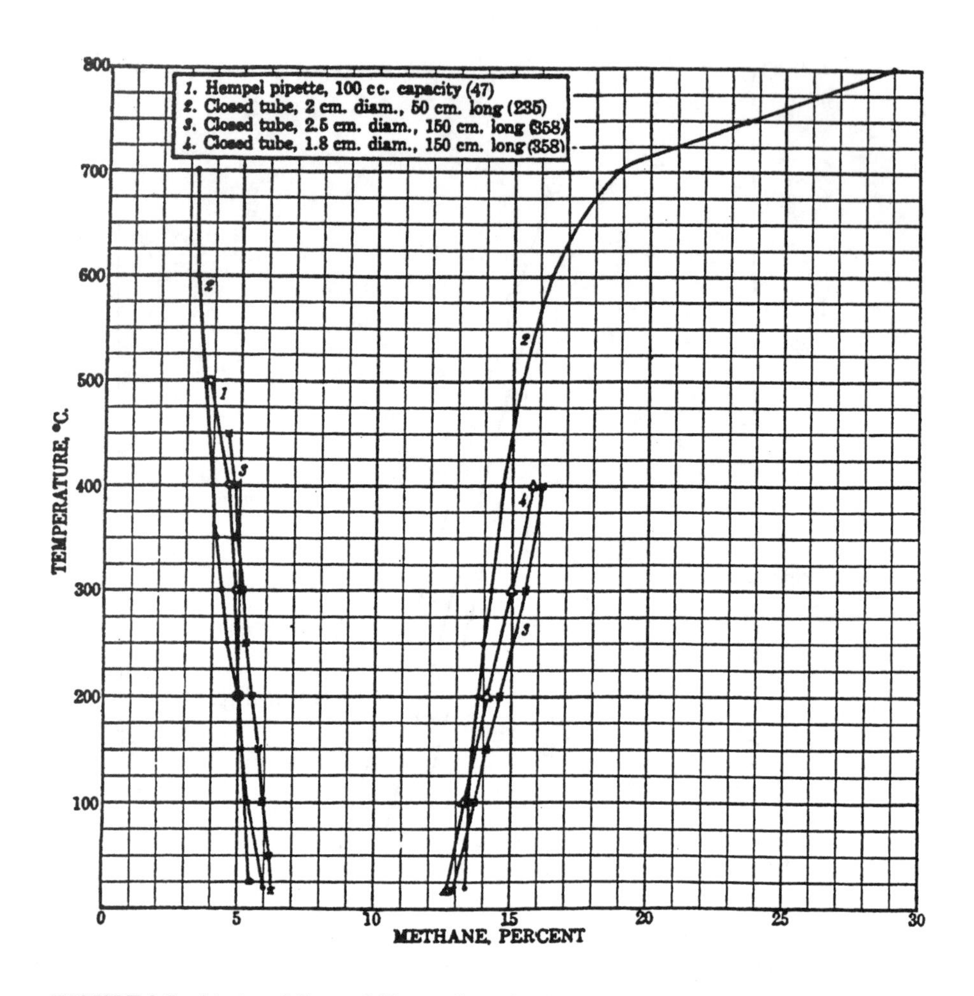

FIGURE 9.5 Limits of flammability methane in air as (downward propagation), showing influence of temperature. (From Coward, H. F. and Jones, G. W., Bulletin 503, U.S. Department of the Interior, Bureau of Mines, 1952.)

pipelines just upstream of the burner to prevent the backflow of any gases into the oxygen or fuel pipelines, which could create a flammable mixture.

Purging the furnace with air, or an inert gas, prior to (and possibly after) an attempt to ignite any burner is a primary technique that is used for avoiding fires or explosions. A combustible gas mixture may have accumulated within the furnace due to a leak of fuel and oxygen into the furnace or due to an unsuccessful burner ignition attempt. Burner ignition is controlled to a specific time limit, and at a specific low firing rate (see Chapter 10 for more details), to minimize the volume of a flammable mixture that may develop as a result of an unsuccessful attempt to ignite the burner (trial for ignition).

Flame supervision devices (including fire eyes and flame rods) are primary safety techniques that are used to avoid a fire or explosion from a flammable mixture caused

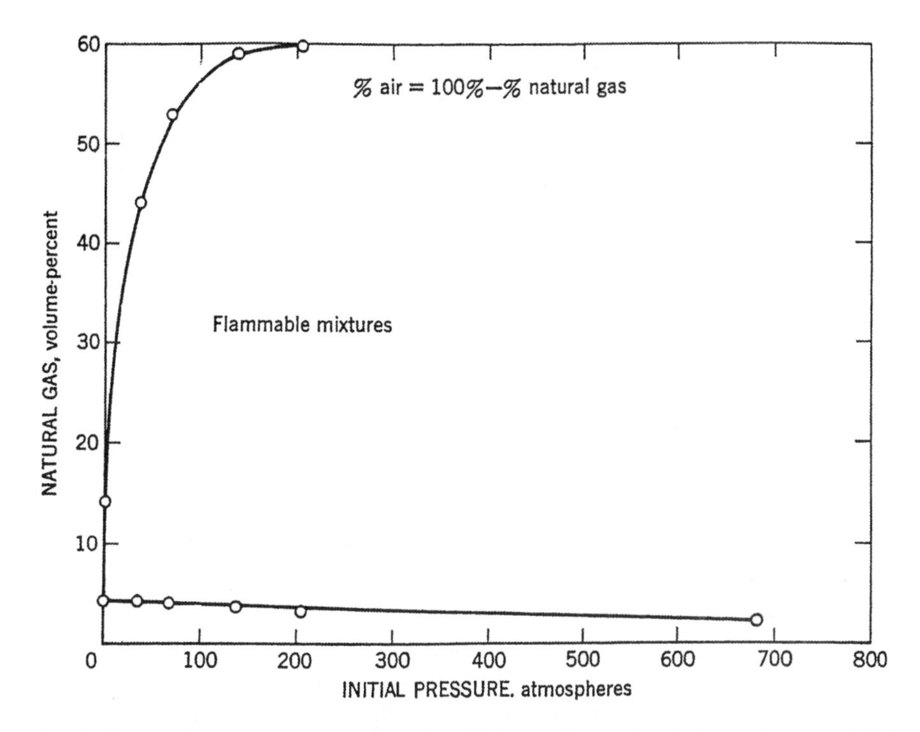

FIGURE 9.6 Effect of pressure on limits of flammability of natural gas in air at 28°C (82°F). (From Zebetakis, M. G., Bulletin 627, U.S. Department of the Interior, Bureau of Mines, 1965.)

by an inadvertent loss of flame. The flame supervision devices are interlocked to the fuel and oxygen safety shutoff valves which automatically stop the flow of fuel and oxygen upon the loss of the flame signal. Some applications (e.g., large glass furnaces) incorporate the use of a furnace low-temperature alarm (interlocked to the safety shutoff valves) to ensure that the furnace temperature is sufficiently above the autoignition temperature of the fuel/oxidant mixture while the burners are in operation.

Oxygen-enriched airstreams frequently incorporate a continuous oxygen analyzer (interlocked to the oxygen safety shutoff valves) to make sure that the percentage of oxygen enrichment is at a safe level. The safe level of oxygen enrichment is a function of the existing materials within the air main pipe, the cleanliness of the air main pipe, and the design of the pipe line and components located downstream of the oxygen injection point. The percentage of oxygen enrichment can also be calculated (and interlocked to the oxygen safety shutoff valves) based on a measured airflow rate and a measured oxygen flow rate.

9.6 IGNITION CONTROL

Ignition of a fuel and an oxidant mixture that is within the flammability limits has many possible causes, including elevated temperature; particle impact; electrical discharge (arc or spark); rapid pressurization; acoustic resonance; spontaneous exothermic reaction; friction; and others. Regardless of the physical effect, an ignition event must introduce a minimum temperature and a minimum energy. In the case of an oxygen-enriched combustion system, the presence of high concentrations of oxygen tends to lower the temperature (see Figure 1.22) and the energy needed to initiate a fire (see Figure 1.21). The introduction of oxygen into a process that formerly used air will tend to lower ignition thresholds; therefore, inoperative ignition mechanisms in the prior air/fuel system may now become important.

Ignition control refers to the attempt to minimize or eliminate any and all sources of ignition that may initiate an unwanted fire or explosion. The total elimination of all ignition sources is generally difficult to obtain. There are so many different ignition possibilities that it may not be practical, or possible, to eliminate all of them. Further, there is the probability that all ignition sources may not be identifiable. For these reasons, ignition control is widely practiced, but it is seldom relied upon as the primary safety technique. However, in general, every ignition source that can be eliminated reduces the probability of a fire or explosion from occurring.

A key issue in designing safe oxygen flow control equipment and piping systems is the oxygen velocity within the system. This issue is directly related to ignition prevention. As any gas moves through a contained space, it tends to carry particles along with it. The source of these particles can be the gas production facility or storage tank, the pipe itself (e.g., scale flaking off of the inner wall), dirt produced or introduced during fabrication, corrosion products, etc. The velocities of these particles approach that of the gas. When such a particle impacts the inner surface of the piping or components, it releases energy; and, in the case of oxygen, this impact energy can be sufficient to cause ignition of the particle and possibly of the piping material. Depending upon the materials of construction of the piping, there will be an upper gas velocity limit (determined by the requirement to remain below the minimum energy release) which can cause ignition. An upper velocity limit of 200 ft/s (60 m/s) for carbon or stainless steel pipe (for pressures below 200 psig) is a common example that is generally considered a safe limit (see Figure 9.7).[7] However, many other oxygen velocity criteria exist that pertain to piping installations in which the oxygen gas impacts surfaces such as inside tees, short-radius elbows, and orifice plates. The expected actual gas velocity in a pipe or space can be readily calculated by standard fluid mechanics theory.

Reduction of the number and size of particles within an oxygen pipeline includes first, and foremost, following proper installation practices. This begins with a thorough cleaning of all components and pipe sections prior to installation. Pipe and equipment installation involves making welded and threaded connections which follow procedures that maintain the cleanliness of the inside of the pipeline. Many common piping components, such as valves, regulators, and gauges, may be purchased as "Cleaned for Oxygen Service." Once the pipeline is complete, it should again be cleaned to remove any particles or other foreign materials that may have been introduced during installation. Most oxygen pipelines are purged with an inert gas, such as nitrogen, prior to commissioning the system. In addition, the oxygen flow control equipment must contain filters and strainers to minimize the number and size of particles that may travel through the pipeline during continued operation

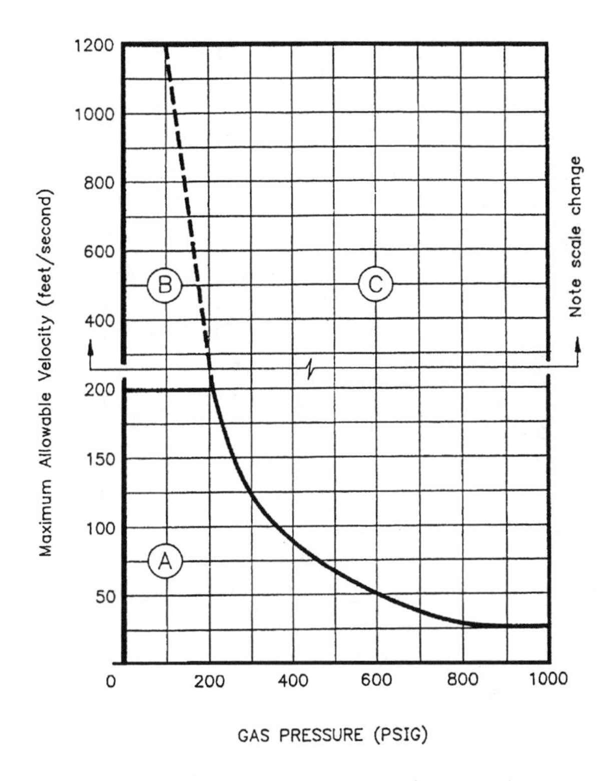

FIGURE 9.7 Material selection for gaseous oxygen (for pipe, fittings, components, and valves). A=Carbon steel, stainless steel, or CU or NI alloys. B=stainless steel (1 in. NPS min. and ⅛ in. thick min.) or CU or NI alloys. C=CU or NI alloys only. *Note:* (1) The material choice for each site is based on the maximum velocity and pressure. (2) Temperature less than 200°F. (3) Consideration must be given to the effect of particle impingement on materials of construction. (*CGA G-4.4,* Compressed Gas Association, 1993, 10. With permission.)

of the oxygen system. Periodic inspection and cleaning is required for removing particles from the filters and strainers. Stringent maintenance procedures that prevent the introduction of particles and incompatible materials are required when working on the oxygen piping and related equipment. This includes the use of clean tools, clean, lint-free rags, and wearing "nonflammable" clothing. The work area and the outside of the oxygen equipment itself may require a thorough cleaning prior to performing any maintenance procedures. Also, of course, smoking by plant personnel must be prohibited in all areas where fuel or oxygen piping systems are present.

9.7 FIRE EXTINGUISHMENT

Fire extinguishment refers to both stopping and limiting the spread of a fire or explosion. This involves removing one or more of the requirements of the combustion triangle (i.e., the fuel, the oxidant, and/or ignition sources) as illustrated in Figure 9.1. There have been systems developed that can sense fire and activate other systems that attempt to extinguish it, but these systems have found limited practical application in industry. When ignition occurs on industrial systems, the events typically happen with such speed that it is relatively difficult to prevent extensive damage. Extinguishment control therefore focuses primarily on limiting the extent of fire damage by design features that are within the combustion system.

An example of an extinguishment design feature as applied to air-based combustion is the "flame arrestor." This device is typically installed in the pipeline close to the burner. The purpose of the flame arrestor is to extinguish a flame front that might move back through the pipeline after a failure in or near the burner. The design is based upon the principle of quenching the flame so that it is cooled below the temperature required for continued propagation. This is normally accomplished by providing a large amount of surface area and mass to accelerate heat conduction away from the flame front. A simple example of this concept in practice can be found in most oxygen/acetylene torches. At the connection of the torch to the fuel hose there is a copper section designed to "absorb" the energy of any flame front moving back through the line, therefore extinguishing the flame.

Other practical fire extinguishment techniques that are used for oxygen-enriched combustion applications involve stopping the flow of the fuel and oxygen to the fire. Burner flow control equipment designs are required to contain automatic shutoff valves that will automatically stop the flow of fuel and oxygen if a high temperature or excess flow condition is sensed, or if an operator presses an "emergency stop" push button. Another example is the incorporation of manual isolation/shutoff valves that are located throughout the pipeline. These valves provide a convenient way for operators or fire safety personnel to shut off the flow of fuel and oxygen that is supporting a downstream fire. A main isolation/shutoff valve is almost always located at the fuel and oxygen supply stations. All valves must have labels that identify their purpose, and all pipelines must have labels that clearly identify the contained fluid.

9.8 OXYGEN COMPATIBILITY

Material and component selection for oxygen-based combustion or oxygen-enriched combustion requires specific knowledge. An existing air delivery system cannot be used safely with even moderate levels of oxygen enrichment without alterations to the piping components, controls, and operating procedures. The term *oxygen compatible* is used to describe a material that has been analyzed to be acceptable for service in an oxygen or oxygen-enriched environment. A material classified as oxygen compatible may have restrictions or limitations that are a function of the intended service.

The American Society for Testing and Materials (ASTM) Committee G-4 is dedicated to working on "Compatibility and Sensitivity of Materials in Oxygen-Enriched Atmospheres." This committee publishes the most comprehensive collection of documents available for understanding oxygen safety issues, analyzing velocity and pressure concerns, incorporating cleaning methods, selecting compatible materials, and designing piping and distribution systems. Committee G-4 is a consensus-based, nonprofit organization dedicated to the advancement of oxygen compatibility knowledge. Committee G-4 references and endorses many of the related practices published by other groups, such as the National Fire Prevention Association

(NFPA), Compressed Gas Association (CGA), European Industrial Gas Association (EIGA), and others. Committee G-4 also conducts seminars and technical training courses; publishes technical books, safety videos, and personal computer software; and continues to promote research in this field.

In general, application of the ASTM procedures to oxygen systems requires a high degree of technical judgment, because there may be alternative ways to achieve the same end result. It is very important to remember that oxygen compatibility is not an exact science and that proper application of common sense safety factors is crucial. Those who are working on a process using oxygen should contact their industrial gas supplier for detailed guidelines and recommendations concerning oxygen compatibility.

In oxygen systems, the oxygen index of a material may be used to rank the materials of construction for their acceptability for service. The oxygen index of a material is defined as the minimum concentration of oxygen, expressed as a volume percent, in a mixture of oxygen and nitrogen that will just support combustion of the material initially at room temperature under the conditions of the test method described in ASTM designation G125-95.[8] Table 9.1 lists the oxygen index for a variety of materials.[8] For example, the rubber compound Buna-N (a material that is typically used for gaskets, seals, and O-rings) has an oxygen index of 18. Silicone rubber compounds are listed as having an oxygen index of 21. Consequently, because both of these materials will burn in air, they are often not suitable for even moderate levels of oxygen enrichment. Figure 9.8 illustrates the oxygen index of carbon steel as a function of pressure and mole percent of oxygen enrichment.[9] Clearly, various equipment made from carbon steel (e.g., pipe, valves, and components) may not be acceptable for oxygen service at certain pressures and levels of oxygen enrichment.

The safety requirements associated with oxygen compatibility consist of meticulous component cleaning and the use of compatible materials. All systems must be thoroughly cleaned to remove any traces of incompatible contaminants such as oils, greases, solvents, and particulates. Any fluids that will be used in the system must be assessed to be compatible for oxygen service. Typical fluid applications include general lubricants, valve greases, thread joining compounds, weld fluxes, and leak check fluids. All rubber compounds and plastic materials must also be assessed to be compatible for oxygen service. Typical applications include valve seals, O-rings, gaskets, flexible hoses, and seals in general. Metal components, such as pipe and valves, must be selected by considering the intended service (i.e., pressure and velocity) and the geometry of the component. Figure 9.7 is a powerful tool for analyzing the acceptance of various metals for oxygen service.[7] In the case of oxygen-based combustion systems operating at up to 250 psig, imposed design practices include scrupulous cleaning, the use of compatible fluids (of the PCTFE and PFE families); compatible polymers (of the PTFE, PCTFE, FKM classes); and compatible metals (such as copper, brass, bronze, monel, inconel, and other copper or nickel alloys). In low-velocity and low-pressure duties, 304 or 316 stainless steels and carbon steels may be used, but specific dimension, pressure, and velocity limitations must be followed (as described in CGA G-4.4).

In the case of oxygen enrichment of an existing air-based system, it is necessary to assess the materials of the existing system and its present cleanliness, and also

TABLE 9.1
Oxygen Index for Selected Materials

Material	Manufacturer	Description	Oxygen Index	Source of Data
Polyacetal	Various		14.2–16.1	B
Polymethylmethacrylate	Various		16.7–17.7	B
Polypropylene (Pure)	Various	Soft plastic	17.4	B
Polyethylene Sheet	Atlas Mineral Prod.	0.140 in. thick white color	17.5	C
Polystyrene (Pure)	Various	Hard plastic	17.8	B
Buna-N		O-ring	18.0	C
Hi Fax Plastic 1900	Hercules Powder	0.128 in. thick polyethylene sheet — white color	18.0	C
Polypropylene Sheet	Atlas Mineral Prod.	0.127 in thick white color	18.0	C
ABS (flame retardant)	Various	Acrylonitrile–Butadiene–Styrene	18.8–33.5	B
Silicone Rubber	Lehigh Rubber Co.	0.030 in thick red color	21.0	C
Flexane 95	Devcon Corp.	Curing urethane gray color	21.5	C
EPT	Various	Ethylene Propylene Terpolymer	21.9	B
Polycarbonate	Various		22.5–39.7	B
Garlock 900	Garlock Mgf. Co.	0.67 in thick tan color	23.0	C
Urethane Foam X-50 Pipe	Triangle Conduit & Cable Co.	Exterior thermal foam insulation factor-foamed on copper tubing	23.5	C
Asbestos Gasket J-M 61	Johns Manville	0.067 in thick asbestos sheet, gray color	24.0	C
Nylon 6	E.I. du Pont de Nemours		24–30.1	B
Hypalon Sheet 0.60 in	E.I. du Pont de Nemours	Chlorosulfonated Polyethylene	25.1	B
Polystyrene (flame retardant)	Koppers	Hard plastic	25.2	B
Nordel Sheet (EPDM)	E.I. du Pont de Nemours	0.121 in thick sheet ethylene propylene rubber — black color	25.5	C
Colma SL Sealant	Sika Chemical Co.	Self leveling, gray color	26.0	C
Melrath 150	Melrath Gasket & Supply	0.066 in thick gray color	26.0	C
Neoprene	E.I. du Pont de Nemours	Chloroprene rubber	26.3	B
Craftsman Silicone Sealant	Sears Roebuck Co.	Curing elastomer	27.0	C
Nomex Nylon	E.I. du Pont de Nemours	Tan cloth	27.0	C
Silicone rubber	Various	Polysiloxane	27.9–39.2	B
Methane Foam FS/25	Owens Corning	Exterior thermal foam insulation	28.5	C
Polypropylene (flame retardant)	Avisun	Soft plastic	29.2	B
Neoprene		Diaphragm nylon reinforced	29.5	C

TABLE 9.1 (continued)
Oxygen Index for Selected Materials

Material	Manufacturer	Description	Oxygen Index	Source of Data
Zytel	E.I. du Pont de Nemours	0.625 in diameter, 0.125 in thick 0.25 in hole, white color	36.0	C
Polyimide Film 0.001 in	Various		36.5	B
Polyvinyl Chloride II High Impact PC	Atlas Mineral Prod.	0.135 in thick sheet gray color	37.0	C
Epoxy Compund	Crest Products Co.	7343 resin, 7139 catalyst	41.0	C
Polyester	Various		41.5	B
Polyvinyl Choride I	Atlas Mineral Prod.	0.129 in thick sheet, dark gray color	42.0	C
Polyvinylidene Fluoride	Various		43.7	B
Scandura 1786	Scandura Ltd.	0.066 in thick red color	45.5	C
Leotite	James Walker Co. Ltd.	0.066 in thick red color	54.0	C
Viton-A		O-Ring black color	57.0	C
Klingerit 661	Richard Klinger Ltd.	0.027 in thick red color	59.0	C
Polyvinylidene chloride	Various		60.0	B
Polyvinylidene chloride	Atlas Mineral Prod.	0.128 in thick sheet, dark gray color	65.0	C
Vespel SP-21	E.I. du Pont de Nemours	0.060 in thick black color polyimide resin with graphite	65.0	C
Alenco Hilyn	Turner Bros. Ltd.	TFE-fluorocarbon tape thread sealant	83.0	C
Gore-Tex Joint Sealant	W. L. Gore, Inc.	0.25 in thick white	91.0	C
PTFE	E.I. du Pont de Nemours	Polytetrafluoroethylene	95.0	B
TFE-fluorocarbon Sheet	E.I. du Pont de Nemours	0.100 in thick white	95.0	C
Klingerit 661	Richard Klinger Ltd.	0.048 in thick red color	100	C
Gore-Tex Packing	W. L. Gore, Inc.	1/8 in rolled string gasket white color	100.0	C
TFE-fluorocarbon		O-Ring, liquid oxygen line seal	100.0	C

A — Measured by method described in Method D2863.

B — Hilado, C. J., Oxygen Index of Materials, *Fire and Flammability Series,* Technomic Publishing Co., Westport, CT, Vol. 4.

C — Lapin, A., Oxygen Compatibility of Materials, *Reliability and Safety of Air Separation Plant,* Bulletin de L'Institut Internationale du Froid, Annexe 1973-1, pp. 79–94.

Source: ASTM G 63-87, Standard Guide for Evaluating Metals for Oxygen Service, American Society for Testing and Materials, Philadelphia, 1987. With permission.

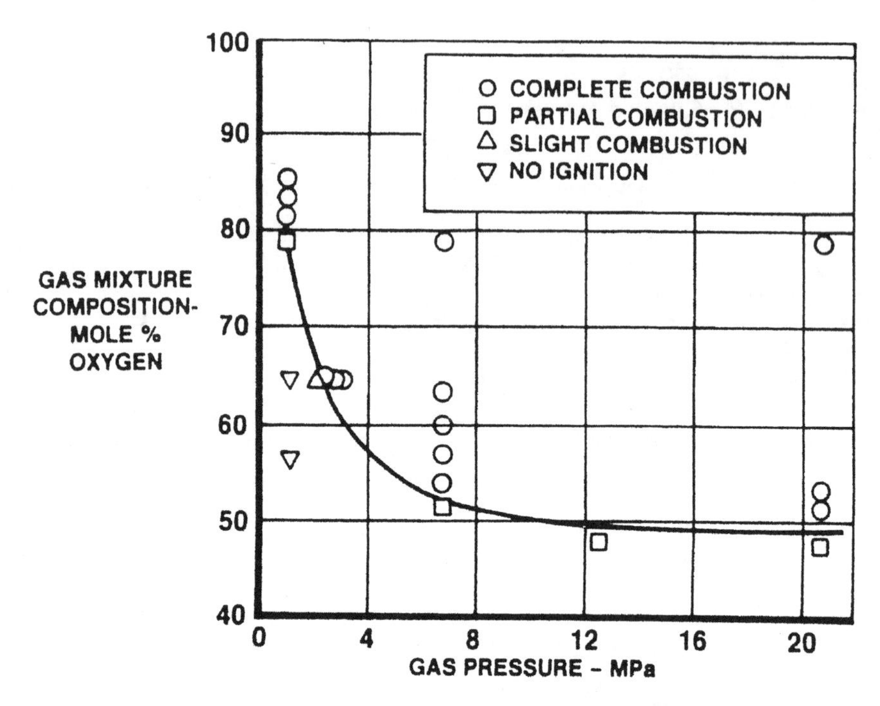

FIGURE 9.8 Oxygen index of carbon steel as a function of pressure and mole percent of oxygen. (From Benning, M. A. and Werley, B. L., *The Flammability of Carbon Steel as Determined by Pressurized Oxygen Index Measurements,* Spec. Tech. Publ. 910, American Society for Testing and Materials, Philadelphia, 1986. With permission.)

its prospect for long-term cleanliness. For example, if the supply of air is from a hydrocarbon-lubricated compressor, then a long-term safety hazard exists because over time oil deposits are likely to be dispersed throughout the air delivery system. This may lead to a fire and/or an explosion during the operation of the oxygen-enrichment system. Some existing air systems may need to be replaced entirely in order to achieve adequate cleanliness. These requirements become more extensive and stringent as the concentration level of oxygen increases.

Every air system that is being oxygen enriched must be carefully hazard-reviewed. This involves examining the burners, the fuel and oxidant delivery systems, the electrical systems, the control systems, the oxygen/air mixing point, the overall heating process, the location of the installation, operating procedures, the personnel who will be operating the equipment, and, any other relevant factors which could impact on the safety of the overall system. HAZOP (Hazard Analysis and Operability), What-If, and Fault Tree analyses are standard methods commonly used to assess the hazards and safeguards of a proposed system.

9.9 REGULATIONS AND STANDARDS

At present there are few statutory regulations in the U.S. that address safety controls for general combustion devices, or for fuel and oxygen control systems. There are regulations, such as the National Electrical Code, which enforces wiring to safe configurations, that do apply to oxygen systems.

There are currently several voluntary standards and guidelines throughout the world that do apply to the design and operation of combustion control systems including specific sections that address oxygen-enriched combustion. These include

NFPA standards,
CEN Standards (European committee for standardization),
CGA Standards,
ASTM Standards, and
EIGA Standards.

In the U.S., the industrial insurance carriers, e.g., Industrial Risk Insurers (IRI), Factory Mutual (FM), etc., may demand conformance to some of the various voluntary standards or to their own requirements. Europe, through CEN, appears to be taking a regulatory approach to safety which may lead to the enforcement of standards in the future.

The design and engineering of oxygen-enriched combustion systems, including the oxygen piping equipment, control systems, and burners, should be performed only by qualified engineers with the proper level of safety training, education, and experience. Oxygen users are encouraged to consult with their industrial gas supplier who has had many years of experience in the safe operation of systems using oxygen. In most cases, the standards developed and followed by the industrial gas suppliers meet or exceed the industrywide voluntary standards. Typically, these industrywide voluntary standards have been written in part by representatives from the various industrial gas suppliers.

REFERENCES

1. Zabetakis, M. G., Flammability Characteristics of Combustible Gases and Vapors, Bulletin 627, U.S. Department of the Interior, Bureau of Mines, 1965.
2. Davy, (Sir) H., Some Researches on Flame, *Philos. Trans. R. Soc. London,* Part 1, January 16, 45–76, 1817.
3. Turin, J. J. and Huebler, J., Gas-Air-Oxygen Combustion Studies, Report no. I.G.R.-61, American Gas Association, 1951.
4. Kuchta, J. M., Investigation of Fire and Explosion Accidents in the Chemical, Mining, and Fuel-Related Industries — A Manual, Bulletin 680, U.S. Department of the Interior, Bureau of Mines, 1985.
5. Coward, H. F. and Jones, G. W., Limits of Flammability of Gases and Vapors, Bulletin 503, U.S. Department of the Interior, Bureau of Mines, 1952.
6. *NFPA 69, Standard on Explosion Prevention Systems,* National Fire and Protection Association, Quincy, MA, 1997.
7. *CGA G-4.4, Industrial Practices for Gaseous Oxygen Transmission and Distribution Piping Systems,* Compressed Gas Association, Arlington, VA, 1993.
8. *ASTM G 125-95, Standard Test Method for Measuring Liquid and Solid Material Fire Limits in Gaseous Oxidants,* American Society for Testing and Materials, Philadelphia, 1995.

9. Benning, M. A. and Werley, B. L., *The Flammability of Carbon Steel as Determined by Pressurized Oxygen Index Measurements,* Spec. Tech. Publ. 910, American Society for Testing and Materials, Philadelphia, 1986.
10. IHEA, *Combustion Technology Manual*, Industrial Heating Equipment Association, Arlington, VA, 1988.
11. Reed, R. J., *North American Combustion Handbook*, Vol. 1, 3rd ed., North American Manufacturing Co., Cleveland, OH, 1986.

10 Equipment Design

Mark A. Niemkiewicz and J. Scott Becker

CONTENTS

10.1 INTRODUCTION

The purpose of this chapter is to provide practical design and engineering guidelines for the control of safe oxygen-enriched combustion systems. This chapter is divided into two categories of combustion systems. The first category includes the integration of an oxygen supply and control system into an existing air/fuel combustion process. The enrichment of the air with oxygen is usually facilitated by an oxygen injection device that is installed into the air supply pipeline. The second category of combustion systems deals with the installation of a new oxygen/fuel combustion process. This type of system usually includes an oxygen and fuel supply and control system, and an oxygen/fuel burner.

0-8493-1695-2/98/$0.00+$.50

This chapter is written from a perspective that treats the National Fire and Protection Agency's (NFPA) standard 86 entitled, "Ovens and Furnaces, 1995 Edition" as the accepted standard for regulating the safe design of combustion equipment. There are many parts of the NFPA 86 standard that may be interpreted in various ways depending upon the design of the oven or furnace, the combustion process, the geometry of the gas delivery system, and so forth. This chapter provides design guidelines that are an interpretation of the NFPA 86 standard as it relates to basic oxygen-enriched air combustion systems and basic oxygen/fuel combustion systems. The American Society for Testing and Materials (ASTM) standards and the Compressed Gas Association (CGA) standards are also referenced throughout the text. The reference list at the end of the chapter provides a variety of relevant literature.

Throughout the chapter, several examples will be provided that incorporate pneumatically actuated valves. It is important to note that the same functionality can be obtained with electric motor–actuated valves or electric solenoid–actuated valves. The selection of appropriate valves for a particular furnace and combustion process is a function of cost, performance, availability of clean instrument air, utility failure analyses, and personal preference.

The design and engineering of oxygen-enriched combustion systems, including the oxygen piping equipment, control systems, and burners, should be performed only by qualified engineers with the proper level of safety training, education, and experience. Oxygen users are encouraged to consult with their industrial gas oxygen supplier who has had many years of experience in the safe operation of systems using oxygen. In all cases, the design, installation, and operation of oxygen-enriched combustion systems must be reviewed and approved by the appropriate authorities.

10.2 DEFINITIONS

The definitions that are listed below include a partial listing of those that are presented in the NFPA 86 standard. Several additional terms have been defined to help clarify the remainder of the text found in this chapter. The definitions presented here are relevant to the design of equipment for oxygen-enriched combustion processes. For more information, the reader should refer to the glossaries and definition lists that are included in the various standards that are referenced at the end of this chapter.

10.2.1 Approved — Acceptable to the *authority having jurisdiction*. Note: The NFPA does not approve, inspect, or certify any installations, procedures, equipment, or materials, nor does it approve or evaluate testing laboratories. In determining the acceptability of installations, procedures, equipment, or materials, the authority having jurisdiction may base acceptance or compliance with NFPA or other appropriate standards. In the absence of such standards, said authority may require evidence of proper installation, procedure, or use. The authority having jurisdiction may also refer to the listings or labeling practices of an organization concerned with product evaluations which is in a position to determine compliance with appropriate standards for the current production of listed items.

10.2.2 Authority having jurisdiction — The organization, office, or individual responsible for approving equipment, an installation, or a procedure. Note: The phrase *authority having jurisdiction* is used in NFPA documents in a broad manner, since jurisdictions and "approval" agencies vary, as do their responsibilities. Where public safety is primary, the *authority having jurisdiction* may be a federal, state, local, or other regional department or individual, such as a fire chief; fire marshal; chief of a fire prevention bureau, labor department, health department; building official; electrical inspector; or others having statutory authority. For insurance purposes, an insurance inspection department, rating bureau, or other insurance company representative may be the authority having jurisdiction. In many circumstances the property owner or his or her designated agent assumes the role of the authority having jurisdiction; at government installations, the commanding officer or departmental official may be the authority having jurisdiction.

10.2.3 Combustion air main pipeline — The main pipeline or duct that supplies air to a burner or combustion process.

10.2.4 Combustion safeguard — A safety control directly responsive to flame properties; it senses the presence and/or absence of flame and correspondingly deenergizes the fuel (and oxygen) safety valves in the event of flame failure within 4 seconds of the loss of the flame signal.

10.2.5 Flame rod — A detector that employs an electrically insulated rod of temperature-resistant material that extends into the flame being supervised, with a voltage impressed between the rod and a ground connected to the nozzle of the burner. The resulting electrical current, which passes through the flame, is rectified, and this rectified current is detected and amplified by the combustion safeguard.

10.2.6 Flow run — A set of components, valves, instruments, and piping assembled together to control the pressure and flow of a particular fluid. Several different flow runs (air, oxygen, natural gas, water, etc.) are typically mounted on a single valve stand.

10.2.7 Fuel gas — Gas used for heating, such as natural gas, manufactured gas, undiluted liquefied petroleum gas (vapor phase only), liquefied petroleum gas–air mixtures, or mixtures of these gases.

10.2.8 Fuel oil — Grades 2, 4, 5, or 6 fuel oils as defined in American Society for Testing and Materials (ASTM) D396, Specifications for Fuel Oils.

10.2.9 Ignition Systems — Automatic-Ignited Burner: a burner ignited by direct electric ignition or by an electric-ignited pilot. Direct Electric Ignition: ignition of flame by an electric-ignition source, such as a high-voltage spark or hot wire, without the use of a separate pilot burner. Manual-Ignited Burner: a burner ignited by a portable torch manually placed in proximity to the burner nozzle. Semiautomatic-Ignited Burner: a burner ignited by direct electric ignition or by an electric-ignited pilot, where the electric ignition is manually activated.

10.2.10 Ignition temperature — The lowest temperature at which an air/oxygen/fuel mixture may ignite and continue to burn when an ignition source is supplied.

Autoignition temperature is defined as the temperature at which an air/oxygen/fuel mixture will self-ignite without the presence of an ignition source. When burners supplied with a gas–air mixture, in the flammable range, are heated above the autoignition temperature, flashbacks may occur. In general, such temperatures range from 870°F (470°C) to 1300°F (700°C). However, higher temperatures are necessary to dependably ignite various air/oxygen/fuel mixtures. For example, the temperature necessary to ignite natural gas is slightly higher than for manufactured gases. For safety reasons, a temperature of about 1200°F (650°C) is required to dependably ignite manufactured gas/oxidant mixtures, and a temperature of about 1400°F (760°C) is required to dependably ignite natural gas/oxidant mixtures.

10.2.11 Operator — An individual responsible for the start-up, operation, shutdown, and emergency handling of the furnace and its associated equipment.

10.2.12 Oxygen — A chemical element of atomic weight 16, which at normal atmospheric temperatures and pressures exists as a colorless, odorless, and tasteless gas. Oxygen comprises about 21% by volume of the Earth's atmosphere. Refer to an industrial gas supplier's Oxygen Material Safety Data Sheet (MSDS) for further details.

10.2.13 Oxygen-enriched air — Air in which the concentration of oxygen exceeds 23% by volume, or the partial pressure of oxygen exceeds 160 torr, or both.

10.2.14 Oxygen enrichment level I — Air in which the concentration of oxygen is greater than 23% and less than or equal to 27% by volume.

10.2.15 Oxygen enrichment level II — Air in which the concentration of oxygen is greater than 27% and less than or equal to 40% by volume.

10.2.16 Oxygen enrichment level III — Air in which the concentration of oxygen is greater than 40% by volume. Note that level III oxygen enrichment must be treated the same as the case of pure oxygen for all purposes including materials selection, piping component selection, and cleaning procedures.

10.2.17 Oxy/fuel — A general term used to describe a combustion process that includes oxygen gas, as the oxidizer and a fuel. The fuel can be a gas (e.g., natural gas), a liquid (e.g., #2 fuel oil), or a solid (e.g., powdered coal).

10.2.18 Pilot — A flame that is used to light the main burner.

10.2.19 Pilot, continuous — A pilot that burns continuously throughout the entire period that the heating equipment is in service whether or not the main burner is firing.

10.2.20 Pilot, interrupted — A pilot that is ignited and burns during the light-off and is automatically shut off at the end of the trial-for-ignition of the main burner(s).

10.2.21 Programmable controller — A digital electronic system designed for use in an industrial environment, that uses a programmable memory for the internal storage of user-oriented instructions for implementing specific functions to control, through digital or analog inputs and outputs, various types of machines or processes.

10.2.22 Purge — The replacement of a flammable, indeterminate, or high-oxygen-bearing atmosphere with another gas that, when complete, results in a nonflammable final state.

10.2.23 Safety Device — An instrument, control, or other equipment that acts, or initiates action, to cause the furnace to revert to a safe condition in the event of equipment failure or other hazardous event. Safety devices are redundant controls, supplementary controls utilized in the normal operation of a furnace system. Safety devices act automatically, either alone or in conjunction with operating controls, when conditions stray outside of design operating ranges and endanger equipment or personnel.

10.2.24 Safety interlock — A device required to ensure safe start-up and safe operation, and to cause safe equipment shutdown.

10.2.25 Shall — Indicates a mandatory response.

10.2.26 Should — Indicates a recommendation or that which is advised but not required.

10.2.27 Safety shutoff valve — A normally closed (closed when de-energized) valve installed in the piping that closes automatically to shut off the fuel or atmosphere gas in the event of abnormal conditions, or during shutdown. The valve can be opened either manually or by a motor-operator, but only after the solenoid coil or other holding mechanism is energized.

10.2.28 Temperature controller — A device that measures the temperature and automatically controls the heat input into the furnace.

10.2.29 Temperature excess limit controller — A device designed to cut off the source of heat if the operating temperature exceeds a predetermined temperature set point.

10.2.30 Trial for ignition period (flame-establishing period) — The interval of time during light-off that a safety-control circuit allows the fuel and oxygen safety shutoff valves to remain open before the combustion safeguard is required to supervise the flame.

10.2.31 Valve stand — A free-standing frame including several fluid flow runs that is used to physically control the fluid flow to a combustion process. A typical oxy/fuel combustion equipment valve stand will include an instrument airflow run, an oxygen flow run, and a fuel flow run. The equipment on the valve stand typically includes the on/off valves, pressure regulators, flow control valves, and associated instrumentation necessary to operate a combustion process.

10.3 OXYGEN-ENRICHMENT GUIDELINES

10.3.1 PURPOSE

This section provides guidelines for the design of safe oxygen-enriched air systems. In general, an oxygen-enriched air system includes an oxygen flow control valve stand, some type of electrical controls, and an oxygen injection device (e.g., a diffuser) that is inserted into an existing air main pipeline.

10.3.2 Scope

This guideline applies specifically to –20 to 200°F (–30 to 90°C) oxygen, or an oxygen-containing gas, with pressures below 250 psig, that is being supplied to an air main pipeline for the purpose of oxygen enriching the air to a maximum of 40% oxygen by volume.

10.3.3 Oxygen Flow Control System

Figure 10.1 illustrates a typical oxygen flow control system that may be used to oxygen enrich a combustion air main pipeline for a variety of industrial applications. All topics below reference the tag numbers as shown in Figure 10.1. The components that are required as well as some of those that are optional are described below. The requirements that are related to the industry standards are based on an interpretation of those standards by Air Products and Chemicals, Inc. All practical system designs and installations must be approved by the authority having jurisdiction.

10.3.3.1 In general, the oxygen piping from the supply, through the valve stand, and to the diffuser, shall comply with the industry standards for oxygen service as summarized in: CGA G-4.1 and 4.4; NFPA 86 (sect. 4-4.3); and ASTM Designation G88-84, G93-88, and G94-88. NFPA 86 (sect. 4-4.3.1) requires that the design, materials of construction, installation, and testing of the oxygen piping shall comply with applicable sections of ANSI B31.3, Code for Pressure Piping, Chemical Plant and Petroleum Refinery Piping.

10.3.3.2 **An Isolation Valve (BV100)** is required by NFPA 86 (sect. 4-4.3.5 and sect. 5-2.2), and design guidelines are provided in CGA G-4.4 (sect. 4.5.1 and sect. 5.3.1). The purpose of the isolation valve is to provide a manual shutoff for maintenance and emergency conditions. Therefore, the valve must be located in a position where a hazardous condition will not impede access to the valve by plant personnel. A quick-opening ball or butterfly valve is recommended, and it must be suitable for oxygen service. It is recommended that the valve be opened slowly to minimize adiabatic compression hazards. The valve should include a locking feature so that the supply of oxygen can be locked-out/tagged-out during maintenance operations.

10.3.3.3 **A Strainer (ST101)** is required by NFPA 86 (sect. 5-14.2), and design guidelines are provided in CGA G-4.4 (sect. 4.8). The purpose of the strainer is to capture particulates and prevent them from proceeding through the oxygen piping and flow components. A brass or monel strainer screen with a 30 to 100 mesh size is recommended.

10.3.3.4 **A "Blowdown" Drain Valve (V101)** is not recommended for oxygen service per CGA G-4.4 (sect. 4.8.3). The purpose of a blowdown drain valve is to provide an easy method for cleaning the strainer. However, this valve could easily be opened at an inappropriate time, which could cause a dangerous localized oxygen-enriched atmosphere when being used to clean out the strainer.

10.3.3.5 **A Pressure Regulator** (not shown) is required by NFPA 86 (sect. 4-4.4.2) whenever the upstream oxygen supply pressure is subject to excessive fluctuations, or exceeds the required pressure for proper burner operation.

10.3.3.6 **A Pressure Relief Device** (not shown) is required by NFPA 86 (sect. 4-4.3.12) whenever it is possible to have an event leading to an oxygen supply pressure that is greater than the MAWP (maximum allowable working pressure) of any of the components or piping within the oxygen pipeline. The pressure relief device must be sized properly per manufacturer and industry standards. The outlet of the pressure relief device must be piped to a safe location per NFPA 86 (sect. 4-4.3.7).

10.3.3.7 **A Safety Shutoff Valve (ABV102)** is required per NFPA 86 (sect. 5-7 and sect. 5-14.1). The safety shutoff valve must have position indication (e.g., "open" and "closed"), it cannot be a flow control valve, and it must fail to a closed position when any safety interlock is not satisfied. An air-operated ball valve is an acceptable safety shutoff valve per CGA G-4.4 (sect. 5.3.4). An instrument air three-way solenoid valve (XV102) with adjustable orifices is recommended to control the opening and closing speed of the ball valve. A second safety shutoff valve is required by NFPA 86 (sect. 5-7 and sect. 5-14.1) when the burner system has a heating capacity greater than 400,000 Btu/h (120 kW). At least one of the two safety shutoff valves must have a limit switch (or similar means) that can be used to prove that the valve is in a "closed" position during the furnace purge interval per NFPA 86 (sect. 5-4.1.2.1 and sect. 5.7.2.2).

10.3.3.8 **A Temperature Element (TE103)** and **Temperature Transmitter (TT103)** are optional components. These components are used to compensate the computed oxygen flow rate for temperature fluctuations of the oxygen supply. The decision to use temperature compensation is a function of the required accuracy of the process.

10.3.3.9 **A Flow Element (FE104)** is recommended to provide a measure of the oxygen flow rate. CGA G-4.4 (sect. 4.7 and sect. 5.4) provides general guidelines for flow element selection. Flow elements may include orifice plates (monel material is recommended); annubars; vortex meters; mass flow meters; and others, provided that all of the materials are oxygen compatible and dimensional requirements are satisfied. Glass tube flowmeters (rotameters) are not recommended. If used, safeguards against personnel injury must be provided per NFPA 86 (sect. 4-4.3.10).

10.3.3.10 **Low and High Oxygen Flow Limit Switches (FSL/H104)** are required per NFPA 86 (sect. 5-14.3 and sect. 5-14.4). The flow limit switches must be interlocked to close the automatic safety shutoff valves in the event of an unsafe oxygen flow rate. In Figure 10.1, the flow transmitter (FT104) generates an electronic flow signal which is monitored by the flow limit switch device (FSL/H104).

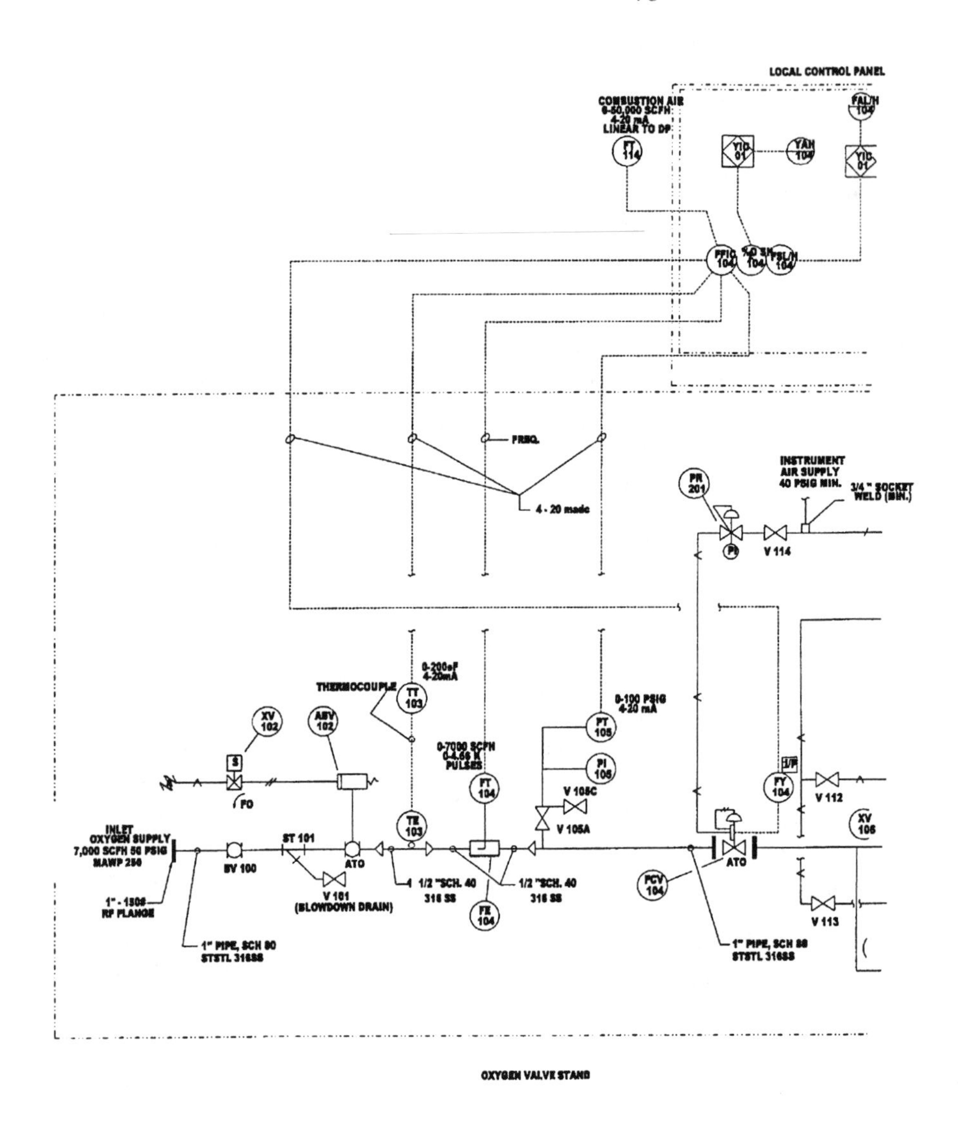

FIGURE 10.1 Typical oxygen flow run used for oxygen enrichment of a combustion air main pipeline.

10.3.3.11 **A Pressure Transmitter (PT105)** is an optional component. This component is used to compensate the computed oxygen flow rate for pressure fluctuations of the oxygen supply. The decision to use pressure compensation is also a function of the required accuracy of the process.

10.3.3.12 **A Pressure Indicator (PI105)** is strongly recommended. This pressure indicator is used to indicate the oxygen metering pressure for the flow element. Many types of flow elements require a precise metering pressure for accurate flow measurement.

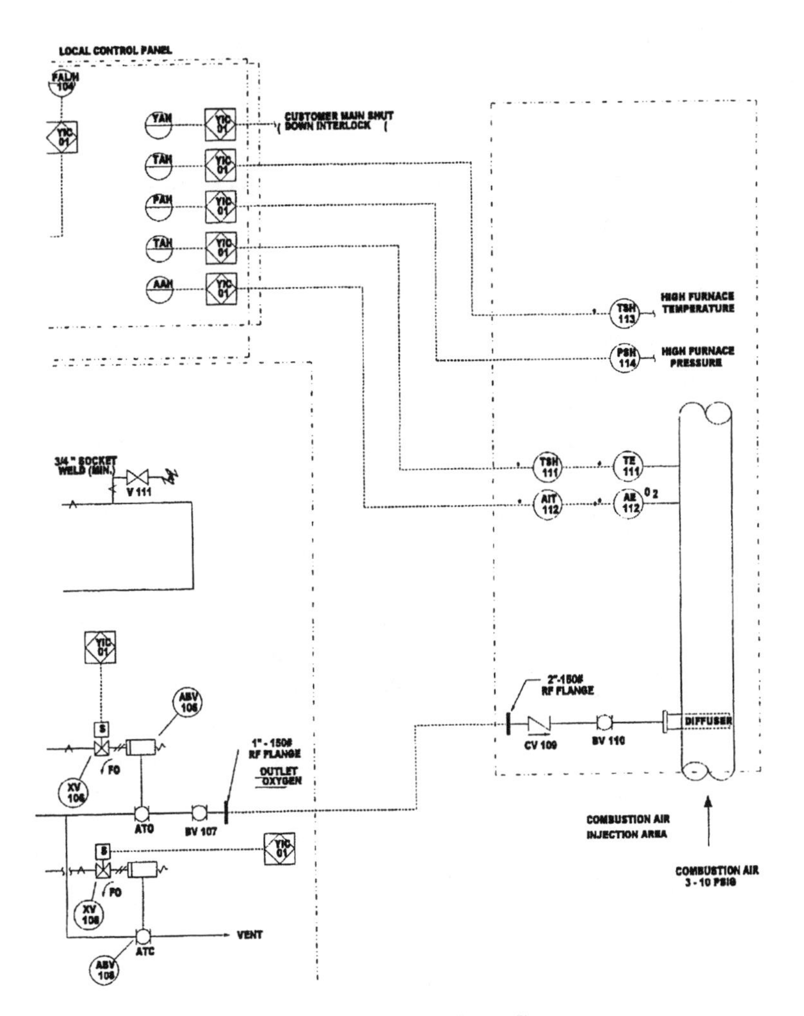

FIGURE 10.1 (continued)

10.3.3.13 **Valves (V105A and V105C)** are optional.

10.3.3.14 **The Flow Control Valve (FCV104)** must be sized properly per NFPA 86 (sect. 4-4.4). Also, CGA G-4.4 (sect. 5.3.2) provides general guidelines for selecting appropriate flow control valves. The purpose of the flow control valve is to control the oxygen flow rate to a value based on the desired oxygen-enrichment percentage. In Figure 10.1, an I/P transducer (FY104) accepts an electrical position signal from the controller (FFIC104). The I/P transducer then converts the electrical signal (I) into a pneumatic signal (P) which is used to position the valve. The controller (FFIC104) accepts an airflow rate signal (FT114) and computes the corresponding oxygen flow rate set point necessary to obtain the desired percentage of oxygen enrichment. The controller (FFIC104) also accepts a measured oxygen flow rate signal from the flow transmitter (FT104) and, correspondingly, tries to position the flow control valve (FCV104) until the measured oxygen flow rate equals the desired set point value. The controller (FFIC104) calculates the actual oxygen-enrichment percentage based on the measured airflow rate and the measured oxygen flow rate, and it will send a signal to shut off the flow of oxygen automatically if an excessive amount of oxygen enrichment exists. A high oxygen-enrichment percentage safety interlock based on the measured airflow rate and measured oxygen flow rates is highly recommended.

10.3.3.15 **A Bleed Valve (ABV108)** and corresponding piping is recommended to allow any oxygen that may be leaking past the upstream safety shutoff valve (ABV102) to vent to a safe location when the oxygen system is not in use. Bleed valve selection and vent piping design guidelines are provided in CGA G-4.4 (sect. 5.3.3). Note that the bleed valve and corresponding pipeline should not be considered a "vent" line. Therefore, the size of the bleed valve and piping should be kept to a minimum (¼ to ½ in. is recommended). The small size will tend to avoid excessive oxygen flow due to a failure of the bleed valve. The bleed valve and piping also provide a permanent and ready means for making tightness checks of the upstream safety shutoff valve (ABV102). An air-actuated ball valve is acceptable per CGA G-4.4 (sect. 5.3.4). An instrument air three-way solenoid valve (XV108) with adjustable orifices is recommended to control the opening and closing speed of the ball valve. The bleed valve pipeline must be routed to a safe location per NFPA 86 (sect. 4-4.3.7).

10.3.3.16 **A Second Safety Shutoff Valve (ABV106)** is required per NFPA 86 (sect. 5-7 and sect. 5-14.1) when the burner system has a heating capacity greater than 400,000 Btu/h (120 kW). The safety shutoff valve must have position indication (e.g., "open" and "closed"), it cannot be a flow control valve, and it must fail to a closed position when any safety interlock is not satisfied. An air-operated ball valve is an acceptable safety shutoff valve per CGA G-4.4 (sect. 5.3.4). An instrument air three-way solenoid valve (XV106) with adjustable orifices is recommended to control the opening and closing speed of the ball valve. At least one of the two safety shutoff valves must have a limit switch (or similar means) that can be used to prove that the valve is in a "closed" position during the furnace purge interval per NFPA 86 (sect. 5-4.1.2.1 and sect. 5-7.2.2).

10.3.3.17 **A Second Isolation Valve (BV107)** is optional. Design guidelines are provided in CGA G-4.4 (sect. 4.5.1 and sect. 5.3.1). The purpose of the second isolation valve is to provide a manual shutoff for maintenance operations. A quick-opening ball or butterfly valve is recommended, and it must be suitable for oxygen service. It is recommended that the valve be opened slowly to minimize adiabatic compression hazards.

The following are general guidelines associated with the oxygen flow control system.

10.3.3.18 Chapter 9 of this book and NFPA 86 (sect. A-4-4.1) provide a general overview of the hazards associated with oxygen-enriched burners and control systems.

10.3.3.19 All electrical wiring, and applicable electrical components, shall be in accordance with the National Electric Code (NFPA 70).

10.3.3.20 Operation and maintenance personnel training is required and is detailed by NFPA 86 (sect. 1-5 and Chapter 10).

10.3.3.21 The use of two different high oxygen-enrichment percentage interlocks is strongly recommended in order to provide a redundant level of safety. Examples of oxygen-enrichment percent interlocks are (1) a calculated oxygen-enrichment percent based on the measured air and oxygen flow rates; (2) an oxygen-enrichment percent from an oxygen analyzer located in the air main pipeline downstream of the oxygen/air mixing point; (3) an oxygen-enrichment percent level based on the inherent maximum oxygen flow rate through the piping, incorporating a high oxygen pressure switch and an airflow rate switch; and (4) an oxygen-enrichment level based on the inherent maximum oxygen flow rate through the piping, coupled with a low air pressure switch. Each proposed system must be reviewed for hazards and safeguards, and must include appropriate methods of ensuring that the percentage of oxygen enrichment remains below the safety limit.

10.3.3.22 Requirements for the use of **Programmable Controllers** are defined by NFPA 86 (sect. 5-3). Combustion safety interlocks, combustion safeguards, high oxygen-enrichment percentage interlocks, high oxygen flow interlocks, and excessive air main or furnace temperature limits shall be wired to directly de-energize the safety shutoff valves, and their operations shall result in a safe system condition.

10.3.3.23 The oxygen safety shutoff valves are required by NFPA 86 (sect. 5-14.5) to shut off the flow of oxygen in the event of any unsatisfied interlock (e.g., low or high flow of the fuel or oxygen, flame failure, high furnace temperature, low burner cooling water flow, high burner cooling water temperature, etc.).

10.3.3.24 When oxygen is added to a combustion air line, the oxygen flow and airflow shall be interlocked to prevent the initiation of oxygen flow prior to establishment of the airflow per NFPA 86 (sect. 5-14.6.1).

10.3.3.25 An audible and/or visible alarm is recommended in the safety circuit to give warning of unsafe conditions or interruption of the safety circuit.

10.3.3.26 A permanent and ready means for making tightness checks of fuel gas shutoff valves is typically required per NFPA 86 (sect. 5-7.2.3). This practice is recommended for the oxygen safety shutoff valves as well.

10.3.3.27 Oxygen shall not be introduced into air-fuel gas mixing piping, fuel gas mixing machines, or into mixing blowers per NFPA 86 (sect. 4-2.6.1).

10.3.3.28 Every air system that is being oxygen enriched must be carefully hazard-reviewed. This involves examining the burner(s), the fuel and oxidant delivery systems, the electrical systems, the control systems, the oxygen/air mixing point, the overall heating process, the location of the installation, operating procedures, the personnel who will be operating the equipment, and any other relevant factors which could impact on the safety of the overall system. HAZOP (hazard analysis and operability), What-If, and Fault Tree analyses are standard methods commonly used to assess the hazards and safeguards of a proposed system.

10.3.4 Oxygen Diffuser

Figure 10.2 illustrates a typical oxygen diffuser. Listed below are general guidelines for the safe design and use of oxygen diffusers.

In general, the diffuser design shall comply with the industry standards for oxygen service as summarized in CGA G-4.1 and 4.4; NFPA 86 (sect. 4-4.3); and ASTM Designation G88-84, G93-88, and G94-88.

10.3.4.1 The diffuser is required by NFPA 86 (sect. 4-4.5.2) to be designed to prevent jet impingement of oxygen onto the interior surfaces of the air main pipeline. This includes engineering the number and size of the holes so that the velocity of the oxygen gas exiting the diffuser remains below a safe limit. In addition, the geometry of the diffuser must be designed to inject the oxygen somewhere near the centerline of the air main pipeline.

10.3.4.2 The diffuser design should promote good air/oxygen mixing in the air main pipeline. The number of holes necessary to maintain a particular oxygen velocity for a particular hole size at a particular oxygen flow rate through the diffuser can be calculated as follows:

$$\text{\# of holes} = \frac{4\cdot\left[Q(\text{acfh})\cdot\left[1(\text{h})/3600(\text{s})\right]\right]}{\pi\cdot\text{oxygen velocity}\ (\text{ft/s})\cdot\left[\text{hole size}\ (\text{in.})\right]^2\cdot\left[1(\text{ft})^2/144(\text{in.})^2\right]} \tag{10.1}$$

The hole size of the diffuser is designed such that the oxygen velocity through each of the holes (at the maximum oxygen operating flow rate through the diffuser) is typically 200 ft/s (60 m/s). Note that the flow rate, Q, must be in actual cubic feet per hour (acfh).

10.3.4.3 The diffuser must be made from an oxygen compatible material. Brass, copper, or monel materials are good choices. Diffusers are frequently made from stainless steel, but in these cases dimensional limits, pressure limits, and velocity limits must be considered.

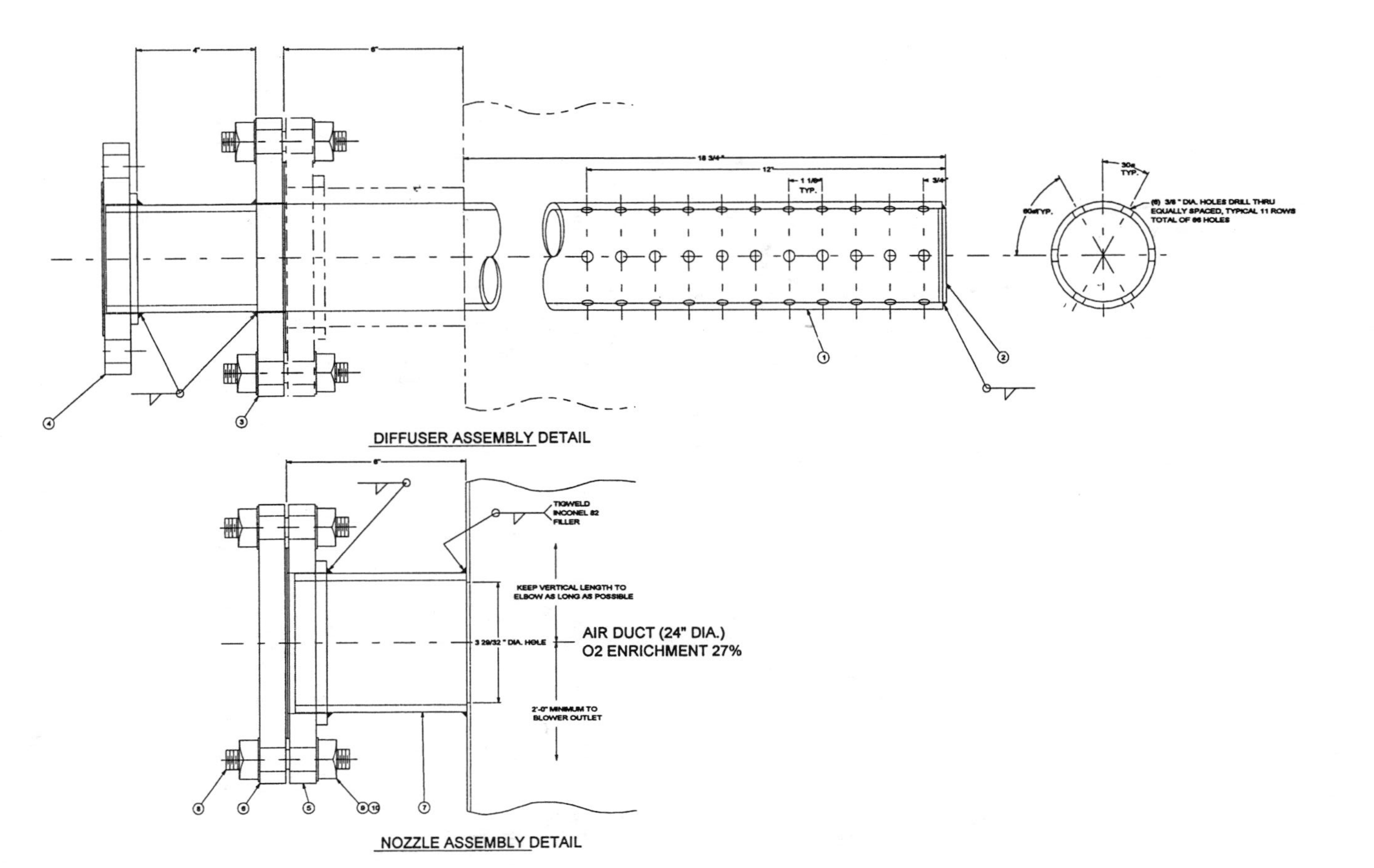

FIGURE 10.2 Typical oxygen diffuser used for oxygen enrichment of a combustion air main pipeline.

10.3.4.4 **A Check Valve** (or similar device) must be installed in the oxygen line immediately upstream of the diffuser per NFPA 86 (sect. 4-4.3.8). The purpose of the check valve is to prevent backflow of any gases into the oxygen piping.

10.3.5 Air Main Pipeline Concerns

NFPA 86 (sections 4-3.2, 4-4.5, and 5-6) provides general requirements for designing oxygen-enriched combustion air pipelines. The following are specific concerns related to the air main pipeline design.

10.3.5.1 Oxygen-enriched air velocities are subject to limitations based on CGA G-4.4. Allowable velocities are a function of the enrichment level, the material of the air main pipeline, the pressure of the air main pipeline, and the cleanliness of the air main pipeline.

10.3.5.2 For level I enrichment, it is recommended that the inside surface of the air main pipeline be cleaned and inspected to be free of hydrocarbons and contaminants for five diameters downstream of the diffuser location.

10.3.5.3 For level II enrichment, it is recommended that the inside surface of the air main pipeline be cleaned and inspected to be free of hydrocarbons and contaminants for all piping downstream of the diffuser location. It is also recommended that the air main pipeline be periodically inspected by qualified personnel, and the results of the inspection should be documented.

10.3.5.4 It is recommended that all contamination of combustible materials from the air main pipeline having oxygen indexes below the high oxygen-enrichment percent interlock settings plus 5% be eliminated. For example, if the high oxygen percent interlocks are set for 31% oxygen enrichment, then all combustible contaminants having oxygen indexes below 36 (31 + 5) must be eliminated from the air main pipeline. The oxygen index of existing contaminants within the air main pipeline can be determined by test methods as described in ASTM Designation G125-95. Approximately 10 oz of contaminants are necessary to perform the test.

10.3.5.5 Branching of the air main pipeline may not occur downstream of the diffuser location until there is a uniform mixture of air and oxygen, as described in NFPA 86 (sect. 4-4.5.4). Oxygen-enriched combustion air must not be introduced to the burner until there is a uniform mixture of air and oxygen as described in NFPA 86 (sect. 4-4.5.3).

10.3.5.6 For level I enrichment, it is recommended that any components in the air main pipeline located within 20 diameters downstream of the diffuser location be suitable for level III oxygen service. It is also recommended that any components in the air main pipeline located more than 20 diameters downstream of the diffuser location be suitable for 27% oxygen enrichment service.

10.3.5.7 For level II enrichment, it is recommended that any components in the air main pipeline located downstream of the diffuser location be suitable for level III oxygen service. This includes gaskets, seals, valve seats, valve packing materials, etc.

10.3.5.8 The air supply compressor or blower must be free of lubricating petroleum oils, greases, or other flammable substances per NFPA 86 (sect. 4-4.2.2). Sealed bearings are recommended. Consider operating the air supply equipment at the maximum output level possible to dislodge any dust or contaminants prior to inspection of the air main pipeline.

10.3.5.9 Filters must be installed on the air supply compressor or blower intake to minimize contamination of the air main pipeline per NFPA 86 (sect. 4-4.5.1).

10.3.5.10 The air main pipeline downstream of the diffuser should be monitored for abnormally high temperatures during commissioning of the oxygen-enrichment system. Abnormally high temperatures in the air main pipeline during the commissioning of the oxygen-enrichment system could be an indication of a fire in the air main pipeline.

10.4 OXY/FUEL GUIDELINES

10.4.1 PURPOSE

This section provides general guidelines for the design of safe oxygen/fuel combustion equipment systems. In general, an oxygen/fuel combustion equipment system includes an oxygen and a fuel flow run mounted on a single valve stand, an electrical control panel, an oxygen/fuel burner inserted into a furnace, furnace purge air controls, ignition equipment, and flame supervision equipment.

10.4.2 SCOPE

This guideline applies specifically to the design of equipment used to control the flow of oxygen gas (–20 to 200°F or –30 to 90°C, with pressures below 250 psig) and a fuel (gas or liquid) to a burner; and, to safely address the related issues involved in a typical oxy/fuel combustion application. The related issues may include furnace purging, ignition, flame supervision, burner concerns, furnace concerns, and miscellaneous concerns.

10.4.3 OXYGEN FLOW RUN

Figure 10.3 illustrates a typical oxygen flow run used to control the flow of oxygen to a burner. All topics below reference the tag numbers as shown in Figure 10.3 except as noted. The components that are required as well as some of those that are optional are described below. The requirements that are related to industry standards are based on an interpretation of those standards by Air Products and Chemicals, Inc. All practical system designs and installations must be approved by the authority having jurisdiction.

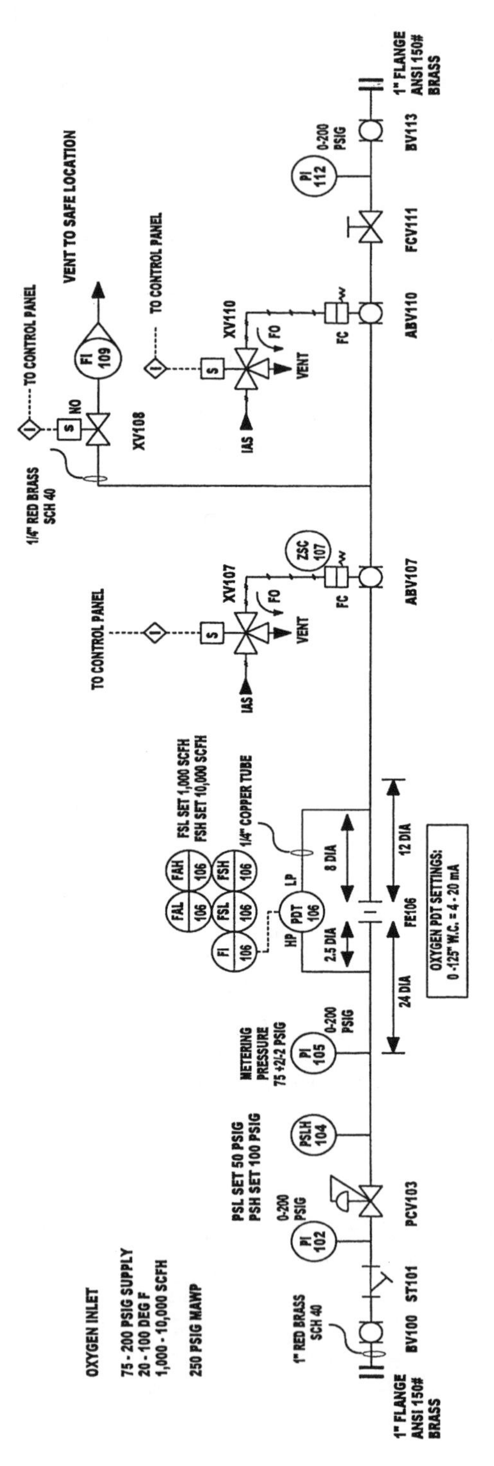

FIGURE 10.3 Typical oxygen flow run for use in an oxygen/fuel combustion system.

10.4.3.1 In general, the oxygen piping from the supply, through the valve stand, and to the burner shall comply with the industry standards for oxygen service as summarized in CGA G-4.1 and 4.4; NFPA 86 (sect. 4-4.3); and ASTM Designation G88-84, G93-88, and G94-88. NFPA 86 (sect. 4-4.3.1) requires that the design, materials of construction, installation, and testing of the oxygen piping shall comply with applicable sections of ANSI B31.3, Code for Pressure Piping, Chemical Plant and Petroleum Refinery Piping.

10.4.3.2 **An Isolation Valve (BV100)** is required by NFPA 86 (sect. 4-4.3.5 and sect. 5-2.2) and design guidelines are provided in CGA G-4.4 (sect. 4.5.1 and sect. 5.3.1). The purpose of the isolation valve is to provide a manual shutoff for maintenance and emergency conditions. Therefore, the valve must be located in a position in which a hazardous condition will not impede access to the valve by plant personnel. A quick-opening ball or butterfly valve is recommended, and it must be suitable for oxygen service. It is recommended that the valve be opened slowly to minimize adiabatic compression hazards. The valve should include a locking feature so that the supply of oxygen can be locked-out/tagged-out during maintenance operations.

10.4.3.3 **A Strainer (ST101)** is required by NFPA 86 (sect. 5-14.2) and design guidelines are provided in CGA G-4.4 (sect. 4.8). The purpose of the strainer is to capture particulates and prevent them from proceeding through the oxygen piping and flow components. A brass or monel strainer screen with a 30 to 100 mesh size is recommended.

10.4.3.4 **A "Blowdown" Drain Valve** off of the strainer (not shown) is not recommended for oxygen service per CGA G-4.4 (sect. 4.8.3). The purpose of a blowdown drain valve is to provide an easy method for cleaning the strainer. However, this valve could easily be opened at an inappropriate time and could cause a dangerous localized oxygen-enriched atmosphere when being used to clean the strainer.

10.4.3.5 **Pressure Indicator (PI102)** is optional. This pressure indicator is used to indicate the oxygen pressure supplied to the valve stand.

10.4.3.6 **A Pressure Regulator (PCV103)** is required by NFPA 86 (sect. 4-4.4.2) whenever the upstream oxygen supply pressure is subject to excessive fluctuations, or exceeds the required pressure for proper burner operation.

10.4.3.7 **A Pressure Relief Device** (not shown) is required by NFPA 86 (sect. 4-4.3.12) whenever it is possible to have an event leading to an oxygen supply pressure that is greater than the MAWP of any of the components or piping within the oxygen pipeline. The pressure relief device must be sized properly per manufacturer and industry standards. The outlet of the pressure relief device must be piped to a safe location per NFPA 86 (sect. 4-4.3.7).

10.4.3.8 **A Pressure Indicator (PI105)** is strongly recommended. This pressure indicator is used to indicate the oxygen metering pressure for the flow element. Many types of flow elements require a precise metering pressure for accurate flow measurement.

10.4.3.9 **A Flow Element (FE106 and PDT106)** is recommended to provide a measure of the oxygen flow rate to the burner. CGA G-4.4 (sect. 4.7 and sect. 5.4) provides general guidelines for flow element selection. Flow elements may include orifice plates (monel material is recommended); annubars; vortex meters; mass flow meters; and others, provided that all of the materials are oxygen compatible and dimensional requirements are satisfied. Glass tube flowmeters (rotameters) are not recommended. If used, safeguards against personnel injury must be provided per NFPA 86 (sect. 4-4.3.10).

10.4.3.10 **Low and High Oxygen Flow Limit Switches (FSL/H106)** are required per NFPA 86 (sect. 5-14.3 and sect. 5-14.4). The flow limit switches must be interlocked to close the automatic safety shutoff valves in the event of an unsafe oxygen flow rate. In Figure 10.3, the flow transmitter (PDT106) generates a differential pressure signal that is monitored by the flow limit switch device (FSL/H106).

10.4.3.11 **A Pressure Transmitter** (not shown) is an optional component. This component is used to compensate the computed oxygen flow rate for pressure fluctuations of the oxygen supply. The decision to use pressure compensation is a function of the required accuracy of the process.

10.4.3.12 **A Temperature Element** and **Temperature Transmitter** (not shown) are optional components. These components are used to compensate the computed oxygen flow rate for temperature fluctuations of the oxygen supply. The decision to use temperature compensation is also a function of the required accuracy of the process.

10.4.3.13 **A Safety Shutoff Valve (ABV107)** is required per NFPA 86 (sect. 5-7 and sect. 5-14.1). The safety shutoff valve must have position indication (e.g., "open" and "closed"), it cannot be a flow control valve, and it must fail to a closed position when any safety interlock is not satisfied. An air-operated ball valve is an acceptable safety shutoff valve per CGA G-4.4 (sect. 5.3.4). An instrument air three-way solenoid valve (XV107) with adjustable orifices is recommended to control the opening and closing speed of the ball valve. A second shutoff valve is required by NFPA 86 (sect. 5-7 and sect. 5-14.1) when the burner system has a heating capacity greater than 400,000 Btu/h (120 kW). At least one of the two safety shutoff valves must have a limit switch (or similar means) that can be used to prove that the valve is in a "closed" position during the furnace purge interval per NFPA 86 (sect. 5-4.1.2.1 and sect. 5.7.2.2).

10.4.3.14 **A Bleed Valve (XV108)** and corresponding piping is recommended to allow any oxygen that may be leaking past the upstream shutoff valve (ABV107) to vent to a safe location when the oxygen/fuel system is not in use. Bleed valve selection and vent piping design guidelines are provided in CGA G-4.4 (sect. 5.3.3). Note that the bleed valve and corresponding pipeline should not be considered a "vent" line. Therefore, the size of the bleed valve and piping should be kept to a minimum (¼ to ½ in. is recommended). The small size will tend to avoid excessive oxygen flow due to a failure of the bleed valve. The bleed valve, piping, and flow indicator (FI109) also provide a permanent and ready means for making a leak tightness check of the upstream safety shutoff valve (ABV107). An air-actuated ball valve (not shown) is acceptable per CGA G-4.4 (sect. 5.3.4). An instrument air three-way solenoid valve (XV107) with adjustable orifices is recommended to control the opening and closing speed of the ball valve. The bleed valve pipeline must be routed to a safe location per NFPA 86 (sect. 4-4.3.7).

10.4.3.15 **A Second Safety Shutoff Valve (ABV110)** is required per NFPA 86 (sect. 5-7 and sect. 5-14.1) when the burner system has a heating capacity equal to or exceeding 400,000 BTU/h (120 kW). The safety shutoff valve must have position indication (e.g., "open" and "closed"), it cannot be a flow control valve, and it must fail to a closed position when any safety interlock is not satisfied. An air-operated ball valve is an acceptable safety shutoff valve per CGA G-4.4 (sect. 5.3.4). An instrument air three-way solenoid valve (XV110) with adjustable orifices is recommended to control the opening and closing speed of the ball valve. At least one of the two safety shutoff valves must have a limit switch (or similar means) that can be used to prove that the valve is in a "closed" position during the furnace purge interval per NFPA 86 (sect. 5-4.1.2.1 and sect. 5.7.2.2).

10.4.3.16 **The Flow Control Valve (FCV111)** must be sized properly per NFPA 86 (sect. 4-4.4). Also, CGA G-4.4 (sect. 5.3.2) provides general guidelines for selecting appropriate flow control valves. The purpose of the flow control valve is to control the oxygen flow rate to a value based on the desired burner firing rate. In Figure 10.3, a hand-operated globe valve (FCV111) is used as the oxygen flow control valve. Other systems may include two parallel solenoid-actuated flow branches, or PID loop control. A typical PID loop control functions as follows. A controller sends an electrical signal to an I/P transducer which produces a pneumatic signal used to position an automatic control valve. The controller also accepts a feedback oxygen flow rate signal and positions the valve until the measured flow rate equals the desired set point value.

10.4.3.17 **Pressure Indicator (PI112)** is optional. This pressure indicator is used to indicate the oxygen pressure delivered from the valve stand.

10.4.3.18 **A Second Isolation Valve (BV113)** is optional. Design guidelines are provided in CGA G-4.4 (sect. 4.5.1 and sect. 5.3.1). The purpose of the second isolation valve is to provide a manual shutoff for maintenance operations. A quick-opening ball or butterfly valve is recommended, and it must be suitable for oxygen service. It is recommended that the valve be opened slowly to minimize adiabatic compression hazards.

The following are general guidelines associated with the oxygen flow control system:

10.4.3.19 Chapter 9 of this book and NFPA 86 (sect. A-4-4.1) provide general overviews of the hazards associated with oxygen-enriched burners and control systems.

10.4.3.20 The oxygen safety shutoff valves are required by NFPA 86 (sect. 5.14.5) to shut off the flow of oxygen in the event of any unsatisfied interlock (e.g., low or high flow of the fuel or oxygen, flame failure, high furnace temperature, low burner cooling water flow, high burner cooling water temperature, loss of furnace ventilation, etc.).

10.4.3.21 A permanent and ready means for making tightness checks of fuel gas shutoff valves is required by NFPA 86 (sect. 5-7.2.3). This practice is recommended for the oxygen safety shutoff valves as well.

10.4.3.22 **A Check Valve** (or similar device) must be installed in the oxygen line immediately upstream of the burner per NFPA 86 (sect. 4-4.3.8). The purpose of the check valve is to prevent backflow of any gases into the oxygen piping.

10.4.4 Fuel Gas Flow Run

Figure 10.4 illustrates a typical natural gas flow run used to control the flow of fuel gas to a burner. All topics below reference the tag numbers as shown in Figure 10.4 except as noted. The components that are required as well as some of those that are optional are described below. The requirements that are related to industry standards are based on an interpretation of those standards by Air Products and Chemicals Inc. All practical system designs and installations must be approved by the authority having jurisdiction.

10.4.4.1 In general, the fuel gas piping from the supply, through the valve stand, and to the burner shall comply with the industry standards for fuel gas service as summarized in the NFPA 54, "The National Fuel Gas Code."

10.4.4.2 **An Isolation Valve (BV200)** is required by NFPA 86 (sect. 4-2.4.1.1 and sect. 5-2.2). The purpose of the isolation valve is to provide a manual shutoff for maintenance and emergency conditions. Therefore, the valve must be located in a position where a hazardous condition will not impede access to the valve by plant personnel. A quarter-turn ball or butterfly valve with usual indication of valve position is required by NFPA 86 (sect. 4-2.4.1.2). The valve should include a locking feature so that the supply of fuel gas can be locked-out/tagged-out during maintenance operations.

10.4.4.3 **A Strainer (ST201)** is required by NFPA 86 (sect. 4-2.4.3). The purpose of the strainer is to capture particulates and prevent them from proceeding through the fuel gas piping and flow components. A stainless steel screen with a 30 to 100 mesh size is recommended.

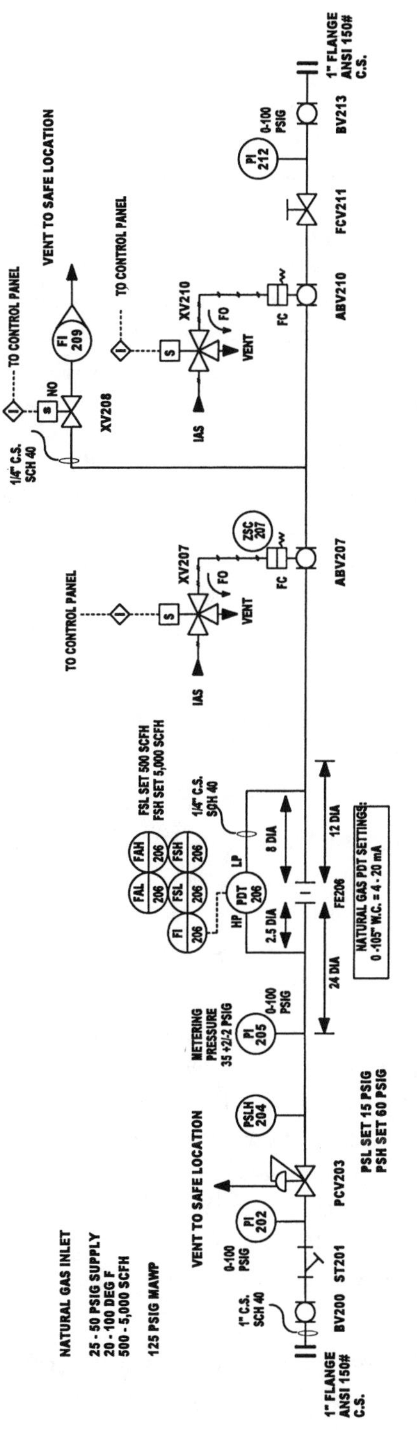

FIGURE 10.4 Typical fuel gas flow run for use in an oxygen/fuel combustion system.

10.4.4.4 **Pressure Indicator (PI202)** is optional. This pressure indicator is used to indicate the fuel gas pressure supplied to the valve stand.

10.4.4.5 **A Pressure Regulator (PCV203)** is required by NFPA 86 (sect. 4-2.4.4.1) whenever the upstream fuel gas supply pressure is subject to excessive fluctuations, or exceeds the required pressure for proper burner operation. Venting of the fuel gas pressure regulator is required under most circumstances as defined by NFPA 86 (sect. 4-2.4.4.2) and NFPA 54 (sect. 2.8.4).

10.4.4.6 **A Pressure Relief Device** (not shown) is required per NFPA 86 (sect. 5-7.1.7) and NFPA 54 (sect. 2.9) whenever it is possible to have an event leading to a fuel gas pressure that is greater than the MAWP of any of the components or piping within the fuel gas pipeline. The pressure relief device must be sized properly per manufacturer and industry standards.

10.4.4.7 **Low and High Fuel Gas Pressure Limit Switches (PSL/H204)** are required per NFPA 86 (sections 5-8.1, 5-8.2, and 5-8.3) to assure that the fuel gas is being supplied at a safe pressure for proper burner operation and flow measurement accuracy. Pressure limit switches can be easily misinterpreted; therefore, the use of low and high flow limit switches (FSL/H206) (which incorporate a direct measure of low and high differential pressure) provides a higher level of safety and therefore are recommended.

10.4.4.8 **Pressure Indicator (PI205)** is strongly recommended. This pressure indicator is used to indicate the fuel gas metering pressure for the flow element. Many types of flow elements require a precise metering pressure for accurate flow measurement.

10.4.4.9 **A Flow Element (FE206 and PDT206)** is recommended to provide a measure of the fuel gas flow rate to the burner. Flow elements may include orifice plates (stainless steel is recommended); annubars; vortex meters; mass flow meters; and others, provided that all of the materials are fuel gas compatible and dimensional requirements are satisfied.

10.4.4.10 **Low and High Fuel Gas Flow Limit Switches (FSL/H206)** have been used to satisfy the reuqirement of NFPA 86 (sections 5-8.1, 5-8.2, and 5-8.3) for low and high fuel gas pressure limit switches (because flow switches incorporate a direct measure of low and high differential pressure). This practice should be reviewed by the authority having jurisdication on a case-by-case basis. The flow limit switches must be interlocked to close the automatic safety shutoff valves in the event of an unsafe fuel gas flow rate. In Figure 10.4, the flow transmitter (PDT206) generates a differential pressure signal that is monitored by the flow switch interlock device (FSL/H206).

10.4.4.11 **A Pressure Transmitter** (not shown) is an optional component. This component is used to compensate the computed fuel gas flow rate for pressure fluctuations of the fuel gas supply. The decision to use pressure compensation is a function of the required accuracy of the process.

10.4.4.12 **A Temperature Element** and **Temperature Transmitter** (not shown) are optional components. These components are used to compensate the computed fuel gas flow rate for temperature fluctuations of the fuel gas supply. The decision to use temperature compensation is also a function of the required accuracy of the process.

10.4.4.13 **A Safety Shutoff Valve (ABV207)** is required per NFPA 86 (sect. 5-7.1 and sect. 5-7.2). The safety shutoff valve must have position indication (e.g., "open" and "closed"), it cannot be a flow control valve, and it must fail to a closed position when any safety interlock is not satisfied. An air-operated ball valve is an acceptable safety shutoff valve. An instrument air three-way solenoid valve (XV207) with adjustable orifices is recommended to control the opening and closing speed of the ball valve. A second safety shutoff valve is required by NFPA 86 (sect. 5-7.2.1) when the burner system has a heating capacity greater than 400,000 Btu/h (120 kW). At least one of the two safety shutoff valves must have a limit switch (or similar means) that can be used to prove that the valve is in a "closed" position during the furnace purge interval per NFPA 86 (sect. 5-4.1.2.1 and sect. 5-7.2.2).

10.4.4.14 **A Bleed Valve (XV208)** and corresponding piping is recommended to allow any fuel gas that may be leaking by the upstream safety shutoff valve (ABV207) to vent to a safe location when the oxygen/fuel system is not in use. Note that the bleed valve and corresponding pipeline should not be considered a "vent" line. Therefore, the size of the bleed valve and piping should be kept to a minimum (¼ to ½ in. is recommended). The small size will tend to avoid excessive fuel gas flow due to a failure of the bleed valve. The bleed valve, piping, and flow indicator (FI209) also provide a permanent and ready means for making a leak tightness check of the upstream safety shutoff valve (ABV207), which is required by NFPA 86 (sect. 4-7.2.3). The bleed valve pipeline must be routed to a safe location per NFPA 86 (sect. 4-2.4.4.2).

10.4.4.15 **A Second Safety Shutoff Valve (ABV210)** is required per NFPA 86 (sect. 5-7.2.1) when the burner system has a heating capacity equal to or exceeding 400,000 Btu/h (120 kW). The safety shutoff valve must have position indication (e.g., "open" and "closed"), it cannot be a flow control valve, and it must fail to a closed position when any safety interlock is not satisfied. An air-operated ball valve is an acceptable safety shutoff valve. An instrument air three-way solenoid valve (XV210) with adjustable orifices is recommended to control the opening and closing speed of the ball valve. At least one of the two safety shutoff valves must have a limit swtich (or similar means) that can be used to prove that the valve is in a "closed" position during the furnace purge interval per NFPA 86 (sect. 5-4.1.2.1 and sect. 5-7.2.2).

10.4.4.16 **The Flow Control Valve (FCV211)** must be sized properly per NFPA 86 (sect. 4-2.5). The purpose of the flow control valve is to control the fuel gas flow rate to a set point value based on the desired burner firing rate. In Figure 10.4, a hand-operated globe valve (FCV211) is used as a fuel gas flow control valve. Other systems may include two parallel solenoid-actuated flow branches, or PID loop control. A typical PID loop control functions as follows. A controller sends an electrical signal to an I/P transducer which produces a pneumatic signal used to position an automatic control valve. The controller accepts a feedback fuel gas flow rate signal and positions the valve until the measured flow rate equals the desired set point value.

10.4.4.17 **Pressure Indicator (PI212)** is optional. This pressure indicator is used to indicate the fuel gas pressure delivered from the valve stand.

10.4.4.18 **A Second Isolation Valve (BV213)** is optional. The purpose of the second isolation valve is to provide a manual shutoff for maintenance operations. A quick-opening ball or butterfly valve is recommended.

The following are general guidelines associated with the fuel gas flow control system:

10.4.4.19 The fuel gas safety shutoff valves are required by NFPA 86 (sect. 5-7.1.2) to shut off the flow of fuel gas in the event of any unsatisfied interlock (e.g., low or high flow of the fuel or oxygen, flame failure, high furnace temperature, low burner cooling water flow, loss of furnace ventilation, etc.).

10.4.4.20 A permanent and ready means for making tightness checks of all shutoff valves shall be provided per NFPA 86 (sect. 5-7.2.3).

10.4.4.21 **A Check Valve** (or similar device) must be installed in the fuel gas line immediately upstream of the burner per NFPA 86 (sect. 4-4.3.8). The purpose of the check valve is to prevent backflow of any gases into the fuel gas piping.

10.4.5 Fuel Oil Flow Run

Figure 10.5 illustrates a typical #6 fuel oil flow run used to control the flow of oil to a burner. All topics below reference the tag numbers as shown in Figure 10.5 except as noted. The components that are required as well as some of those that are optional are described below. The requirements that are related to industry standards are based on an interpretation of those standards by Air Products and Chemicals, Inc. All practical system designs and installations must be approved by the authority having jurisdiction.

10.4.5.1 In general, the oil piping from the supply, through the valve stand, and to the burner shall comply with the industry standards for oil service as summarized in NFPA 31, "Oil Burning Equipment."

10.4.5.2 **An Isolation Valve (BV400)** is required by NFPA 86 (sections 4-3.4.11, 4-3.4.12, and 5-2.2). The purpose of the isolation valve is to provide a manual shutoff

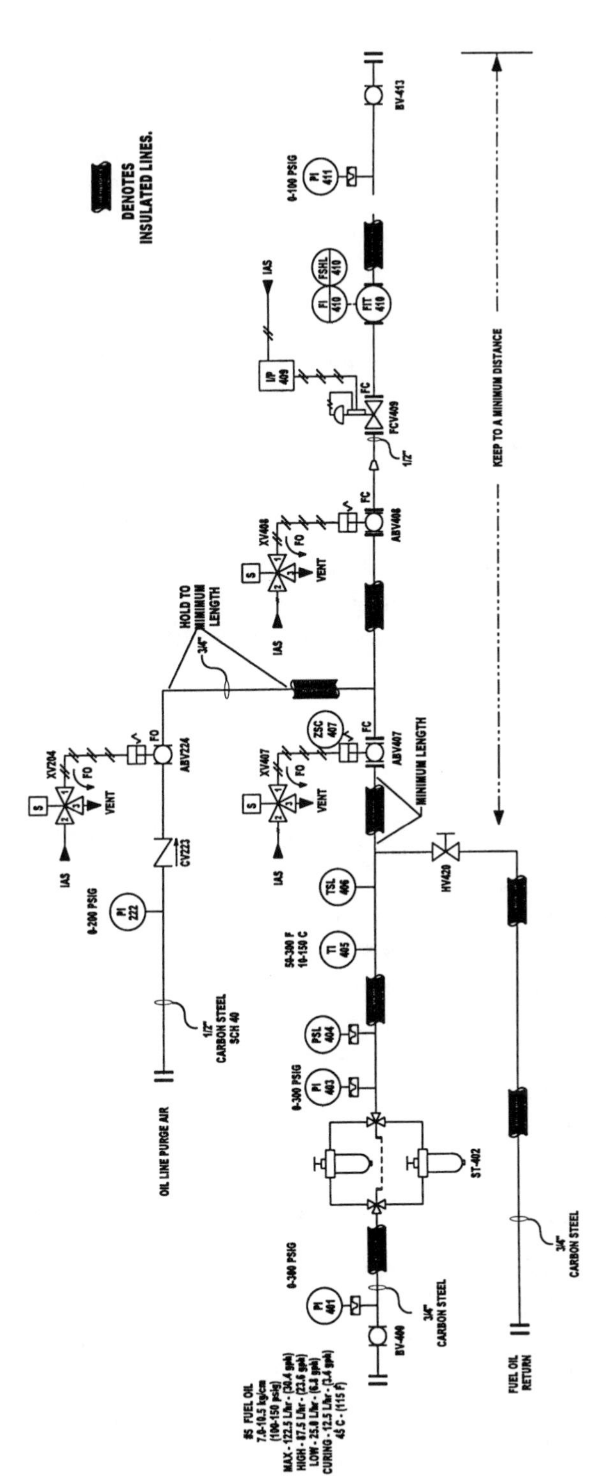

FIGURE 10.5 Typical heavy oil flow run for use in an oxygen/fuel combustion system.

for maintenance and emergency conditions. Therefore, the valve must be located in a position in which a hazardous condition will not impede access to the valve by plant personnel. A quarter-turn valve with visual indication of valve position is required by NFPA 86 (sect. 4.3.4.1.3). The valve should include a locking feature so that the supply of fuel oil can be locked-out/tagged-out during maintenance operations.

10.4.5.3 **Pressure Indicator (PI401)** is optional. This pressure indicator is used to indicate the oil pressure supplied to the valve stand.

10.4.5.4 **A Strainer/Filter (ST402)** is required by NFPA 86 (sect. 4-3.4.3). The purpose of the strainer/filter is to minimize any contaminants from proceeding through the fuel oil piping and flow components.

10.4.5.5 **Pressure Indicator (PI403)** is optional. This pressure indicator is used to indicate the fuel oil pressure drop across the strainer/filter (ST402). The pressure drop across the strainer/filter can indicate if cleaning maintenance is necessary.

10.4.5.6 **A Pressure Regulator** (not shown) is required by NFPA 86 (sect. 4-3.4.4) whenever the upstream fuel oil supply pressure is subject to excessive fluctuations, or exceeds the required pressure for proper burner operation.

10.4.5.7 **A Pressure Relief Device** and/or **Expansion Chamber** (not shown) is/are required per NFPA 86 (sect. 4-3.3.3 and sect. 4-3.3.6) whenever it is possible to have an event leading to a fuel oil supply pressure that is greater than the MAWP of any of the components or piping within the fuel oil pipeline or whenever the fuel oil can be trapped in the pipeline between two valves. The pressure relief device must be sized properly per manufacturer and industry standards.

10.4.5.8 **Low and High Fuel Oil Pressure Switches (PSL/H404)** are required per NFPA 86 (sections 5-8.1, 5-8.2, and 5-8.3) to assure that the fuel oil is being supplied at a safe pressure for proper burner operation and flow measurement accuracy. Pressure limit switches can be easily misinterpreted; therefore, the use of low and high flow limit switches (FSL/H306) (which incorporate a direct measure of low and high differential pressure) provides a higher level of safety and therefore are recommended.

10.4.5.9 **Temperature Indicator (TI405)** is optional. This temperature indicator is used to indicate the fuel oil temperature being supplied to the valve stand. Many types of flow elements require a precise oil temperature for accurate flow measurement.

10.4.5.10 **A Low Fuel Oil Temperature Limit Switch (TSL406)** is required per NFPA 86 (sect. 5-11) to assure that the temperature of the fuel oil is within the acceptable limits for fuel oil flow measurement instrumentation viscosity concerns, and for proper burner operation.

10.4.5.11 **A Safety Shutoff Valve (ABV407)** is required per NFPA 86 (sect. 5-7.1 and sect. 5-7.3). The safety shutoff valve must have position indication (e.g., "open" and "closed"), it cannot be a flow control valve, and it must fail to a closed position when any safety interlock is not satisfied. An air-operated ball valve is an acceptable safety shutoff valve. An instrument air three-way solenoid valve (XV407) with adjustable orifices is recommended to control the opening and closing speed of the ball valve.

10.4.5.12 **A Second Safety Shutoff Valve (ABV408)** is required per NFPA 86 (sect. 5-7.3.1) whenever (a) the oil pressure is greater than 125 psig; (b) the oil pump operates when the burner is not firing, regardless of the oil pressure; or (c) the oil pump operates during fuel gas burner operation. The safety shutoff valve must have position indication (e.g., "open" and "closed"), it cannot be a flow control valve, and it must fail to a closed position when any safety interlock is not satisfied. An air-operated ball valve is an acceptable safety shutoff valve. An instrument air three-way solenoid valve (XV408) with adjustable orifices is recommended to control the opening and closing speed of the ball valve. At least one of the two safety shutoff valves must have a limit switch (or similar means) that can be used to prove that the valve is in a "closed" position during the furnace purge interval per NFPA 86 (sect. 5-4.1.2.1 and sect. 5-7.3.2).

10.4.5.13 **The Flow Control Valve (FCV409)** must be sized properly per NFPA 86 (sect. 4-3.5). The purpose of the flow control valve is to control the fuel oil flow rate to a set point value based on the desired burner firing rate. In Figure 10.5, an automatic control valve (FCV409) is used as a fuel oil flow control valve. Other systems may include two parallel solenoid-actuated flow branches, or a manual hand valve. A typical PID loop control functions as follows. A controller sends an electrical signal to an I/P transducer which produces a pneumatic signal used to position an automatic control valve. The controller accepts a feedback fuel oil flow rate signal and positions the valve until the measured flow rate equals the desired set point value.

10.4.5.14 **A Flow Element (FIT410)** is recommended to provide a measure of the fuel oil flow rate to the burner. Flow elements may include positive displacement meters, magnetic meters, orifice plates (stainless steel is recommended), vortex meters, mass flow meters, and others, provided that all of the materials are oil compatible and dimensional requirements are satisfied.

10.4.5.15 **Low and High Fuel Oil Flow Limit Switches (FSL/H410)** have been used to satisfy the requirement of NFPA 86 (sections 5-8.1, 5-8.2, and 5-8.3) for low and high fuel oil pressure limit switches (because flow switches incorporate a direct measure of low and high differential pressure). This practice should be reviewed by the Authority having jurisdiction on a case-by-case basis. The flow limit switches must be interlocked to close the automatic safety shutoff valves in the event of an unsafe fuel oil flow rate. In Figure 10.5, the flow transmitter (FIT410) generates a signal, based upon a differential pressure, that is monitored by the flow switch interlock device (FSL/H410).

10.4.5.16 **A Temperature Element** and **Temperature Transmitter** (not shown) are optional components. These component are used to compensate the computed fuel oil flow rate for temperature fluctuations of the fuel gas supply. The decision to use temperature compensation is a function of the required accuracy of the process.

10.4.5.17 **Pressure Indicator (PI411)** is optional. This pressure indicator is used to indicate the fuel oil pressure delivered from the valve stand.

10.4.5.18 **A Second Isolation Valve (BV413)** is optional. The purpose of the second isolation valve is to provide a manual shutoff for maintenance operations. A quick-opening ball valve is recommended.

10.4.5.19 **A Fuel Oil Return Line** and **Hand Valve (HV420)** is recommended by NFPA 86 (sect. 4-3.3.4) as a design practice for eliminating air entrainment in the fuel oil line. The fuel oil return line is also used in heavy oil applications to ensure that the fluid is being supplied at an adequate temperature and viscosity.

The following are general guidelines associated with the fuel oil flow control system:

10.4.5.20 The fuel oil safety shutoff valves are required by NFPA 86 (sect. 5-7.1.2) to shut off the flow of fuel oil in the event of any unsatisfied interlock (e.g., low or high flow of the fuel or oxygen, flame failure, high furnace temperature, low burner cooling water flow, low fuel temperature, low atomization gas pressure, loss of furnace ventilation, etc.).

10.4.5.21 **An Atomizing Gas Low Pressure/Flow Limit Switch** is required by NFPA (sect. 5-10.1 and sect. 5-10.2). The purpose of an atomizing gas low pressure/flow limit switch is to inhibit the flow of fuel oil to the burner if the appropriate amount of atomizing gas is not available.

10.4.5.22 **A Check Valve** (or similar device) must be installed in both the fuel oil line and atomizing gas line immediately upstream of the burner per NFPA 86 (sect. 4-4.3.8). The purpose of the check valves is to prevent backflow of any gases into the fuel oil and atomizing gas piping.

10.4.6 Burner Equipment

Figure 10.6 illustrates a typical oxy/oil burner and associated equipment. All topics below reference the tag numbers as shown in Figure 10.6 except as noted. The components that are required are described below. The requirements that are related to industry standards are based on an interpretation of those standards by Air Products and Chemicals Inc.

10.4.6.1 In general, per NFPA 86 (sect. 4-2.7, 4-3.7, and 5-14.6), the oxy/fuel burner should be selected to be suitable for the type of fuel and oxidizer to be used, and for the pressures and temperatures to which they will be subjected.

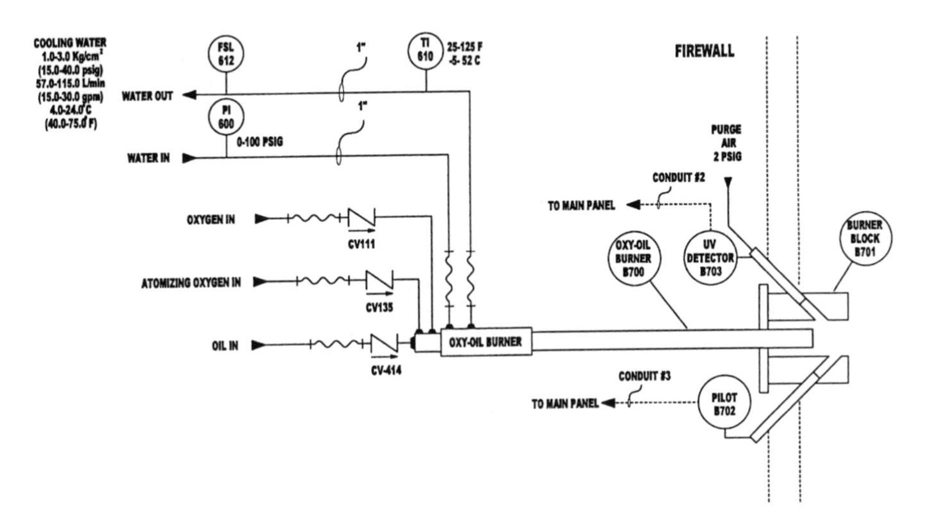

FIGURE 10.6 Typical oxy/oil burner equipment for use in an oxygen/fuel combustion system.

10.4.6.2 **A Check Valve (CV111)** (or similar device) must be installed in the oxygen line immediately upstream of the burner per NFPA 86 (sect. 4-4.3.8). The purpose of the check valve is to prevent backflow of fuel, air, or furnace gases through the oxygen piping.

10.4.6.3 **Check Valves (CV414 and CV135)** (or similar devices) must be installed in the fuel (oil) line and atomizing gas line immediately upstream of the burner per NFPA 86 (sect. 4-4.3.8). The purpose of the check valve is to prevent backflow of oxygen, air, or furnace gases through the fuel and atomizing gas piping.

10.4.6.4 **A Cooling Water Low Flow Interlock (FSL612)** is required by NFPA 86 (sect. 5-14.7) if the burner requires cooling water. The purpose of a cooling water low flow interlock is to protect the burner from damage due to overheating. The interlock must shut off the flow of fuel and oxygen to the burner whenever the water flow is below an acceptable minimum limit. The cooling water low flow interlock device must be located on the water exit line from the burner.

10.4.6.5 **A Cooling Water High Temperature Interlock** is recommended to protect the burner from damage due to overheating. This interlock should shut off the flow of fuel and oxygen to the burner whenever the water temperature is above an acceptable maximum limit. The cooling water high-temperature interlock device should be located on the water exit line from the burner.

10.4.7 Furnace Purge Requirements

Figure 10.7 illustrates a typical furnace purge forced-air equipment system. All topics below reference the tag numbers as shown in Figure 10.7 except as noted. The

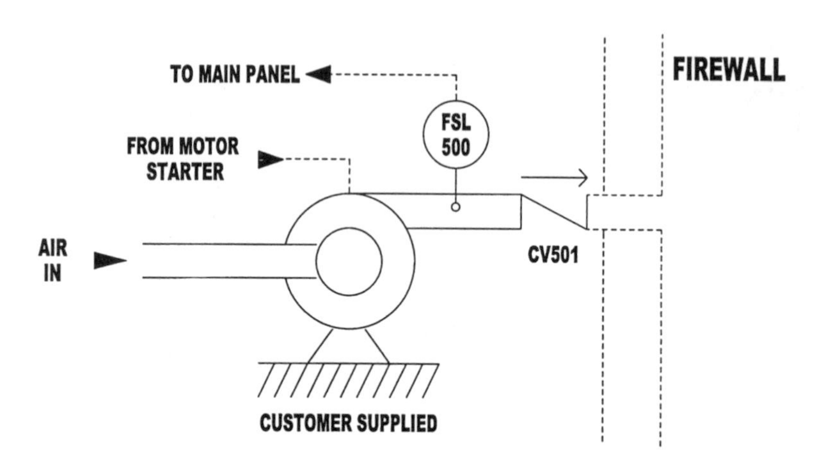

FIGURE 10.7 Typical forced-air purge equipment for use in an oxygen/fuel combustion system.

components that are required are described below. The requirements that are related to industry standards are based on an interpretation of those standards by Air Products and Chemicals, Inc.

10.4.7.1 In general, prior to each furnace ignition trial, provision shall be made for the removal of flammable vapors and/or gases that may have entered during the shut-down period, per NFPA 86 (sect. 5-4.1.1). In some cases, purging the furnace with a forced-air purge is unacceptable; therefore, an inert gas may be required.

10.4.7.2 At least four volume changes of purge gas shall be introduced during the purging cycle, per NFPA 86 (sect. 5-4.1.2).

10.4.7.3 The purging cycle must be timed, and must be interlocked to prevent the start-up of the oxygen/fuel combustion equipment, per NFPA 86 (sect. 5-4.1.2). The purging process is considered a critical safety interlock and therefore must be hard-wired to de-energize the safety shutoff valves per NFPA 86 (sect. 5-3.3).

10.4.7.4 **A Purge Gas Low Flow Limit Switch (FSL500)** or similar device is required by NFPA 86 (sect. 5-4.1.2) to prove that the purge gas is flowing at the minimum requirement during the timing of the purge cycle.

10.4.7.5 **A Check Valve (CV501)** must be installed in the purge gas line immediately upstream of the fire wall per NFPA 86 (sect. 4-4.3.8). The purpose of the check valve is to prevent backflow of any gases into the purge gas pipeline.

10.4.7.6 At least one of the two safety shutoff valves (on the fuel gas flow run, and on the oxygen flow run) must have a **Limit Switch (ZSC)** (or similar means) that can be used to prove that the valve is in a "closed" position during the furnace purge interval per NFPA 86 (sect. 5-4.1.2.1 and 5.7.2.2) when the burner system has a heating capacity equal to or exceeding 400,000 Btu/hr (120 kW). At least one of the two safety shutoff valves (on the fuel oil flow run) must have a **Limit Switch (ZSC)** (or similar means) that can be used to prove that the valve is in a "closed" position during the furnace purge interval per NFPA 86 (sect.5-4.1.2.1 and 5.7.3.2) whenever: (a) the fuel oil pressure is greater than 125 psig; (b) the fuel oil pump operates when the burner is not firing, regardless of the oil pressure; or (c) the fuel oil pump operates during fuel gas burner operation.

10.4.8 Burner Ignition

10.4.8.1 For burners that cannot be safely ignited at all firing rates encountered, positive provisions shall be made to reduce the burner firing rates during light-off to a level that will assure a smooth and reliable ignition of the main flame (forced low-fire start), per NFPA 86 (sections 4-2.7.3.2, 4-3.7.4.2, and 5-15.1).

10.4.8.2 In general, burners can be ignited by a manual torch; by a continuous, intermittent, or interrupted pilot burner; or by direct electrical means per NFPA 86 (sect. 4-2.8.1 and sect. 4-3.8.1). Burners may also be ignited by autoignition due to a furnace temperature that is above the autoignition temperature per NFPA 86 (sect. 5-9 and sect. 5-17). See Section 10.4.9 for more details.

10.4.8.3 Electrical ignition energy for direct spark ignition systems shall be terminated after the main burner trial-for-ignition period, per NFPA 86 (sect. 5-15.2).

10.4.8.4 Trial for ignition of pilots and main burners shall not exceed 15 seconds, per NFPA 86 (sect. 5-4.2).

10.4.8.5 Automatic relight after unintentional flame failure shall be prohibited, per NFPA 86 (sect. 5-2.3). Manual intervention by an operator is required to relight the burner system.

10.4.8.6 In general, prior to each furnace ignition trial, provisions shall be made for the removal of flammable vapors and/or gases that may have entered during the shutdown period per NFPA 86 (sect.5-4.1). See Section 10.4.7 for more details.

10.4.9 Flame Supervision

10.4.9.1 In general, each burner flame shall be supervised by a combustion safeguard, having a nominal flame response timing of 4 seconds or less, interlocked with the safety circuitry, per NFPA 86 (sect. 5-9.1).

10.4.9.2 Under certain circumstances, multiple burners in a single furnace zone may not require individual flame supervision, per NFPA 86 (sect. 5-9.2.2).

10.4.9.3 It is permissible to supervise flames at the intersection of the main burner flame and the pilot flame rather than providing supervision equipment for both, as long as the pilot flame is interrupted after ignition trial of the burner, per NFPA 86 (sect. 5-9.2.1).

10.4.9.4 It is permissible to switch the flame supervision out of the safety circuitry for a furnace zone when the zone temperature is at or above 1400°F (760°C), per NFPA 86 (sect. 5-9.1). Burners without flame supervision shall be interlocked to prevent their operation unless the furnace is at or above 1400°F (760°C). A 1400°F (760°C) bypass controller must be used for this purpose per NFPA 86 (sect. 5-17). The bypass controller and temperature-sensing element must be independent from any other controller or element. Failure of the element must cause the bypass controller to sense a temperature below 1400°F (760°C) and therefore shut off the combustion system. Visual indication must be provided to indicate that the bypass controller action is in effect.

10.4.9.5 Trial for ignition of pilots and main burners shall not exceed 15 seconds, per NFPA 86 (sect. 5-4.2).

10.4.9.6 Automatic relight after unintentional flame failure shall be prohibited, per NFPA 86 (sect. 5-2.3).

10.4.9.7 In general, prior to each furnace ignition trial, provision shall be made for the removal of flammable vapors and/or gases that may have entered during the shutdown period, per NFPA 86 (sect. 5-4.1).

10.4.9.8 Electric arc furnaces have different requirements for flame supervision (see sect. 9-2.6 and 9-2.7).

10.4.10 Electrical Equipment and Wiring

10.4.10.1 In general, all electrical wiring, and applicable electrical components, shall be in accordance with NFPA 70, the National Electric Code.

10.4.10.2 An audible and/or visible alarm is recommended in the safety circuit to give warning of unsafe conditions or interruption of the safety circuit.

10.4.10.3 A safety shutdown of the oxy/fuel combustion equipment system by any of the prescribed safety features or devices shall require manual intervention of an operator for reestablishment of normal operation of the system, per NFPA 86 (sect. 5-2.3).

10.4.10.4 Jumpering of any safety interlock circuitry or improper adjustment of any safety interlock circuitry for sustaining production is prohibited, per NFPA 86 (sect. 5-2.7 and sect. 5-2.8).

10.4.10.5 Requirements for the use of programmable logic controllers (PLCs) are defined in NFPA 86 (sect. 5-3). All critical protection interlocks (e.g., combustion safety interlocks, flame supervision, high and low oxygen and fuel flow interlocks, excessive temperature limits, etc.) shall be wired to de-energize the safety shutoff valves directly, and their operations shall result in a safe system condition.

10.4.10.6 Typically, an oxy/fuel combustion equipment system is considered to be a Class 1, Division 2, Group D electrical installation per NFPA 70 (articles 500, 501, and 502). Note that the Group is defined as a function of the flammable fuel being handled. See NFPA 497A for help in determining the classification of a particular oxy/fuel combustion application. All components, enclosures, conduits, and wires must be suited for the particular classification of the oxy/fuel equipment system electrical installation.

10.4.10.7 For Class 1, Division 2 electrical installations, any enclosure and/or component (within 5 ft of the flammable fuel pipeline) that contains an internal spark ignition source (relay contact, switch contact, motor, etc.) must be explosion proof, be constructed using hermetically sealed switches, or must include a type Z purge. Note that solenoid valve coils, transmitters, and I/P transducers are typically not considered spark ignition sources. A type Z purge (see NFPA 496 for details) includes an uninterrupted air supply (or an inert gas supply), a positive pressure or flow indication device, blowout plugs, and poured fittings for all wire conduits attached to the enclosure.

10.4.11 Miscellaneous Concerns

10.4.11.1 All safety devices (e.g., safety shutoff valves, pressure switches, flame supervision equipment) shall be listed for the service intended per NFPA 86 (sect. 5-2.1). Safety devices shall be applied and installed with established safe practices and manufacturer's instructions. When listed devices are not available for the service intended, the selected device shall require approval by the authority having jurisdiction.

10.4.11.2 **A High Furnace Temperature Interlock** is required by NFPA 86 (sect. 5-16) whenever it is possible for the controlled temperature to exceed a safe limit. The high-temperature interlock device and associated equipment including thermocouple must be independent of the equipment used to control the temperature of the furnace. The high furnace temperature interlock is considered a critical safety interlock.

10.4.11.3 Every oxygen/fuel combustion system must be carefully hazard-reviewed. This involves examining the burner(s), the fuel and oxygen delivery systems, the electrical systems, the control systems, the overall heating process, the location of the installation, operating procedures, the interactions of the personnel who will be operating the equipment, and any other relevant factors which could impact on the safety of the overall system. HAZOP, what-if, and fault tree analyses are standard methods commonly used to assess the hazards and safeguards of a proposed system.

10.4.11.4 **An Operating and Maintenance Manual** is required for every oxy/fuel combustion equipment system per NFPA 86 (sect. 1-5.6). As a minimum, the manual must include oxygen and fuel safety literature; appropriate approved drawings of the system; ORI (operation readiness inspection) procedures; start-up and shutdown procedures; emergency procedures; and maintenance and inspection procedures.

10.4.11.5 Recommended maintenance and inspection procedures for the safety devices must be provided per NFPA 86 (sect. 5-2.4 and Chapter 10). It shall be the responsibility of the user to establish, schedule, and enforce the frequency of, and the extent of, the maintenance and inspection program. Most of the critical safety devices are passive devices (they only react under abnormal conditions); therefore, they must be inspected frequently to prove that they are operational. Appendix B of NFPA 86 entitled, "Example of Operational and Maintenance Checklist," provides examples of various checklists and inspection/maintenance procedures.

10.4.11.6 **Operator and Maintenance Personnel Training** is required by NFPA 86 (sect. 1-5). A training program must include an initial training session and regularly scheduled retraining and testing sessions. Operator and maintenance personnel must have access to the operating and maintenance manuals at all times.

10.4.11.7 Chapter 9 and Appendix A-4-4.1 of NFPA 86 discuss special considerations for oxy/fuel combustion equipment systems associated with electric arc furnaces used to make steel. In summary, the oxy/fuel combustion equipment must be interlocked to the arc furnace controls to prevent operation of the burners until the proper time in the sequence of operation of the furnace or when an unsafe furnace condition occurs. The oxy/fuel combustion equipment cannot be operated until there is proof that enough electrical current is flowing to maintain a strong arc in the furnace or until there is proof that the furnace temperature is above 1400°F (760°C). Per NFPA 86, "operation of a burner shall not be required to be halted in the event of a momentary interruption of the arc, nor after arc heating has been intentionally discontinued, provided the contents of the furnace are incandescent or determined to be at a temperature in excess of 1400°F (760°C). The arc, hot furnace walls, and molten metal close to the burner outlets may be considered dependable ignition sources." Oxy/fuel burners installed on an electric arc furnace are exempt from the requirements of burner ignition pilots and ignitors and flame supervision.

REFERENCES

1. CGA Pamphlet P-14, Accident Prevention in Oxygen-Rich and Oxygen-Deficient Atmospheres, Compressed Gas Association, Arlington, VA, 1992.
2. CGA G-4, Oxygen, Compressed Gas Association, Arlington, VA, 1996.
3. CGA G-4.1, Cleaning Equipment for Oxygen Service, Compressed Gas Association, Arlington, VA, 1996.
4. CGA G-4.3, Commodity Specification for Oxygen, Compressed Gas Association, Arlington, VA, 1994.
5. CGA G-4.4, Industrial Practices for Gaseous Oxygen Transmission and Distribution Piping Systems, Compressed Gas Association, Arlington, VA, 1993.

6. CGA O_2-DIR, 1997 Directory of Cleaning Agents for Oxygen Service, Compressed Gas Association, Arlington, VA, 1996.
7. NFPA 31, Standard for the Installation of Oil Burning Equipment, National Fire and Protection Agency, Quincy, MA, 1997.
8. NFPA 53, Fire Hazards in Oxygen-Enriched Atmospheres, National Fire and Protection Agency, Quincy, MA, 1994.
9. NFPA 54, National Fuel Gas Code, National Fire and Protection Agency, Quincy, MA, 1992.
10. NFPA 69, Explosion Prevention Systems, National Fire and Protection Agency, Quincy, MA, 1997.
11. NFPA 70, National Electric Code, National Fire and Protection Agency, Quincy, MA, 1996.
12. NFPA 86, Ovens and Furnaces, National Fire and Protection Agency, Quincy, MA, 1995.
13. NFPA 496, Purged and Pressurized Enclosures for Electrical Equipment, National Fire and Protection Agency, Quincy, MA, 1993.
14. NFPA 497A, Classification of Class I Hazardous (Classified) Locations for Electrical Installations in Chemical Process Areas, National Fire and Protection Agency, Quincy, MA, 1997.
15. ASTM Designation D396, Specifications for Fuel Oils, American Society for Testing and Materials, Philadelphia, 1989.
16. ASTM Designation D2863-87, Standard Test Method for Measuring the Minimum Oxygen Concentration to Support Candle-like Combustion of Plastics (Oxygen Index), American Society for Testing and Materials, Philadelphia, 1987.
17. ASTM Designation D4809-90, Standard Test Method for Heat of Combustion of Liquid Hydrocarbon Fuels by Bomb Calorimeter (Intermediate Precision Method), American Society for Testing and Materials, Philadelphia, 1990.
18. ASTM Designation G63-87, Standard Guide for Evaluating Nonmetallic Materials for Oxygen Service, American Society for Testing and Materials, Philadelphia, 1987.
19. ASTM Designation G88-84, Standard Guide for Designing Systems for Oxygen Service, American Society for Testing and Materials, Philadelphia, 1984.
20. ASTM Designation G93-88, Standard Practice for Cleaning Methods for Material and Equipment Used in Oxygen-Enriched Environments, American Society for Testing and Materials, Philadelphia, 1988.
21. ASTM Designation G94-88, Standard Guide for Evaluating Metals for Oxygen Service, American Society for Testing and Materials, Philadelphia, 1988.
22. ASTM Designation G125-95, Standard Test Method for Measuring Liquid and Solid Material Fire Limits in Gaseous Oxidants, American Society for Testing and Materials, Philadelphia, 1995.
23. ANSI/ASME B31.3, Code for Pressure Piping, Chemical Plant and Petroleum Refinery Piping, American National Standards Institute, New York, 1990.
24. IHEA, *Combustion Technology Manual*, 4th ed., Industrial Heating Equipment Association, New York, 1988.
25. Matheson, *Gas Data Book*, 6th ed., Lyndhurst, NJ, 1980.
26. Reed, R. J., *North American Combustion Handbook*, 3rd ed., Vol. 1, North American Manufacturing Company, Cleveland, OH, 1986.

11 Fuels

Yanping Zhang

CONTENTS

11.1 INTRODUCTION

Fossil fuels are commonly used to react with oxygen to produce high-temperature combustion flue gases to heat loads in most industrial heating processes, such as metal heating and melting, glass melting, calcining, and coal and rubber gasification. The hot flue gases transfer their sensible energy to heat sinks by radiation, convection, and conduction. Of these, radiation heat transfer, which is dependent on the fourth power of the temperature, is dominant in a high-temperature furnace (>2700°F or 1500°C). Detailed discussions of the heat transfer between a flame and the charge are given in Chapter 4. There are three types of fossil fuels: gaseous (natural gas and synthetic fuels), liquid (gasoline, light and heavy fuel oil, synthetic fuels, and waste liquids), and solid (coals and wastes), which are used in the high-temperature processes. A fuel could also consist of a mixture of more than one type.

0-8493-1695-2/98/$0.00+$.50

The chemical compositions of fossil fuels mainly consist of carbon, hydrogen, nitrogen, sulfur, oxygen, and other trace elements. When fossil fuels are combusted with an adequate amount of oxygen, the carbon is converted to carbon dioxide, the hydrogen is converted to water, the nitrogen partly forms nitrogen oxides, the sulfur is converted into sulfur dioxide, and the oxygen reacts with the carbon, sulfur, and hydrogen in the fuels to reduce the amount of oxygen to be supplied. Thus, we have

$$C + O_2 \rightarrow CO_2$$

$$H_2 + O_2 \rightarrow 2H_2O$$

$$N_2 + O_2 \rightarrow 2NO$$

$$N_2 + 2O_2 \rightarrow 2NO_2$$

$$S + O_2 \rightarrow SO_2$$

These are the primary combustion products when fossil fuels are burned in a high-temperature furnace. Reactions of carbon and hydrogen with oxygen provide a high-temperature process with heat. Nitrogen and sulfur, however, react with oxygen to form undesired nitrogen oxide and sulfur oxide compounds. Inorganic minerals and some organic compounds will generate fly ash (particulate) and other organic materials in the flue gases if solid fuels are involved in the combustion.

There are, conventionally, three ways to control emissions in a high-temperature furnace. One is to clean fuels by removal of the unwanted elements. The second is to control the combustion process properly by designing appropriate burners to minimize pollutant generation. The third is combustion flue gas cleanup. Oxygen-enhanced combustion (OEC) is a very effective way to decrease the amount of nitrogen molecules involved in a combustion process. Environmental regulations and economics are driving forces when deciding which methods or combinations of methods to use.

It is very important to understand the fuel properties when OEC is involved in a high-temperature process since the characteristics of OEC are very different in terms of the reaction intensity, the equilibrium temperature, and flame patterns, as well as pollutant emissions. Therefore, the properties of the fuels commonly fired in different industries and some basic concepts applied to combustion of each fuel will be presented in this chapter. Understanding them will enable combustion engineers and scientists to optimize a process with a desirable flame and heat release pattern, low NOx, CO, and other pollutant emissions, and safe and stable operation with minimum requirement of maintenance.

11.2 COMBUSTION REACTIONS

Combustion is an intensive oxidation reaction of combustible compounds, resulting in high temperatures, heat release, and strong radiation. As discussed in Chapter 1, the rate of the OEC reaction increases significantly due to the higher partial pressures for both oxygen and fuel which leads to higher equilibrium temperature.

11.2.1 Enthalpy of Formation of a Compound

The enthalpy of formation of a compound is the energy released (or needed) by a reaction of natural elements to form the compound. It is conventionally defined as the energy change of the reaction by generating 1 mol of the compound as the standard enthalpy of formation of the compound when the reaction takes place at 1 atm, and a standard temperature of 298 K (77°F). The standard enthalpy of formation is commonly written as Δh^0_{f298}, where "0" and "298" refer to 1 atm and the standard temperature. The enthalpies of formation of some compounds commonly encountered in combustion processes are shown in Table 11.1.[1,2]

The standard enthalpy of formation of carbon dioxide is described by the following formula:

$$C + O_2 \rightarrow CO_2, \quad \Delta h^0_{f298} = -94.05 \text{ kcal}/[\text{molCO}_2]$$

It is important to note that the energy released by the reaction, $CO + O_2 \rightarrow CO_2$, $\Delta h^0_{R298} = -67.63$ kcal/[mol CO], of carbon monoxide and oxygen to form carbon dioxide at 1 atm and the standard temperature is not the standard enthalpy of formation of carbon dioxide, since the reactant, CO, is not a natural element.

11.2.2 Reaction Enthalpy

The enthalpy of a reaction is defined as heat released or adsorbed when several reactants (compounds or elements) react to form products. If the reaction involving components 1 through j occurs, the general form may be expressed as

$$\sum_{j=1}^{i} \nu_j C_j = 0 \tag{11.1}$$

Equation 11.1 is an atom balance; C_j could be considered the chemical formula for j and ν_j the molar stoichiometric coefficient. For reactants, ν_j is always defined as a negative number and, for products, a positive number.

If the reaction is stoichiometrically balanced, there is no net change in mass before and after reaction; therefore,

$$\Sigma m_j \nu_j = 0 \tag{11.2}$$

where m_j is the molecular weight of j. If the extent, $d\psi$, of reaction is introduced,

$$d\psi = \frac{dN_j}{\nu_j}, \quad j = 1, 2, \ldots, 1$$

TABLE 11.1
Enthalpy of Formation (Δh^0_{f298}) and Combustion Heat ($-\Delta h^0_{R298}$) for Some Compounds (kcal/g mol at 1 atm and 25°C); H_2O (Liquid) and CO_2 (Gas) as Combustion Products

Compound	Formula	State	Δh^0_{f298}	$-\Delta h^0_{R298}$	Compound	Formula	State	Δh^0_{f298}	$-\Delta h^0_{R298}$
Carbon monoxide	CO	Gas	–26.42	67.64	Hexene	C_6H_{12}	Gas	–9.96	964.24
Carbon dioxide	CO_2	Gas	–94.05	—	One-carbon increase	—	Gas	Adds –4.93	Adds 157.44
Methane	CH_4	Gas	–17.89	212.8	Oxygen	O_2	Gas	0	—
Ethane	C_2H_6	Gas	–20.24	372.82	Nitrogen	N_2	Gas	0	—
Propane	C_3H_8	Gas	–24.82	530.60	Carbon	C	Graphite	0	94.05
Butane	C_4H_{10}	Gas	–30.15	687.64	Carbon	C	Diamond	0.45	—
Pentane	C_5H_{12}	Gas	–35.00	845.16	Water	H_2O	Gas	–57.80	—
Hexane	C_6H_{14}	Gas	–39.96	1002.57	Water	H_2O	Liquid	–68.32	—
One-carbon increases	C_nH_{2n+2}	Gas	Adds –4.93	Adds 157.44	Sulfur dioxide	SO_2	Gas	–70.96	—
Ethylene	C_2H_4	Gas	12.50	337.15	Sulfuric oxide	SO_3	Gas	–104.80	—
Propene	C_3H_6	Gas	4.88	491.99	Nitric oxide	NO	Gas	21.57	—
Butene	C_4H_8	Gas	–0.03	649.38	Nitrogen oxide	NO_2	Gas	7.93	—
Pentene	C_5H_{10}	Gas	–5.00	806.70	Nitrous oxide	N_2O	Gas	19.51	—

where N_j is the molar number of compound j. Therefore, all dN_j for reacting species may be expressed in terms of $d\psi$. The remaining reactant, j, is $N_j = N_{j0} + \nu_j\, d\psi$ after the reaction.

The reaction enthalpy of the reaction is calculated by the difference of enthalpies of the reactants and products,

$$\Delta H^0_{RT} = \sum_P m_P \Delta h^0_{fT_P} - \sum_R m_R \Delta h^0_{fT_R} \tag{11.3}$$

where ΔH^0_{RT} is the reaction enthalpy at temperature, T, and 1 atm, m_P and m_R are the molar numbers of the products, P, and reactants, R. The combustion heats for some typical hydrocarbons and CO are shown in Table 11.1. The following examples show how to use the above concepts to compute the reaction enthalpy, or combustion heat.

Example 11.1: Calculate the enthalpy for the reaction: $CH_4 + 2O_2 \rightarrow CO_2 + 2H_2O$ at 25°C (77°F) and 1 atm.

Total enthalpy of formation of the reactants: $1 \times (-17.9) + 2 \times (0.0) = -17.9$ kcal;
Total enthalpy of formation of the products: $1 \times (-94.0) + 2 \times (-68.3) = -230.6$ kcal;
The reaction enthalpy is $-230.6 - (-17.9) = -212.7$ kcal.

It is not uncommon that the enthalpies of formation of some reactants or products are unknown or not readily available; therefore, the enthalpy of reaction cannot be calculated using the above method. However, we can estimate it based on the bonding energies that a compound contains. A certain amount of energy, which is called as bonding energy, is needed to break a chemical bond between two atoms in a molecule. Conversely, a certain amount of energy will be released when two atoms combine to form a chemical bond. The enthalpy of the reaction is roughly equal to the difference of the energy needed to break the bond and the energy released to form the bond. Table 11.2 shows the bonding energies for chemical compounds commonly found in combustion engineering.[3]

TABLE 11.2
Bonding Energy for Typical Bonds (kcal/mol)

Bond	Energy	Bond	Energy	Bond	Energy	Bond	Energy
C–C	85	C–O	86	C–Cl	78	N=N	60
C=C	143	C=O	173	C–S	64	N≡N	225
C≡C	198	C–N	81	O–O	33	H–H	103
C–H	98	C≡N	210	O=O	117	O–H	109
O–N	150	N–H	88	S–S	50	H–S	81

Example 11.2: Calculate enthalpy of reaction for CH_4 + $2O_2$ to form CO_2 + $2H_2O$ based on bonding energy data shown in Table 11.2.

In the reaction, four C–H bonds and two O–O bonds are broken; then the atoms form two C–O bonds and four O–H bonds. The energy needed to break the C–H and O–O bonds is $(4 \times 98 + 2 \times 33) = 458$ kcal/mol, and the total energy released to form the C–O and O–H bonds $(2 \times 86 + 4 \times 109) = 608$ kcal/mol. Hence, the enthalpy of the reaction is $458 - 608 = -150$ kcal/mol. Comparing this result with that in Example 11.1, we find that the value is an approximate estimation.

It is very useful to adopt the concept of the bonding energy to calculate the enthalpy of reaction when no information on enthalpy of formation for a compound is available, even though the method is approximate.

In the above example, the reaction enthalpy is calculated at 25°C (77°F) and 1 atm. It is usually called the standard enthalpy of reaction. In the most cases, the reaction is not at standard temperature and pressure (STP: 25°C and 1 atm). The reaction enthalpy of a reaction at any temperature and pressure for a reaction is still calculated by the enthalpy difference before and after reaction,

$$\Delta H_{PR} = \sum_P H_P - \sum_R H_R \tag{11.4}$$

where P and R represent products and reactants. The change of ΔH_{PR} with temperature at constant pressure is

$$\left(\frac{d\Delta H_{PR}}{dT}\right)_p = \left(\sum_P \frac{dH_P}{dT} - \sum_R \frac{dH_R}{dT}\right)_p \tag{11.5}$$

where subscript p presents a constant pressure. From the definition of heat capacity at constant pressure, we have the following equation:

$$\left(\frac{d\Delta H_{PR}}{dT}\right)_p = \left(\sum_P m_P C_{pP} - \sum_R m_R C_{pR}\right) \tag{11.6}$$

By integrating the above equation from 298 K (25°C) to a temperature, T, a conveniently used formula is obtained:

$$\Delta H_{PR} = \int_{298}^{T} \left(\sum_P m_P C_{pP} - \sum_R m_R C_{pR}\right) dT + \Delta H_{298}^0 \tag{11.7}$$

The above relationship is usually called Kirchoff's law. For an ideal gas, its heat capacity, C_p, is not a function of pressure and only changes with temperature. The dependence of the heat capacity on temperature is empirically written as

$$C_p = a + \beta T + \gamma T^2 \quad \text{or} \quad C_p = a + bT + cT^{-2} \tag{11.8}$$

TABLE 11.3A
Specific Heat for Some Organic Compounds

Compound	Formula	α	β × 10³	γ × 10⁶
Methane	CH_4	3.381	18.044	−4.300
Ethane	C_2H_6	2.247	38.201	−11.049
Propane	C_3H_8	2.410	57.195	−17.533
Butane	C_4H_{10}	3.844	73.350	−22.655
Pentane	C_5H_{12}	4.895	90.113	−28.039
Hexane	C_6H_{14}	6.011	106.746	−33.363
Ethylene	C_2H_4	2.830	28.601	−8.726
Propene	C_3H_6	3.253	45.116	−13.74
Butene	C_4H_8	3.909	62.848	−19.617
Acetylene	C_2H_2	7.331	12.622	−3.889

Parameters in $C_P = \alpha + \beta T + \gamma T^2$, where units of C_p and T are Btu/(lb mol °F) and K. Equation is good for T from 298 to 1500 K (77 to 2200°F).

TABLE 11.3B
Specific Heat for Some Inorganic Compounds

Compounds	Formula	Approx. Temperature Range (K)	*a*	*b* × 10³	*c* × 10⁻⁵
Carbon monoxide	CO	298–2500	6.79	0.98	−0.11
Carbon dioxide	CO_2	298–2500	10.57	2.10	−2.06
Hydrogen	H_2	298–3000	6.52	0.78	+0.12
Hydrogen sulfate	H_2S	298–2300	7.81	2.69	−0.46
Nitrogen	N_2	298–3000	6.83	0.90	−0.12
Oxygen	O_2	298–3000	7.16	1.00	−0.40
Nitrous oxide	N_2O	298–2000	10.92	2.06	−2.04
Nitric oxide	NO	298–2500	7.03	0.92	−0.14
Nitrogen oxide	NO_2	298–2000	10.07	2.28	−1.67
Sulfur oxide	SO_2	298–2000	11.04	1.88	−1.84
Water	H_2O	298–2750	7.30	2.46	0.00

Parameters in $C_P = a + bT + cT^{-2}$, where units of C_p and T are Btu/(lb mol °F) and K.

The coefficients for some gaseous compound in the above equations are listed in Table 11.3.[4,5] It is found that the heat capacity of a real gas is almost the same as its ideal state if the pressure is not too high. Therefore, it is proper to use the heat capacity of an ideal gas for a practical gas when the pressure is low. The approximate method to calculate the heat capacity can also be used if empirical data is not readily available.[6]

Example 11.3: Methane is delivered at 298 K (77°F) to a glass plant that operates a melting furnace at 1600 K (2420°F). The fuel is mixed with a quantity of air, which is 10% in excess of the amount theoretically needed for complete combustion, at 500 K (440°F). Assume that air is approximately 21% O_2 and 79% N_2 and that the products are CO_2 and H_2O vapor.

1. Assuming complete combustion, what is the composition of the flue gas?
2. The furnace processes 2000 kg (4400 lb) of glass per hour and its heat losses to the surroundings average 400,000 kJ/h (380,000 Btu/h). Calculate the fuel consumption at STP (in m^3/h) assuming that for the glass $h_{1600K} - h^0_{298K} = 1200$ kJ/kg.
3. If oxygen is used to replace air, recalculate the fuel consumption.

Data:

$$\Delta h^0_{R298}(CH_4) = -74.9 \text{ kJ/mol}$$
$$\Delta h^0_{P298}(CO_2) = -393.5 \text{ kJ/mol}$$
$$\Delta h^0_{P298}(H_2O) = -241.8 \text{ kJ/mol}$$
$$C_p(CH_4) = 42 \text{ J/K-mol}$$
$$C_p(CO_2) = 57.3 \text{ J/K-mol}$$
$$C_p(H_2O) = 49.8 \text{ J/K-mol}$$
$$C_p(O_2) = 34.3 \text{ J/K-mol}$$
$$C_p(N_2) = 34.3 \text{ J/K-mol}$$

From the reaction $CH_4 + 2O_2 = CO_2 + 2H_2O$, we know that a complete combustion of 1 mol of methane needs 2 mol of oxygen and produces 1 mol of carbon dioxide and 2 mol of water. In the problem, 10% excess air is supplied to the furnace; therefore, the above reaction can be written as

$$CH_4 + (1.0 + 10\%)\left(2O_2 + 2x\frac{79}{21}N_2\right) = CO_2 + 2H_2O + 0.2O_2 + 8.276N_2$$

1. The composition of flue gas is

$$CO_2\text{: } 1/(1 + 2 + 0.2 + 8.276) \times 100\% = 8.71\%$$

$$H_2O\text{: } 2/(1 + 2 + 0.2 + 8.276) \times 100\% = 17.42\%$$

$$O_2\text{: } 0.2/(1 + 2 + 0.2 + 8.276) \times 100\% = 1.74\%$$

$$N_2\text{: } 8.276/(1 + 2 + 0.2 + 8.276) \times 100\% = 72.11\%$$

2. Assuming the flue gas temperature is 1550 K (2330°F), that is, 50 K (90°F) lower than the furnace temperature. If X moles of fuel are burned every hour, $2.2X$ moles of O_2 and $8.276X$ moles of N_2 are needed, and X moles of CO_2, $2X$ moles of H_2O are produced. There are $0.2X$ moles

of oxygen and $8.276X$ moles of N_2 in the flue gas. By applying the first law of thermodynamics to the furnace system, the energy balance is

$$\sum_R H_R - \sum_P H_P + Q = 0$$

The total enthalpy to the furnace is

$$\sum_R H_R = X\left\{\Delta H^0_{R298}(CH_4) + 2.2 \int_{298}^{500} C_p(O_2)\,dT + 8.276 \int_{298}^{500} C_p(N_2)\,dT\right\}$$

$$= X\{-74.9 + 2.2 \times 34.3 \times 10^{-3}(500-298) + 8.276 \times 34.3 \times 10^{-3}(500-298)\}$$

$$= -2.316X \text{ kJ/h}$$

The total enthalpy out of the furnace is

$$\sum_P H_P = X\left\{\Delta H^0_{P298}(CO_2) + \int_{298}^{1550} C_p(CO_2)\,dT + 2\left[\Delta h^0_{P298}(H_2O) + \int_{298}^{1550} C_p(H_2O)\,dT\right]\right.$$

$$\left. + 0.2\int_{298}^{1550} C_p(O_2)\,dT + 8.276\int_{298}^{1550} C_p(N_2)\,dT\right\}$$

$$= X\{-393.5 + 2(-241.8) + [57.3 + 2 \times 49.8 + (0.2 + 8.276) \times 34.3] \times 10^{-3}$$

$$\times (1550-298)\}$$

$$= -318.749X \text{ kJ/h}$$

The total heat transfer to the glass and the surroundings is

$$Q = -\left[2000 \times \left(h_{1600K} - h^0_{298K}\right)_{glass} + 400000\right] = -[2000 \times 1200 + 400000]$$

$$= -2.8 \times 10^6 \text{ kJ/h}$$

Substituting the above data into the energy balance equation, we obtain the fuel flow rate:

$$X = \frac{2.8 \times 10^6}{-2.316 - (-318.749)} = 8848.6 \text{ mol/h} = 198.2 \text{ m}^3\text{/h (STP)}$$

For gaseous fuels, the volumetric flow rate is more readily measured.

3. If pure oxygen is used, the same procedure can be used to determine the fuel flow rate. For oxy/fuel combustion, no excess oxygen is needed. The fuel flow is 4487.1 mol/h, which is only 51% of the fuel required compared with that when air is used for combustion.

If the reaction in Equation 11.1 is in equilibrium, the equilibrium constant for the reaction is written as

$$K_a = \exp\left(\frac{-\Delta G^0}{RT}\right) \tag{11.9}$$

where ΔG^0 is the standard Gibbs energy for the reaction. It can be calculated using the following equation:

$$\Delta G^0 = \sum_{j=1}^{i} \nu_j G_j^0 \tag{11.10}$$

where ΔG_j^0 is the standard Gibbs free energy, which is the free energy of a compound at standard pressure and temperature. An excellent reference for values of ΔH_j^0 and G_j^0 can be found in tabulations over a wide range of temperatures.[7,8]

For ideal gases, the equilibrium constant is written as

$$K_a = \prod_{j=1}^{i} P_j^{\nu_j} \tag{11.11}$$

where $P_j^{\nu_j}$ is the partial pressure of gas j. If the total pressure and molar fraction for a component j are known, the gas partial pressure is the product of the total pressure and its molar fraction.

Van't Hoff, from thermodynamic reasoning, examined the variation of the equilibrium constant K_a with temperature T and concluded:

$$\frac{d \ln K_a}{dT} = \frac{d(-\Delta G^0/T)}{RdT} = -\frac{1}{RT}\frac{d\Delta G^0}{dT} + \frac{\Delta G^0}{RT^2} = \frac{\Delta S^0}{RT} + \frac{\Delta G^0}{RT^2} = \frac{\Delta H^0}{RT^2} \tag{11.12}$$

where ΔS^0 is the change in standard entropy. By using this equation, ΔG^0 can be found at any other temperature if ΔG^0 and ΔH^0 are available at one temperature.

Equation 11.12 may be integrated if ΔH^0 is known at one temperature T_1, since

$$\Delta H_T^0 = \Delta H_{T_1}^0 + \int_{T_1}^{T}\left(\sum_P m_P C_{pP} - \sum_R m_R C_{pR}\right) dT \tag{11.13}$$

Equation 11.13 is the same as Equation 11.6.

11.2.3 Reaction Rate

The reaction rate of Equation 11.1 is often expressed as

$$r = k' \prod_{jR} C_j^{\nu_j} = k e^{-\frac{E}{RT}} \prod_{jR} C_j^{\nu_j} \tag{11.14}$$

where C_j is the concentration of reactant j, k is the pre-exponential factor, and E is the activation energy. In most cases, the reaction order, ν_j, is not equal to the molar stoichiometric coefficient. Therefore, Equation 11.14 should be written in the following form:

$$r = k' \prod_{jR} C_j^{\nu'_j} = k e^{-E/RT} \prod_{jR} C_j^{\nu'_j} \tag{11.15}$$

where the reaction order, ν'_j, often is an empirical constant for a reactant.

For a high-temperature combustion process in practical engineering, no generalized reaction rate can theoretically be formulated since combustion involves complicated chain reactions. In fact, the combustion rate is often expressed in an empirical formula in terms of the concentrations of the fuel and oxidant:

$$r = k\sqrt{T}\rho^n f_F^r f_{\text{ox}}^{n-r} \exp\left(-\frac{E}{RT}\right) \tag{11.16}$$

where ρ is the gas density, f_F and f_{ox} are the molar fraction of fuel and oxidant. Activation energy, E, ranges from 20,000 to 40,000 kcal/k-mol, and the reaction order, n, from 1.5 to 2 for gaseous hydrocarbon fuels.

11.3 GASEOUS FUELS

Gaseous fuel refers to any fuel in the form of a gas state at normal temperature and pressure. Natural gas and, sometimes, synthetic gas are used most often in high-temperature processes. The use of natural gas as a fuel depends on the availability and its economics. In North America and Europe, natural gas is widely used in heating and melting processes. In some regions, however, synthetic gas is more common. In this section, chemical compositions are summarized and basic concepts of gaseous fuel combustion are briefly discussed.

11.3.1 PROPERTIES

Gaseous fuels are measured volumetrically by a calibrated flowmeter, and their compositions are analyzed by gas chromatography. The heating value can be calculated based on fuel composition or determined experimentally using a bomb calorimeter.

When the bomb calorimeter is used, a small of quantity of material is ignited and burned in a heavy-walled, sealed container immersed in water. The resulting heat release is measured from the final temperature of the bomb, water, and associated container. The final temperature is in the range 20 to 35°C (36 to 63°F), such that the measured heat effect includes the latent heat of condensation of water vapor formed in combustion. This is the so-called higher heating value (HHV), which is substantially greater than the heat appearing in the combustion flue gas stream as sensible heat (LHV). In the literature, most reported heats of combustion data are for the higher heating values. For gaseous fuels, the heating value is measured volumetrically.

The regulations on pollutant emissions from a combustion source have been getting more and more stringent worldwide. Therefore, the content of sulfur and nitrogen in gaseous fuels should be examined more carefully. The use of the terms *sweet* and *sour*, which traditionally indicate a small or large proportion of hydrogen sulfide or other sulfur compounds, is not enough since they do not quantitatively give how much sulfur a fuel contains. In oxy/fuel combustion, a small amount of nitrogen in fuels could lead to high NOx emissions.

If natural gases contain more than 0.1 gal of condensible hydrocarbons per 1000 ft^3 of gas, they are called wet gases. Otherwise, they are dry gases.

The compositions of natural gases vary geographically. Properties of gaseous fuels from different areas are listed in Table 11.4.[9–12]

TABLE 11.4
Composition of Typical Gaseous Fuels (vol.%)

Type of Gas	CH_4	C_2H_6	C_3H_8	C_4H_{10}	CO	CO_2	H_2	N_2	O_2
Blast furnace, B	—	—	—	—	22.7	19.3	2.3	55.0	0.7
Blast furnace, U	—	—	—	—	27.5	11.5	1.0	60.0	—
Blast furnace, China	—	—	—	—	27.0	11.0	2.0	60	—
Coke oven, R	28.3	3.4	0.2	—	4.2	0.9	50.6	10.8	1.6
Coke oven, U	29	3.3	0.6 H_2S	—	5.6	1.4	55.4	4.3	0.4
Coke oven, China	22–28	1.5–3.0	—	—	5–7	2–3.5	45–60	2–5	0.1–1.0
Natural, AK	99.6	—	—	—	—	—	—	0.4	—
Natural, Birmingham, AL	90.0	5.0	—	—	—	—	—	5.0	—
Natural, Pittsburgh, PA	83.4	15.8	—	—	—	—	—	0.8	—
Natural, Kansas City, MO	84.1	6.7	—	—	—	0.8	—	8.4	—
Natural, Kuwait	86.7	8.5	1.7	0.7	—	1.8	—	0.6	—
Natural, Libya LNG[a]	70.0	15.0	10.0	3.5	—	—	—	0.90	—
Natural, North Sea gas	94.4	3.0	0.5	0.4	—	0.2	—	1.5	—
Natural gas L, Germany	81.4	2.9	0.6	0.3	—	0.9	—	13.9	—
Natural gas H, Germany	92.5	3.1	1.2	0.5	—	1.2	—	1.5	—
Natural, Naxi, China	95.0	—	—	2.4	0.1	0.5	1.0	—	—
Natural, Luzhou, China	97.8	1.1	—	0.2	0.2	—	0.5	0.2	—
Natural, China	30–85	C_mH_n: 10–50	—	—	—	0.1–1.0	—	2.0	0.5
Pipeline, Gascor, South Africa	87–88	≤2.0	—	—	≤4.0	—	≤3.0	≤16.0	—
Propane, refinery gas	—	2.0	72.9	0.8	24.3 C_3H_6	—	—	—	—
Refinery gas, China	29–52	40–50 C_mH_n	—	—	—	1– 2	5– 10	2.0	0.5
Syngas, oxygen-blown, Koppers-Totzek	—	—	—	—	58.7	7.0	32.9	1.1	—

[a] Liquified natural gas.

11.3.2 Ignition

The start-up of combustion equipment commonly requires a stable and quick ignition of the fuel, although the equipment may vary in configuration for various applications. Further, it can maintain a stable flame when its operation condition is changed. These are the two major steps — ignition and stabilization of a flame — for a combustion process. In this section, some simplified ignition theories will be briefly discussed.

There are many methods to ignite a fuel and oxidant mixture; however, they are traditionally classified in three types: chemical, spontaneous, and forced ignitions. Chemical ignition refers to a chemical reaction–induced combustion, e.g., the reaction of pure sodium and oxygen in air at room temperature and pressure. When a mixture of fuel and oxygen is uniformly heated, ignition will take place at a certain temperature. This method is traditionally called spontaneous ignition. Forced ignition is a process in which the intense oxidation reaction is induced by a "hot" source.

The theory of spontaneous ignition can easily be understood by examining how combustion starts in a closed system, while forced ignition is discussed in an open system.[13]

A premixed combustible mixture fills a closed vessel, and the ambient temperature outside the vessel is T_∞. It is assumed that

(a) The volume and surface of the vessel are V and S, respectively. The temperature of wall, which is the same as the ambient temperature initially, will equal the temperature of the mixture during reaction.
(b) The gas temperature, T, and composition, F, are uniformly distributed, and no natural and forced convection exists in the vessel. The initial gas temperature is T_∞.
(c) The heat transfer coefficient, α, between the vessel and the ambient does not change as the vessel temperature changes.
(d) The change in gas composition is very small prior to ignition.

The simplified reaction system and process are shown in Figure 11.1. The energy equation for the system is expressed as

$$V\rho_\infty C_v \frac{dT}{dt} = VQ_s r - \alpha S(T - T_\infty) \tag{11.17}$$

where ρ_∞, C_v, and Q_s are the density, specific heat, and reaction heat of the combustible gas, respectively. In addition, r is the reaction rate expressed as Equation 11.16, and t is time. Rewriting Equation 11.17, we have

$$\rho_\infty C_v \frac{dT}{dt} = Q_s r - \frac{\alpha S(T - T_\infty)}{V} = q_g - q_l \tag{11.18}$$

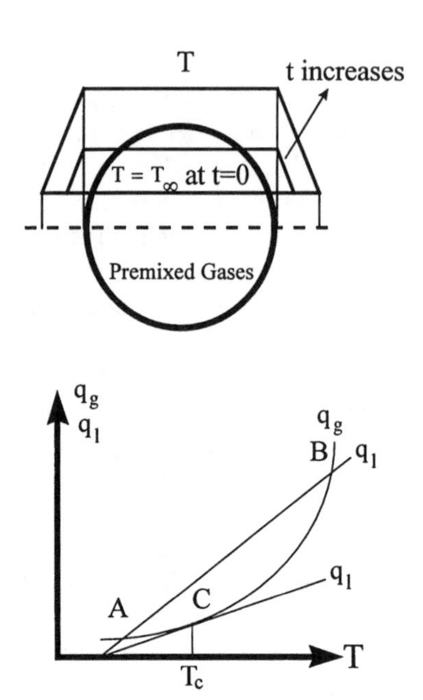

FIGURE 11.1 Spontaneous ignition of premixed fuel and oxidant in a sealed vessel. (Courtesy of Fu and Wei.)

where q_g is the rate of heat release per unit of the gas volume by the reaction and q_l is the rate of heat loss per unit gas volume from the vessel. The rates of heat release and loss are shown in Figure 11.1. By analyzing the physical and chemical processes, one can conclude that points A and B in Figure 11.1 are not the ignition point. In the vicinity of point C, the rate of heat generation always is higher than the rate of heat loss. The transition of the reaction from slow to fast takes place at point C, which is defined as the ignition point. The corresponding temperature, T_c, is called the ignition temperature. The ignition conditions can mathematically be expressed as

$$q_g = q_l \tag{11.19}$$

$$\frac{\partial q_g}{\partial T} = \frac{\partial q_l}{\partial T} \tag{11.20}$$

Substituting Equation 11.18 into Equations 11.19 and 11.20, one obtains the ignition temperature and ignition condition. The ignition temperature is expressed as

$$T_c = T_\infty + \frac{RT_c^2}{E} \tag{11.21}$$

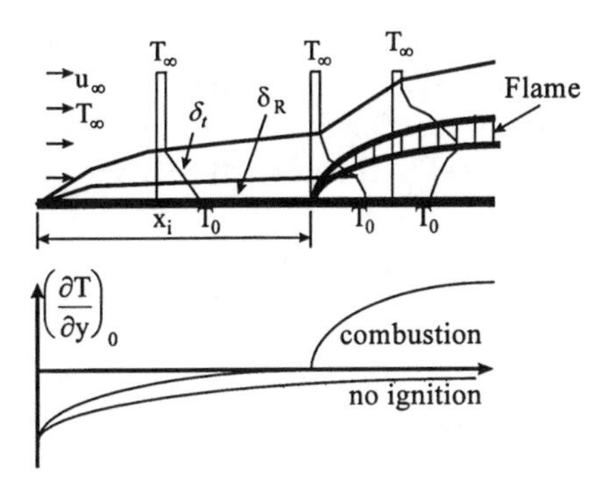

FIGURE 11.2 Ignition of premixed fuel and oxidant by a hot plate (T_0: plate temperature). (Courtesy of Fu and Wei.)

Solving the above equation, one has

$$T_c = \frac{E}{2R}\left(1 - \sqrt{1 - \frac{4RT_\infty}{E}}\right)$$

The ignition condition can be expressed as

$$\frac{E}{RT_c^2} Q_s r = \frac{\alpha F}{V} \tag{11.22}$$

Spontaneous ignition is illustrated by the above sample. The forced ignition model will be explained by the ignition of a cold premixed combustible stream passing along the surface of a hot inert plate, as shown in Figure 11.2. As the gas stream travels along the plate, the temperature of the gas stream will increase first due to heat transfer from the hot surface. It accelerates the reaction rate of the fuel and oxygen in the region near the plate, and the heat released from the reaction will raise the temperature of the gas stream. If the plate is long enough, ignition will take place. In the region where the flame exists, the gas temperature is higher than that of the plate. Therefore, the temperature gradient of the gas stream goes from negative to positive in the direction normal to the plate. The distance, x_i, where the temperature gradient is zero, is usually considered as the ignition distance, and x_i is called the ignition point. After passing the ignition point, the gas stream no longer gets heat from the plate, but transfers energy to the plate. The ignition point is characterized by a zero temperature gradient, i.e.,

$$\left(\frac{\partial T}{\partial y}\right)_{x=x_i} = 0 \tag{11.23}$$

If $x_i \leq L$ (the plate length), a flame will be formed. Otherwise, the ignition failed.

The thermal boundary of thickness δ_t is divided into two regions: region I is a small layer near the plate and region II is the rest of the thermal boundary. In region I with thickness δ_r, the rate of reaction is much higher than that in region II, while the gas velocity is very small (~0). Therefore, the energy equation is simplified to

$$\frac{\partial}{\partial y}\left(\lambda \frac{\partial T}{\partial y}\right) = -Q_s r \tag{11.24}$$

In the region II, the reaction is negligible; hence, the energy equation can be written

$$\rho u \frac{\partial (C_p T)}{\partial x} + \rho v \frac{\partial (C_p T)}{\partial y} = \frac{\partial}{\partial y}\left(\frac{\lambda}{C_p} \frac{\partial (C_p T)}{\partial y}\right) \tag{11.25}$$

The boundary conditions are

$$T = T_0 \quad \text{at} \quad y = 0$$

$$\left(\frac{\partial T}{\partial y}\right) = 0 \quad \text{at} \quad y = 0 \quad \text{and} \quad x = x_i$$

$$u = u_\infty, \quad T = T_\infty \quad \text{at} \quad y = \infty$$

On the boundary between region I and region II,

$$\left(\frac{\partial T}{\partial y}\right)_{\text{I}} = \left(\frac{\partial T}{\partial y}\right)_{\text{II}}$$

Since the thickness of region I is very small, we can assume that the gas heat conductivity $\lambda = \lambda_0$. Equation 11.24 can be further simplified,

$$\left(\frac{\partial^2 T}{\partial y^2}\right) = -\frac{Q_s r}{\lambda_0}$$

Integrating the above equation from $y = 0$ to $y = \delta_r$, the boundary between regions I and II, at $x = x_i$, and considering the temperature gradient at $y = 0$ as zero, we have

$$\left(\frac{\partial T}{\partial y}\right)_{\text{I}} = -\sqrt{\frac{2Q_s}{\lambda_0} \int_{T_0}^{T^*} r dT} \tag{11.26}$$

where T^* is the temperature at $y = \delta_r$. If we assume that $(\partial T/\partial y)_{II}$ is approximately equal to the temperature gradient of the boundary in which no reaction occurs, we have

$$\left(\frac{\partial T}{\partial y}\right)_{II} = \left(\frac{\partial T}{\partial y}\right)_{*} = \frac{-\alpha_*(T_0 - T_\infty)}{\lambda_0} = -\frac{Nu_*(x)}{x}\left(T_0 - T_\infty\right) \qquad (11.27)$$

where subscript * refers to the case of no reaction and α_* is the local coefficient of heat transfer and is expressed by

$$\alpha_* = 0.332\lambda_0 \sqrt{\frac{u_\infty}{\nu_0 x}}\left(T_0 - T_\infty\right)$$

Based on the continuity of temperature gradient at $y = \delta_r$, we have the condition of ignition of a cold combustible stream:

$$\frac{Nu_*(x)}{x}\left(T_0 - T_\infty\right) = \sqrt{\frac{2Q_s}{\lambda_0}\int_{T_0}^{T^*} r dT} \qquad (11.28)$$

If we further simplify the problem by assuming that gas composition is constant before the ignition, the reaction rate is expressed by Equation 11.16, $T_0 - T \ll T_0$, and $T_0 - T^* \cong RT_0^2/E$, a very simple formula is obtained:

$$\tau_i = \frac{x_i}{u_\infty} = k\left(\frac{T_0 - T_\infty}{T_0}\right)^2 e^{-E/RT_0} \qquad (11.29)$$

where τ_i is the induction time of the ignition and k is a constant. Experimental data have confirmed that Equation 11.29 is qualitatively correct.

11.4 LIQUID FUELS

A liquid fuel refers to any fuel in form of a liquid state at normal temperature and pressure. Most liquid fuels are petroleum derivatives, the others coal, oil shale, and tar sands derivatives. The latter sometimes are called synthetic fuels.

11.4.1 Chemical Composition

All liquid fuels mainly consist of mixtures of hydrocarbon compounds with a higher ratio of carbon to hydrogen than that of gaseous fuels, fuel nitrogen (~0.1 to 2.2% by weight), sulfur (~0.2 to 3% by weight), and ash (minerals). The hydrocarbons are complex materials, which is difficult to analyze; therefore, it is common practice to measure an ultimate chemical analysis and several physical properties, rather than to find their chemical structures.

An ultimate chemical analysis refers to an analysis routine that includes carbon, hydrogen, nitrogen, sulfur, and ash. It may be appropriate to request analyses for chlorine and metals that are important to air or water pollution aspects. The nitrogen in fuels exists in pyrroles, indoles, isoquinolines, acridines, and porphyrines, which are complex heterocyclic compounds.[14] The sulfur compounds mainly are hydrogen sulfide, alkyl sulfate, sulfur oxides, sulfones, thiphene, etc.[15] From the ultimate analysis of a fuel, one can estimate the heating values, combustion oxygen requirement, and flue gas analysis, as discussed in previous sections. It provides the combustion engineer with information to adopt an appropriate strategy to minimize the generation of nitrogen oxides and sulfur oxides when the burner and furnace are designed. The compositions of ash consist of minerals, light compounds which vaporize to form small particulates emitted as aerosols in the flue gas, and heavy compounds which could lead to equipment fouling. The elemental analyses for some typical oils and synthetic fuels are summarized in Table 11.5.[10–12,14,16,17] If the chemical composition of the fuel changes, the appropriate action should be taken to achieve an optimal furnace operation, which can often be found by monitoring the flue composition in the stack.

11.4.2 Physical Properties

The physical properties for liquid fuels that interest combustion engineers most are gravity, both specific gravity and API (American Petroleum Institute) degree; density; flash and fire temperatures; viscosity; carbon residue; ash; water; and sediment. These properties will be discussed below.

Specific gravity and API gravity are expressions of the density or weight of a unit volume of liquid fuels. The specific gravity is defined as the ratio of the weight of liquid fuel to the weight of the same volume of water (62.3 lb/ft^3 or 1000 kg/m^3) at a standard temperature, 60°F, while the density is the weight of a unit volume of materials. The specific gravity (spgr) of a liquid fuel is correlated by the following equation

$$\text{spgr} = \frac{\text{fuel density}\left(\text{lb/ft}^3\right)}{\text{water density}\left(62.3\ \text{lb/ft}^3\right)} = \frac{\text{fuel density}\left(\text{kg/m}^3\right)}{\text{water density}\left(1000\ \text{kg/m}^3\right)} \tag{11.30}$$

The relation of API gravity (°API) and specific gravity is expressed as

$$^\circ\text{API} = \frac{141.5}{\text{spgr}} - 131.5 \tag{11.31}$$

Specific gravity or API gravity is measured by means of hydrometers, pycnometers, or the chainomatic specific gravity balance if a small amount of sample is tested. Compared with the other properties, specific gravity is the property most easily measured. Therefore, much of the information for the other physical properties is empirically correlated with the specific gravity or API gravity. Fuel suppliers usually provide their customers with these physical data for their products. If this is not case, however, the following empirical formula can be used to estimate some of the physical properties after the specific gravity is measured.[9]

TABLE 11.5
Elemental Analysis and Characteristics of Typical Fuel Oils and Synthetic Fuels (ultimate analysis in wt.%; heating value in Btu/lb)

Fuel	C	H	N	S	O_2	Water	Ash	API at 60°F	Specific Gravity	Viscosity, SSU	Heating Value
Distillate, Alaska[a]	86.99	12.07	0.007	0.31	0.62	—	<0.001	33.1	—	33.0 at 140°F	—
Distillate, California[a]	86.8	12.52	0.053	0.27	0.36	—	<0.001	32.6	—	30.8 at 140°F	19,330 gross
Distillate, Texas[a]	88.09	9.76	0.026	1.88	0.24	—	<0.001	18.3	—	32.0 at 140°F	—
Residual, Alaska[a]	86.04	11.18	0.51	1.63	0.61	—	0.034	15.6	—	1071 at 140°F	18,470 gross
Residual, California[a]	86.66	10.44	0.86	0.99	0.85	—	0.20	12.6	—	720 at 140°F	18,230 gross
Residual, Gulf of Mexico[a]	84.62	10.77	0.36	2.44	1.78	—	0.027	13.2	—	835 at 140°F	18,240 gross
Residual, Daqing, China[b]	86.5	12.56	—	0.17	—	0.03	0.017	—	0.9326	600 at 212°F	18,144 LHV
Residual, Shengli, China[b]	80.45	10.89	0.7–0.8	0.9–1.2	0.70	—	<0.1	—	0.945	170–700 at 212°F	17,550 LHV
Gas oil (L), Germany[c]	—	—	—	—	—	—	—	—	—	46 at 212°F	17,627 LHV
Heavy oil, Germany[c]	—	—	—	—	—	—	—	—	—	180 at 212°F	17,154 LHV
Sasol waxy oil 1, South Africa[d]	—	—	—	<0.03	—	<0.3	<0.01	—	0.815 at 68°F	<42 at 104°F	19,647
Sasol waxy oil 20, South Africa[d]	—	—	—	<0.2	—	<1.3	<0.4	—	0.893 at 68°F	65 at 212°F	19,561
Middle East, Exxon[a]	86.78	11.95	0.18	0.67	0.41	—	0.012	19.8	—	490 at 140°F	19,070 gross
Indo/Malaysia[a]	86.53	11.93	0.24	0.22	1.04	—	0.036	21.8	—	199 at 140°F	19,070 gross
Venezuela[a]	85.24	10.96	0.40	2.22	1.10	—	0.081	14.1	—	742 at 140°F	18,400 gross
Shale derived[e]	86.18	13.0	0.24	0.51	1.07	—	0.003	33.1	—	—	19,270 gross
Raw											
Paraho shale oil[f]	83.55	11.69	2.15	0.74	1.65	—	0.09	37	—	27 at 70°F	19,400 gross

[a] Data from Reference 16.
[b] Data from Reference 10.
[c] Data from Reference 11.
[d] Data from Reference 12.
[e] Data from Reference 14.
[f] Data from Reference 17.

Gross heating value (GHV)

$$
\begin{aligned}
&= 17{,}887 + (57.5 \times {}^\circ\text{API}) - (102.2 \times \%\ \text{S}),\ \text{in Btu/lb} \\
&= 5738 + 4521/\text{spgr} - 56.8 \times \%\ \text{S},\ \text{in kgal/kg}
\end{aligned}
\tag{11.32}
$$

Weight percentage of hydrogen

$$
= A - 2122.5/({}^\circ\text{API} + 131.5) \tag{11.33}
$$

where $A = 24.50$ for $0 \le {}^\circ\text{API} \le 9$
$= 25.00$ for $9 < {}^\circ\text{API} \le 20$
$= 25.20$ for $20 < {}^\circ\text{API} \le 30$
$= 25.45$ for $30 < {}^\circ\text{API} \le 40$

Net heating value (NHV)

$$
\begin{aligned}
&= \text{GHV (Btu/lb)} - 91.23 \times \%\ \text{H},\ \text{in Btu/lb} \\
&= \text{GHV (kcal/kg)} - 50.70 \times \%\ \text{H},\ \text{in kcal/kg}
\end{aligned}
\tag{11.34}
$$

Specific heat (C_p)

$$
= \frac{0.388 + [0.00045 + \text{temperature, }{}^\circ\text{F}]}{\sqrt{\text{spgr}}},\ \text{in Btu/lb }{}^\circ\text{F or kcal/kg }{}^\circ\text{C} \tag{11.35}
$$

Latent heat of vaporization

$$
\begin{aligned}
&= \frac{110.9 - (0.09 \times \text{temperature, }{}^\circ\text{F})}{\text{spgr}},\ \text{in Btu/lb} \\
&= \frac{60.02 - (0.09 \times \text{temperature, }{}^\circ\text{C})}{\text{spgr}},\ \text{in kcal/kg}
\end{aligned}
\tag{11.36}
$$

If oxygen enrichment is $x\%$, Equation of 2.14 of Reference 9 is modified as:
The volume of pure oxygen needed

$$
\begin{aligned}
&= \left(\frac{x\%}{0.79} \times \frac{0.21}{x\% + 0.21}\right)(25.1 \times \text{spgr} \times \%\ \text{H} + 1260 \times \text{spgr}),\ \text{in ft}^3/\text{gal oil} \\
&= \left(\frac{x\%}{0.79} \times \frac{0.21}{x\% + 0.21}\right)(0.188 \times \%\ \text{H} + 9.43),\ \text{in m}^3/\text{kg oil}
\end{aligned}
\tag{11.37}
$$

The volume of air required

$$= \left(\frac{0.79 - x\%}{0.79} \times \frac{0.21}{x\% + 0.21} \right) (25.1 \times \text{spgr} \times \%\,\text{H} + 1260 \times \text{spgr}), \text{ in ft}^3/\text{gal oil}$$

$$= \left(\frac{0.79 - x\%}{0.79} \times \frac{0.21}{x\% + 0.21} \right) (0.188 \times \%\,\text{H} + 9.43), \text{ in m}^3/\text{kg oil} \tag{11.38}$$

The total volume of oxygen-enriched air will be the sum of the volumes of the pure oxygen and the air calculated from Equations 11.37 and 11.38. The 0% and 79% oxygen enrichments ($x\%$ in Equations 11.37 and 11.38) correspond to no oxygen enrichment and pure oxygen, respectively.

Example 11.4 An analysis of Gulf #6 petroleum is as follows: API gravity of 13.2 at 60°F (289 K), a gross heating value of 18,400 Btu/lb (42,800 kJ/kg), 88.29% carbon, 12.31% hydrogen, 0.44% nitrogen, and negligible sulfur.[14]

1. Compare the estimated gross heat value based on Equation 11.32 with the experimental data, and
2. Compute how much oxygen and air are needed to burn 1 gal of fuel if an oxygen enrichment of 25% is used.

Solution:

1. Using Equation 11.32, the estimated gross heating value is 18,646 Btu/lb (43,400 kJ/kg). The relative error is only 1.3%, compared with the experimental data.
2. First calculate the specific gravity, which is 0.978, by means of Equation 11.18. Substituting the data into Equations 11.37 and 11.38, we find out that 221.7 ft^3 (6.28 m^3) of oxygen and 478.8 ft^3 (13.6 m^3) of air are needed to completely burn a gallon of this fuel.

The above relationships, Equations 11.32 through 11.38, are empirical generalizations at a temperature of 60°F (289 K) for a pure hydrocarbon. Specific and API gravity of typical fuel oils are given in Table 11.6. It was found that a small percentage of an impurity, which is true for most commercial fuels, in the oils does not affect the application of the formula. If the specific gravity (or API gravity) is given at another temperature, one needs to convert the data into the value at 60°F (289 K).

Experimental data show that specific gravity linearly decreases as temperature increases up to 400 for the oils with the specific gravity ranging from ~0.88 to ~1.04, and up to 200°F (366 K) for the oils with the specific gravity ranging from ~0.624 to ~0.88.[15] This implies that their volumes linearly expand with temperature. The specific gravity at 60°F (289 K) can be found by means of

$$\text{spgr@60°F} = \text{spgr@T}_2 \times \left[1 + \alpha_p (T_2 - 60)\right] \tag{11.39}$$

TABLE 11.6
API and Specific Gravity of Typical Fuel Oils (60°F or 289 K)

Typical Fuel (ASTM)	API	Specific
No. 1	36–40	0.845–0.825
No. 2	30–34	0.876–0.855
No. 4	20–28	0.934–0.887
No. 5	18–22	0.986–0.922
No. 6	12–16	0.986–0.959

TABLE 11.7
Coefficient of Oil Expansion

API Gravity	Mean Coefficient of Expansion	Valid for Temperature (F) Up To
0–14.9	0.00035	400
15–34.9	0.00040	300
35–50.9	0.00050	200
51–63.9	0.00060	200

where α_p is a coefficient of expansion and is expressed as

$$\alpha_p = \left(\frac{dV}{dT}\right)_p \tag{11.40}$$

The mean approximate coefficients of expansion for materials of different gravity up to about 400°F (480 K) are shown in Table 11.7.[15]

Flash and fire points are two very important physical properties, which are of interest from the standpoint of safety and ignition characteristics. The flash point is the temperature at which the vapor above an oil will momentarily flash when in the presence of a flame, while the fire point is the temperature at which the vapors are vaporized quickly enough to have a sustained flame. The flash point is determined by using an open cup and Pensky–Martens closed tester for heavy oils and a tag closed tester for light oils. The vapor pressure–temperature relationship can be written as

$$\frac{d \ln P^0}{dT} = \frac{\Delta H_{l-v}}{RT^2} \tag{11.41}$$

where $\Delta H_{l\text{-}v}$ is heat of vaporization, P^0 is the vapor pressure when the equilibrium is achieved, R is the gas constant, and T is the temperature. When the temperature is below the flash and fire points, an oil can be handled without danger of fire. The safety issues are discussed in Chapter 9.

The primary difference between liquid and gaseous fuel burners is that each liquid fuel burner has an atomizer, which is used to facilitate the distribution of the liquid into a spray of droplets. This is due to the higher viscosity of liquid fuels compared with gaseous fuels. Therefore, combustion engineers examine the viscosity of oils closely since it is the most important factor for the design of the flow train and burner. The viscosity of a liquid fuel is a measure of its resistance to internal flow. For liquid fuels, the viscosity increases as the ratio of carbon to hydrogen increases (from light to heavy oils), and decreases as the temperature of oil is raised. The higher the viscosity is, the more difficult it is for the oil to be pumped and atomized. It has been found that certain ranges of viscosity of liquid fuels are good for pumping and for atomizing. The ranges of viscosity are approximately the following: 5000 to 10,000 for pumpable; 2000 to 5000 for easy pumping; 100 to 150 for atomizing; and 70 to 100 SSU (Saybolt Second Universal) for easy atomizing.

Due to high viscosity, heavy oils are often heated by an electrical heater or steam before they are pumped to delivery lines. The resulting temperature is required to be high enough so that the oil can be easily atomized. Otherwise, a spray of undesirably large oil droplets will be formed. It leads to an unstable flame and carbon buildup on the atomizer and burner block. The American Society for Testing and Materials (ASTM) reported the relationship between the viscosity and temperature for different fuel oils, as shown in Figure 11.3.[18]

The other physical properties are carbon residue, ash, water, and sediment in the oils. The carbon residue is the leftover after an oil is vaporized, which is an indicator of how easy an oil tends to carbonize. Ash is the materials left after an oil totally combusts. The ash consists of the inorganic mineral materials. The effect of water on combustion is to make the flame unstable. Sediments clog the tip of the burner atomizer and equipment.

11.4.3 Vaporization and Combustion

The combustion of sprays in a high-temperature furnace is a complex physical and chemical process that involves simultaneous heat, mass and momentum transfer, phase transition, and chemical reactions. The droplet size, composition of the fuel, ambient temperature and pressure, and oxygen concentration are major factors that affect the combustion process. Owing to the complexity of the process, it is very difficult to obtain accurate information on the combustion of the spray. However, the evaporation and combustion of a single droplet of oil have been well studied since it is relative easy to carry out an experiment for the measurement of combustion. Furthermore, it has been theoretically investigated due to its simplicity.

If a droplet of oil is immersed in a high-temperature oxidant stream, the droplet temperature will increase. The droplet surface will vaporize. The fuel vapor and oxygen diffuse from opposite directions, and a flame forms at their contact surface

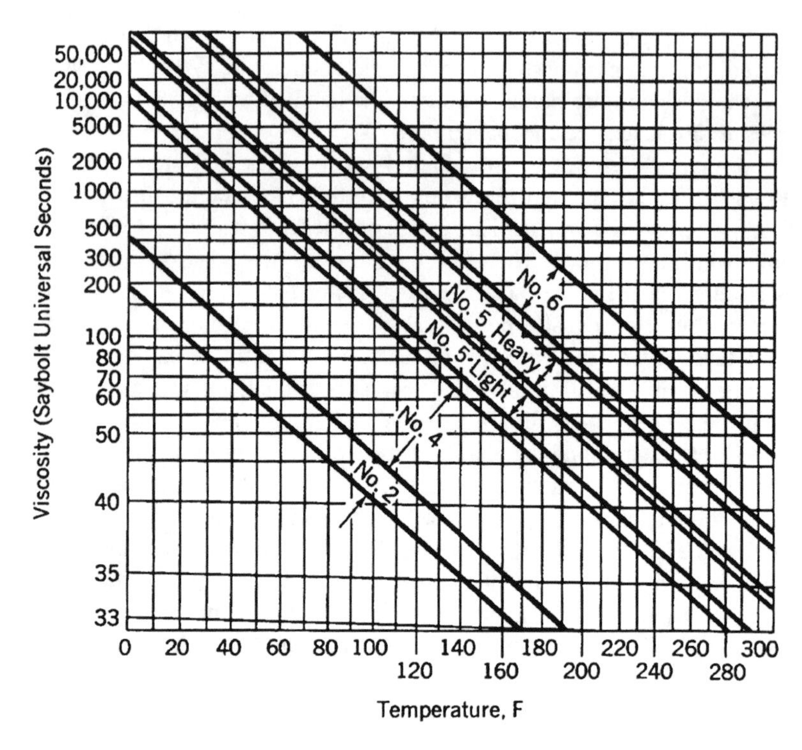

FIGURE 11.3 Viscosity–temperature relationship of typical fuel oils. (From ASTM. With permission.)

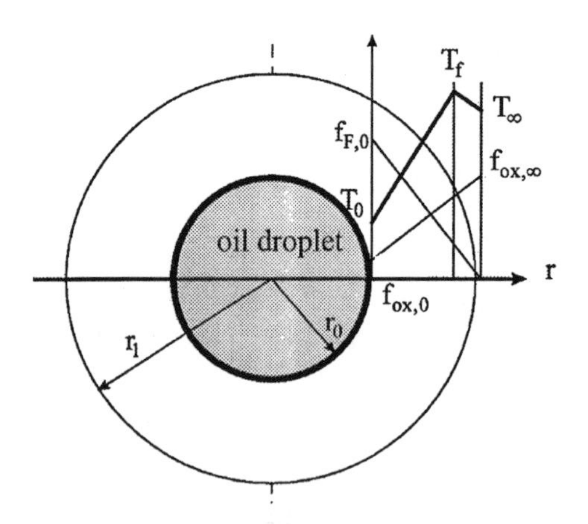

FIGURE 11.4 Schematic of evaporation and combustion of an oil droplet (r_0 and r_1 are the droplet and flow boundary radii, respectively; f_{ox} and f_F are oxidant and fuel distribution around the droplet; T is the temperature distribution).

some distance away from the droplet. The phenomenon is illustrated in Figure 11.4. Modeling the combustion process is beyond the scope of this section because of its complexity, as mentioned above. Here we only give the final results.

TABLE 11.8
Burning Rate of an Oil Droplet for Typical Fuels

Fuel	k_c (mm²/s)
Kerosene	0.96
Octane	0.95
Alcohol	0.81
Diesel	0.79
n-Hexadecane	0.72

The process of vaporization is simply expressed as

$$d_0^2 - d^2 = k_v t \tag{11.42}$$

and the process of combustion is written as

$$d_0^2 - d^2 = k_c t \tag{11.43}$$

where k_V and k_c are vaporization and combustion constants, d is the diameter of the droplet at time, t, and d_0 is the initial diameter of the droplet. The combustion constants for some fuels are listed in Table 11.8 when a droplet of fuel burns in the atmosphere.[13]

11.4.4 Several Points on Synthetic Fuels

Synthetic fuels derived from coal, shale, and tar sands are different from petroleum fuels in chemical and physical properties. Those different properties lead to substantial differences in combustion characteristics and emission. Relative to conventional oils, the synthetic fuels contain high aromatic, fuel nitrogen, and ash and trace mineral; and low hydrogen, volatility, and heat of combustion; and more corrosive in nature. Therefore, the combustion of synthetic fuels will generate more soot and more NOx and SO_2 emissions, and is more difficult to be used than that of petroleum fuels in a practical system. This is a challenge for combustion engineers to utilize synthetic fuels in full-scale equipment.[19]

11.5 SOLID FUELS

Solid fuels refer to any fuels, such as coals, wastes, biomass, etc., that are in a solid state when they are burned in a high-temperature furnace. Coals are the most abundant fossil fuel on Earth, and are the most commonly used solid fuels for power generation, in the metallurgical industry, for gasification, and for heat supply. The wastes sometimes are burned in incinerators, as discussed in Chapter 8, and the generated heat is recovered to produce steam and electricity. Biomass mainly refers to plant materials, which are rarely used as fuels in industry. In this section, we will review the chemical properties, devolatilization, and combustion of coals and wastes, as well as coal gasification by oxygen in the Integrated Coal Gasification Combined Cycle (ICGC).

11.5.1 Properties of Coals

Coals are heterogeneous organic materials and are formed by the decomposition and metamorphism of plants underground under conditions of high pressure and temperature for a long time. The composition of coals varies greatly as a result of the different plants and the different degrees of their changes. In order to correlate coal properties with combustion and gasification behaviors, great efforts have been made to classify coals based on their chemical and physical properties. The most commonly used category is one developed by ASTM. In this method, the coals are classified into 12 ranks — lignite, subbituminous C, B, and A, high volatile bituminous C, B, A, medium volatile bituminous, low volatile bituminous, semianthracite, anthracite, and subanthracite — based on their carbon and heating values. The fixed carbon contents and heating values increase from 35 to 95% and from 8100 to 15,000 Btu/lb, respectively; the volatile matters increase from 25 to 45%, then decrease to 5%; and the water contents decrease from 42 to 5% with coal ranks (from lignite to anthracite).

The chemical composition of a coal usually is expressed by either an ultimate or proximate analysis or by both. Ultimate analysis refers to an analysis routine that includes moisture, combined water, carbon, hydrogen, sulfur, nitrogen, and ash.

The proximate analysis is to measure water content, volatile matter, fixed carbon, and ash. The procedures include the following steps: (1) Heat the sample up to between 104 and 110°C for 1 h; the weight loss is due to moisture. (2) Ignite in a covered crucible for 7 min at 950°C (1740°F); the weight loss is considered as volatile matter. (3) Combust in an open crucible at 725°C (1340°F) until the sample weight no longer changes; the weight loss is reported as fixed carbon, and the residue is reported as ash.

Knowing the contents of nitrogen and sulfur in coals, combustion engineers are in a position to take appropriate approaches to control nitrogen and sulfur dioxide formations. Another useful application of the ultimate analysis is to estimate the heat of combustion by assuming a negligible heat of formation of the organic matter relative to the heat of combustion of the elements:[20]

$$\begin{aligned}\text{Heating value of a coal (dry basis)} &= 14{,}554 \times (\text{weight fraction carbon}) \\ &+ 62{,}077 \times (\text{wt\% hydrogen}) \\ &- 0.225\,(\text{wt. fraction oxygen}) \\ &+ 4{,}053 \times (\text{wt. fraction sulfur}), \text{ in Btu/lb}\end{aligned} \tag{11.44}$$

The combustible volatile matter is very important to coal combustion and gasification. The higher the volatile matter is, the easier it is to ignite coal particles and stabilize the resulting flame, especially for an entrained flow burner in large-scale furnaces. The amount of volatile matter released from a coal increases as temperature increases. A useful application of approximate analysis is to estimate the volatile matter at any temperature, T. This is expressed as[21]

$$V = 1.2 V_{daf}^{0.8} \exp\left(-\frac{2 \times 10^6}{R(T - T_0)^2} \right) \tag{11.45}$$

where V is the weight percentage of volatile matter in the dry and ash-free basis at a temperature, T; V_{daf} is the weight percentage of volatile matter in the dry and ash-free basis obtained by the approximate analysis; T_0 is the ambient temperature; and R is the universal gas constant.

Ash consists of mineral material compounds, which include clays, silicates, carbonates, sulfides, sulfates, oxides, and phosphates. Major elements are Al, Si, Ca, and Fe; minor elements are K, Na, Mg, and others; trace elements are As, Be, Hg, etc. The mineral matter influences fouling, slagging, and heat transfer in high-temperature furnaces; the performance of particulate control equipment; and the health and ecological effect of particles escaping to the atmosphere.[22]

More volatile matters, such as those of alkali metals, are commonly found in fume which is about 0.05 μm in diameter on average. Less volatile elements form the flying ash in a larger size, ranging from 1 to 20 μm. Deposition on relatively cool surfaces, such as water-cooled walls, mainly consists of iron compounds.

Table 11.9 shows the properties of some coals from different areas.[10,11,23-25] The properties vary greatly from coal to coal. Therefore, it is a common practice to design a burner for specific kinds of coals, which have similar chemical composition to each other.

11.5.2 Properties of Wastes

The composition of wastes refers to the category of material (paper, metal, plastics, etc.) in the waste stream. It is very different from one source to another and is not well sampled in many instances. Generally, the composition of the wastes is classified based on the waste sources: residential, commercial, and industrial. The average compositions of commercial and residential wastes are summarized in Table 11.10.[26] The composition of wastes from industries (chemicals, petroleum, rubber and plastics, wood products, food, textile, etc.) varies greatly since different industries use different raw materials to produce various products. The content of paper, however, is the highest among the categories of materials in every industry (except for leather manufacturing), ranging from 20 to 55% by weight.

Some wastes, which are combustible, are burned in a high-temperature furnace, like a boiler or incinerator. The heat released by the combustion process is recovered by producing high-temperature steam to supply heat or to drive a turbine to generate electricity. The heating values of some combustible materials are shown in Table 11.11.[26] The proximate analyses and heating values of municipal refuse, wood, and peat are listed in Table 11.12.[26] If the ultimate analysis is known for a waste, the heating values can be estimated by means of Equation 11.44.

TABLE 11.9
Proximate and Elemental Analyses of Typical Coals

Coals	Proximate (wt.%)	Moisture	Volatile	Fixed C	Ash	Elemental (wt.%)	H	C	N	O	S	Ash
Montana Rosebud[a]	ASTM	2.8	39.0	49.0	9.2	Dry	5.1	70.1	1.3	15.4	0.6	7.5
Utah King[a]	ASTM	4.0	34.9	54.8	6.3	Dry	6.1	74.2	1.4	11.0	0.7	6.6
Wyodak[a]	ASTM	6.0	47.3	40.3	6.4	Dry	5.8	67.5	1.3	19.8	0.3	5.3
Pittsburgh[a]	ASTM	2.0	36.6	55.4	6.0	Dry	5.3	77.5	1.5	8.5	1.2	6.0
Kentucky #9[a]	ASTM	5.1	38.2	47.6	9.1	Dry	5.1	73.5	1.3	8.5	2.9	8.8
Wyoming Monarch[a]	ASTM	11.6	37.0	46.4	5.0	Dry	5.0	69.1	1.3	18.2	0.7	5.7
Bulli, Australia[b]	Air dried	1.4	19.4	67.3	11.9	Air-dried	3.7	75.3	1.4	5.8	0.5	—
Greta, Australia[b]	Air dried	2.2	43.5	48.9	5.4	Air-dried	5.0	72.0	1.5	12.8	1.1	—
Portuguese anthracite[c]	As received	4.1	6.1	41.2	48.6	Dry, ash-free	1.57	84.57	1.14	10.18	2.45	—
Fushun, China[d]	As received	13.0	33.2	39.0	14.79	Dry, ash-free	6.1	78.8	1.7	12.6	0.8	—
Datong, China[d]	As received	7.5	20.6	46.0	25.9	Dry, ash-free	4.4	83.0	1.0	11.1	0.5	—
Kailuan, China[d]	As received	7.0	23.72	46.0	23.25	Dry, ash-free	5.2	83.5	1.5	8.4	1.4	—
Jiaozuo, China[d]	As received	7.0	5.08	67.46	20.46	Dry, ash-free	3.1	92.2	1.4	2.8	0.5	—
Huaibei, China[d]	As received	9.2	12.9	38.9	39.04	Dry, ash-free	6.09	79.9	1.92	10.92	0.77	—
Brown coal, Germany[e]	—	—	—	—	—	Dry, ash-free	5–8	65–75	0.2–2.0	15–26	0.5–4.0	—
Hard coal, Germany[e]	—	—	—	—	—	Dry, ash-free	4–9	80–90	0.6–2.0	4–12	0.7–1.4	—
Anthracite, Germany[e]	—	—	—	—	—	Dry, ash-free	3–4	90–94	1–1.5	0.5–4	0.7–1.0	—

[a] Data selected from Reference 23.
[b] Data from Reference 24.
[c] Data from Reference 25.
[d] Data from Reference 10.
[e] Data from Reference 11.

TABLE 11.10
Composition of Typical Wastes

Component (wt.%)	Commercial Wastes	Residential Wastes
Metal	8.3	8.6
Paper	58.1	44
Plastics	5.2	1.4
Food wastes	~15.6	17.1
Yard waste	~0	9.4
Glass	8.7	8.8
Others	3.5	10.7

TABLE 11.11
Heating Values for Typical Wastes (Btu/lb)

Waste	Average Heating Value (as received)
Paper	7,371
Food	6,000
Wood	4,600
Domestic waste	11,400
Plastics	10,000–22,700

TABLE 11.12
Proximate Analysis of Some Wastes

Waste	Water (wt.%)	Ash (wt.%)	Volatile (dry, wt.%)	HHV (dry, ash-free, Btu/lb)	LHV (as-fired, Btu/lb)
Refuse	28.2	20.8	62.3	8,721	3,933
Peat	64.3	10.0	67.3	12,591	2,394
Wood	46.9	1.5	78.1	9,090	3,906

Detailed information on the chemical composition, physical properties, and handling of wastes will not be presented here. The interested readers should read an excellent book[26] in which a great deal of information has been well documented.

11.5.3 Devolatilization and Combustion

When solid fuels are introduced into a high temperature furnace, they are heated by the high-temperature gases and walls of the furnace. The moisture in the wastes is first released at about 100°C (212°F); the volatile matter begins to decompose at about 200°C (392°F) and transports to the outside of the particles. The volatile materials and solid combustibles (mainly carbon) ignite if their temperatures are high enough. Whether the volatiles or the solid materials first ignite depends on whose ignition condition is met first.

Devolatilization is very complicated, involving multiple chemical reactions coupled with transport processes, and it is very difficult to describe the process theoretically. In practice we resort to the use of a simplified global devolatilization model, which can be an effective approach if properly used.[27]

For waste decomposition, studies have extensively investigated wood and synthetic polymers. Surface pyrolysis of woods may be assumed to be a first-order reaction:[28]

$$\frac{d\rho}{dt} = -k_1\left(\rho - \rho_{c;v}\right)\exp\left(-\frac{E}{RT}\right) \tag{11.46}$$

where ρ (g/cm) is the density at time t; subscripts c and v refer to char or virgin solid; pre-exponential frequency factor $k_1 = 10^6\ \text{min}^{-1}$; and activation energy $E =$ 19,000 kcal/kg.

Thermal degradation[29] of paper is expressed as

$$\frac{dm}{dt} = -k\exp\left(-\frac{E}{RT}\right)\ \text{g}/\left(\text{cm}^2\ \text{s}\right) \tag{11.47}$$

where $k = 5.9 \times 10^6$ and $E = 26{,}000$ for temperatures below 382°C (720°F) and $k = 1.9 \times 10^{16}$ and $E = 54{,}000$ for temperatures above 382°C (720°F).

The content of volatile matter in coal is very important to ignition, combustion, and flame stabilization in a practical flame. Hence, coal devolatilization was extensively studied, and many models were proposed. The rate of devolatilization is described by single-step, two-step, and multiple-step reaction models and a functional group decomposition model. In the single-step model, the rate of decomposition is a first-order reaction, and is proportional to the content of volatiles, v, in coals[30]

$$\frac{dV}{dt} = k\left(V_\infty - V\right)\exp\left(-\frac{E}{RT}\right) \tag{11.48}$$

where V_∞ is the overall content of devolatilization, and k and E are pre-exponential frequency factor and activation energy, respectively. Although these parameters are coal dependent, the model is very simple to use if the parameters are known.

The two-step model simulates the decomposition by two single first-order reactions, in which one reaction is dominant at low temperatures and the other at high temperatures.[31] The rate of decomposition is expressed as

$$\frac{dV}{dt} = \left[Y_1k_1\exp\left(-\frac{E_1}{RT}\right) + Y_2k_2\exp\left(-\frac{E_2}{RT}\right)\right]C \tag{11.49}$$

where k_1, k_2, and E_1, E_2 are the exponential frequency factors and activation energies for two reactions, respectively; Y_1 and Y_2 are roughly equal to the percentages of volatile matter released from a coal under condition of proximate analysis and at high temperature. Dependence of the parameters on coal types limits the model application.

The multiple reaction model consists of a large number of independent parallel first-order reactions, all having the same pre-exponential factor, k_0, and activation energies in Gaussian distribution with mean E_0 and standard deviation σ,[32] which gives

$$k = k_0 \left[\frac{\displaystyle\int_0^\infty \exp\left\{ -\frac{k_0}{m} \int_{T_0}^{T} \exp\left(-\frac{E}{RT} \right) dT - \frac{(E-E_0)^2}{2\sigma^2} - \frac{E}{RT} \right\} dE}{\displaystyle\int_0^\infty \exp\left\{ -\frac{k_0}{m} \int_{T_0}^{T} \exp\left(-\frac{E}{RT} \right) dT - \frac{(E-E_0)^2}{2\sigma^2} \right\} dE} \right] \tag{11.50}$$

Replacing the product of $k \exp(-E/RT)$ in Equation 11.48 by Equation 11.50, we have the overall reaction rate that would be observed at temperature T, to which the coal is assumed to be heated at constant rate m from the initial temperature T_0. For devolatilization of Montana lignite, $k_0 = 1.07 \times 10^{10}$ s^{-1}, $E_0 = 48.72$ kcal/mol, and $\sigma = 9.38$ kcal/mol.

In the above models, the parameters are different for different coals. A model was proposed in which the equivalent activation energy, E, and the equivalent frequency factor, k, are independent of coal types and depend only on the final temperature, T, of the coal particles.[33] The model is the same as Equation 11.48 in math form. The relationships among E, k, and T were given based on the experimental results. The V_∞ in the model can be determined by Equation 11.45.

More-complex models have been proposed to link the coal structure with volatile decomposition. The model, which considers the changes of functional groups and reactions of char-forming repolymerization (cross-linking), can predict gas species and tar in volatiles.[34-36] Another model applies concepts of polymer decomposition to describe the release of tar fragments from the coal macromolecule.[37] The models will not be summarized here since they are more complex than the single-step, two-step, and multiple-step models. Interested readers should consult the appropriate literature for more details.

The volatile combustion usually refers to oxidation of CH_4, C_2H_6, … C_9H_{20}, C_2H_4, C_3H_6, etc. in high temperature. The process can be described by different models. For some applications, a single pseudo reaction step may be sufficient. For each composition in volatile matter, this is expressed by the general equation for reaction,

$$\text{Fuel} + n_1 O_2 \rightarrow n_2 CO_2 + n_3 H_2O \tag{11.51}$$

The n_i are determined from the choice of fuel. The rate law of reaction is Equation 11.16.

The two-step reaction mechanism may be represented by

$$C_nH_m + \left[\frac{n}{2} + \frac{m}{4}\right] O_2 \rightarrow nCO + \frac{m}{2} H_2O \tag{11.52}$$

followed by

$$CO + \tfrac{1}{2} O_2 \rightarrow CO_2 \tag{11.53}$$

This corrects the single-step reaction that overpredicts the heat of combustion by assuming that all the carbon in the fuel is burned to CO_2 and all the hydrogen in the fuel is converted to H_2O.

The porous carbon, or char, is left after the volatile matter is released from the coal. The combustion of carbon in the gaseous oxygen phase is a complex set of heterogeneous reactions for both the gas and the solid. It may occur on the surface or inside the pores of the char and is controlled by gas diffusion and kinetics, reaction temperature, char structure. Generally, the reactions are expressed by three heterogeneous reactions and one gas-phase reaction.

The heterogeneous reactions include

$$C + O_2 \rightarrow CO_2 \tag{11.54}$$

$$2\,C + O_2 \rightarrow 2CO \tag{11.55}$$

$$C + CO_2 \rightarrow 2CO \tag{11.56}$$

The gas-phase reaction is

$$2CO + O_2 \rightarrow CO_2 \tag{11.57}$$

For heterogeneous reactions, the rates are described by

$$r = k_0 A C_o^n \exp\left(-\frac{E}{RT}\right) \tag{11.58}$$

where k_0 is pre-exponential factor, A is the surface area where the reaction takes place, C_o^n is the oxygen concentration on the char surface, and E and R are activation energy and the universal gas constant, respectively.

The reaction rate in the gas phase can be expressed by Equation 11.16. To solve the above equations, the parameters in the rates of reactions need to be determined experimentally. Modeling the above reaction is beyond the scope of this chapter. A simple case will be discussed in the following paragraph.

If there only is one heterogeneous reaction, $C + O_2 \rightarrow CO_2$, on the surface of the char and the reaction is controlled by gas diffusion and surface kinetics, the relationship between the rate, g_C, of carbon combustion and the rate, g_{O_2} of oxygen consumption can be expressed as

$$\beta g_C = g_{O_2} = \alpha\left(\rho_{O_2,\infty} - \rho_{O_2,0}\right) \tag{11.59}$$

and the reaction rate on the surface of char is written as

$$\beta g_C = g_{O_2} = k_{O_2}\rho_{O_2,0} = k_{O_2,0}\exp\left(-\frac{E}{RT}\right)\rho_{O_2,0} \tag{11.60}$$

where α is "convection diffusion coefficient," $\rho_{O_2,\infty}$ and $\rho_{O_2,0}$ are the concentration of bulk oxygen and the concentration of oxygen at char surface, respectively, and β is the stoichiometric coefficient. Solving Equations 11.59 and 11.60, we have

$$g_{O_2} = \frac{\rho_{O_2,\infty}}{\frac{1}{\alpha} + \frac{1}{k_{O_2}}} \tag{11.61}$$

If further assuming that oxygen diffusion to a char particle is similar to oxygen diffusion to a solid sphere, we have

$$\alpha = \frac{N_u D}{d_0} = \frac{\left(2 + 0.6R_e^{0.5}S_c^{0.33}\right)D}{d_0} \tag{11.62}$$

where N_u, R_e, and S_c are the Nusselt, Reynolds, and Schmidt numbers, respectively; D is the diffusion coefficient; and d_0 is the diameter of the char particle.

When the reaction temperature (or pressure) is very high, the reaction is controlled by oxygen diffusion. The reaction rate is

$$\beta g_c = g_{O_2} = \frac{N_u D}{d_0}\rho_{O_2,\infty} \tag{11.63}$$

and the change of the char particle diameter with time is

$$d_{0,0}^2 - d_0^2 = \left(\frac{4}{\beta\rho_c}N_u D\rho_{O_2,\infty}\right)t = k_c t \tag{11.64}$$

where ρ_c is the char density, $d_{0,0}$ is the initial diameter of the char particle, and d_0 is the char diameter at any time, t.

When the reaction temperature is not high and α is large, the reaction rate is controlled by

$$\beta g_C = g_{O_2} = k_{O_2,0} \exp\left(-\frac{E}{RT}\right) \rho_{O_2,\infty} \tag{11.65}$$

and the change of the char diameter is

$$d_{0,0} - d_0 = \frac{2k_{O_2}\rho_{O_2,\infty}}{\beta\rho_c} \exp\left(-\frac{E}{RT}\right) = k_c' t \tag{11.66}$$

The above example is the simplest case in char combustion. Detailed discussions on coal combustion are given elsewhere.[38-40]

11.5.4 Integrated Coal Gasification Cycle with an Air Separation Unit

There are a lot of public concerns regarding the environmental impacts of coal applications in high-temperature processes, such as pollutant and greenhouse gas emissions. To address the environmental concerns, great efforts have been made to increase the thermodynamic efficiency of coal use for electric power generation. ICGC coupled with an air separation unit (ASU) is among them. This cycle has achieved more than 46% efficiency, which is substantially higher than the conventional coal-fired power plant efficiency of 35%. Coal gasification is partial combustion; the final global reactions are expressed as

$$C + O_2 = CO, \ \Delta h^0_{f298} = -110.5 \text{ kJ/kmol} \tag{11.67}$$

$$C + H_2O = CO + H_2, \ \Delta h^0_{f298} = +131.4 \text{ kJ/kmol} \tag{11.68}$$

$$C + CO_2 = 2CO, \ \Delta h^0_{f298} = +172.0 \text{ kJ/kmol} \tag{11.69}$$

In the gas phase, the following reactions are the most significant to the process of coal gasification,

$$CO + 0.5O_2 = CO_2, \ \Delta h^0_{f298} = -283.1 \text{ kJ/kmol} \tag{11.70}$$

$$CO + H_2O = CO_2 + H_2, \ \Delta h^0_{f298} = -41.0 \text{ kJ/kmol} \tag{11.71}$$

In a gasifier, the temperature and pressure are controlled to obtain optimal combustible gases; CO and H_2 and some hydrocarbons are formed by the reaction of $C + 2H_2 = CH_4$. The products from coal gasification are carbon monoxide and hydrogen and a small amount of hydrocarbon.

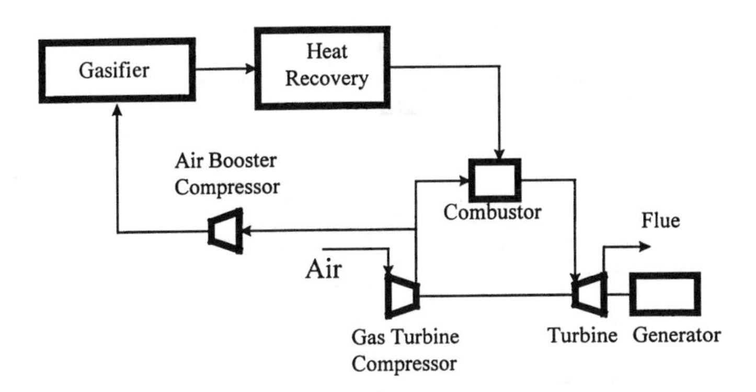

FIGURE 11.5 Basic coal gasification cycle (CGC) flow sheet arrangement: air-blown configuration. (Courtesy of Klosek, P. et al.)

There are three types of gasifiers: entrained flow, fluidized-bed, and fixed-bed gasifiers. The entrained flow gasifier is characterized by using pulverized coal and operating at high temperature (1500 to 2500°C or 2700 to 4500°F) and high pressure (~360 to 515 psia). The fluidized-bed gasifier is operated at a lower temperature with larger coal particles. The fixed-bed gasifier is run at lower temperatures with large coal particles.

Figures 11.5 and 11.6 show the basic flow sheet arrangements for air-blown and oxygen-blown ICGC power production.[41-43] In an air-blown CGC, air is compressed to high pressure and split into two streams. One air stream is further compressed and supplied to the gasifier. Another stream goes to the combustor for oxygen to react with syngas to produce high-pressure and -temperature combustion products to drive a turbine. The turbine is coupled with a generator to produce electricity.

In contrast, oxygen-blown ICGC using a cryogenic ASU (as discussed in Chapter 3) has several configurations: (1) the traditional stand-alone ASU in which there is no integration with the gas turbine; (2) a partially integrated ASU in which part or none of the air fed to the ASU is extracted from the gas turbine (the balance is made up by a separate ASU air compressor) and part or all of the ASU "waste" nitrogen is sent to the gas turbine; and (3) a fully integrated ASU, where full air extraction and full nitrogen returned are used. Each of the cryogenic configuration options has been demonstrated. Air Products and Chemicals, Inc. (Allentown, PA) has designed a 1360 MT/D O_2 stand-alone ASU supplying oxygen to Destec at Plaquemine, LA, an 1800 MT/D O_2 partially integrated ASU for Tampa Electric at Polk County, FL, and a 1760 MT/D O_2 fully integrated ASU for Demkolec at Buggenum, The Netherlands, to meet different customer needs.

At the present time, the overall economical analysis shows that oxygen-blown type is favored when the gasifier is operated at high temperatures.

The effects of air vs. oxygen can be appreciated by considering the key technical parameters in (or influencing) the ICGC flow sheet. This is done below for a reference ICGC design based on entrained flow gasification. The results apply generally to any high-temperature, entrained-flow gasifier (e.g., Texaco, Shell, Destec, Prenflo) and directionally to any type of gasifier. The key technical parameters are compared below.

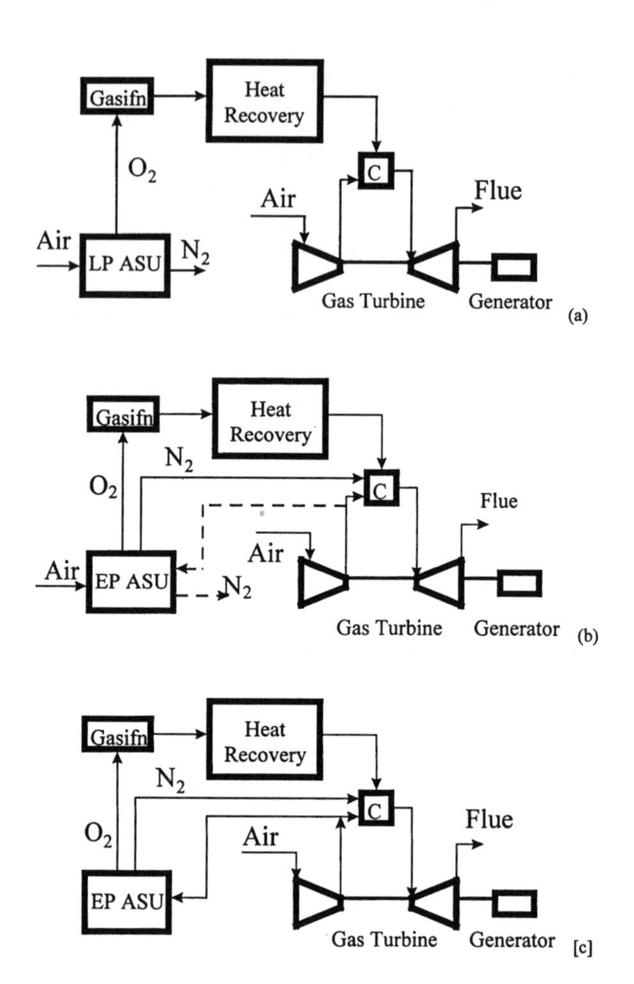

FIGURE 11.6 Basic oxygen-blown flow sheet arrangements: (a) stand-alone ASU; (b) partially integrated ASU; (c) fully integrated ASU (Gasifn — gasifier; C — combustor; LP — low pressure; EP — elevated pressure. Courtesy of Klosek, P. et al.).

The preheat temperature of air for the air-blown case is required to be much higher than for the oxygen-blown case. For example, the oxygen delivery temperature to gasifier is only 150°C (300°F) while air needs to be heated to 1200°C (2200°F) if the ratio of oxygen to dry, ash-free coal (O_2/DAF coal) is 0.99 (wt/wt). As discussed earlier, coal gasification involves endothermic reactions (Equations 11.68 and 11.69). To sustain the gasifier temperature, the preheat temperature of the oxidant must be increased as the oxygen concentration decreases in the oxidant stream. This is because the nitrogen in the oxidant stream acts as a quench to the gasification reactions. If the preheat temperature is fixed, the O_2/DAF coal ratio must be increased. Therefore, more coal can be burned to release more heat to maintain the furnace temperature. For a rational preheat temperature range of 260 to 540°C (500 to 1000°F), the required O_2/DAF coal ratio with air is 27 to 42% higher than that with the oxygen-blown case.

Oxygen-blown gasification has a carbon conversion of 98 to 99.9+%, which is higher than that (90 to 95% carbon conversion) for the air-blown case. This leads to less generation of solid waste (ash or slag).

Indirectly related to gasifier temperature and the type of gasifier is the type of the coal. High-rank coals (from bituminous to anthracite) and petroleum coke that have high fixed carbon and low-to-moderate volatiles contents are generally more difficult to gasify. Higher temperature is helpful, and the demand of oxygen for every pound of coal is high. Hence, these feedstocks are excellent candidates for high-temperature, entrained flow, oxygen-blown gasifiers.

The cold gas efficiency (heating value of clean fuel gas/heating value of feed coal) is 79.2% on a low-heating-value basis (80.7% on a high-heating-value basis) for oxygen-blown gasification, which is higher than that for air-blown gasification. This is in part because more coal completely combusts to convert to carbon dioxide for the air-blown case. Further, oxygen-blown gasification produces the gas with a heating value of 5250 Btu/lb (12,200 kJ/kg) LHV, while the air-blown gasification generates gas with a heating value of 1470 Btu/lb (3420 kJ/kg) LHV.

Clearly, oxygen-blown gasification has many advantages over air-blown gasification in high-temperature gasification. However, the decision on the selection of oxygen-blown or air-blown gasification is dictated by the overall economics and by environmental considerations.

REFERENCES

1. Selected Values of Physical and Thermodynamics of Hydrocarbon and Related Compounds, American Petroleum Institute Research Project 44, Texas A&M University, College Station, Texas.
2. Selected Values of Chemical Thermodynamic Properties, *Natl. Bur. Standa. U.S. Circ.* 500, 1952.
3. Kanury, A. M., *Introduction to Combustion Phenomena*, Gordon and Breach, New York, 1975.
4. Spencer, H. M., *Ind. Eng. Chem.*, 40, 2152, 1948.
5. Kelley, K. K., *U.S. Bur. Mines Bull.*, 584, 1960.
6. Reid, R.C. and Sherwood, T. K., *The Properties of Gases and Liquids*, 2nd ed., McGraw-Hill, New York, 1966, Chap. 5.
7. Stull, D. R., Westrum, E. F., Jr., and Sinke, G. C., *The Chemical Thermodynamics of Organic Compounds*, Wiley, New York, 1969.
8. Stull, D. R. and Prophet, H., *JANAF Thermochemical Tables*, 2nd ed., NSRDS-NBS 37, June 1971.
9. *North American Combustion Handbook*, 3rd ed., North American Manufacturing Co., Cleveland, OH, 1986.
10. Chang, L., Air Products China, Inc., Shanghai, China, personal communication, 1997.
11. Niehoff, T. B., Air Products GmbH, Hattingan, Germany, personal communication, 1997.
12. Wilson, R., Air Products South Africa (PTY) Limited, Sandton, South Africa, personal communication, 1997.
13. Fu, W. and Wei, J., *Fundamentals of Combustion Physics*, Press of Mechanical Industries, Beijing, 1984, 55, 76, and 107.

14. Hanson, S., Beer, J. M., and Sarofim, A. F., *Combustion of Synthetic Fuels*, ACS Symposium Series 217, Washington, D.C., 1983, 95.
15. Nelson, W. L., *Petroleum Refinery Engineering*, McGraw-Hill, New York, 1969, 189.
16. Schmidt, P. F., *Fuel Oil Manual*, 3rd ed., Industrial Press, Inc., New York, 1969.
17. Pepper, M. W., Panzer, J., Maaser, and Ryan, D. F., Combustion of coal-derived fuel oils, in *Combustion of Synthetic Fuels*, ACS Symposium Series 217, Washington, D.C., 1983, 173.
18. American Society for Testing and Materials, Designation: D 341-89 and D 396-92. *1996 Annual Book of ASTM Standards,* West Coushohockeu, PA.
19. Boni, A. A., Edelman, R. B., Bienstock, D., and Fischer, J., Research issues and technology — an overview, in Combustion of Coal-Derived Fuel Oils, in *Combustion of Synthetic Fuels,* ACS Symposium Series 217, Washington, D.C., 1983, 1.
20. Selvig, W. S. and Gibson, F. H., *Chemistry of Coal Utilization*, Lowry, H. H., Ed., John Wiley and Sons, New York, 1945, Chap. 4.
21. Zhang, Y. P., Fu, W., and Mou, J., *Fuel*, 69, 401, March 1989.
22. Sarofim, A. F., Pollutant formation and destruction in *Fundamentals of the Physical-Chemistry of Pulverized Coal Combustion*, J. Lahaye and G. Prado, Eds., Martinus Nijhoff, Dordrecht, 1987, 245.
23. Wells, W. F., Kramer, S. K., Smoot, L. D., and Blackham, A. U., *20th Symposium (International) on Combustion*, The Combustion Institute, Pittsburgh, PA, 1984, 1539.
24. Stubington, J. F., *Fuel*, 63, 1013, July 1984.
25. Seixas, J. P. S. and Essenhigh, R. H., *Combust. Flame*, 66, 215, 1986.
26. Niessen W. R., *Combustion and Incineration Processes*, Marcel Dekker, New York, 1978, 200, 210, and 223.
27. Howard, J. B., Fong, W. S., and Peters, W. A., Kinetics of devolatilization, in *Fundamentals of the Physical-Chemistry of Pulverized Coal*, J. Lahaye and G. Prado, Eds., Martinus Nijhoff, Dordrecht, 1987, 77.
28. Kanury, M. A., Thermal decomposition kinetics of wood pyrolysis, *Combust. Flame*, 18, 75, 1972.
29. Shivadev, U. K. and Emmons, H. W., Thermal degradation and spontaneous ignition of paper shell in air by irradiation, *Combust. Flame*, 22, 223, 1974.
30. Badzioch, S. and Hawksley, P. G. W., Kinetics of thermal decomposition of pulverized coal particles, *Ind. Eng. Chem. Process Des. Dev.* 9, 521, 1970.
31. Kobayashi, H., Howard, J. B., and Sarofim, A. F., Coal devolatilization at high temperatures, in *18th Symposium (International) on Combustion*, The Combustion Institute, Pittsburgh, PA, 1977, 411.
32. Howard, J. B., Fundamentals of Coal Pyrolysis and Hydropyrolysis, in *Chemistry of Coal Utilization*, 2nd Suppl. Vol., M. A. Elliott, Ed., John Wiley & Sons, New York, 1981, 665.
33. Fu, W., Zhang, Y.-P., Han, H., and Wang, D., *Fuel*, 68, 505, April 1989.
34. Solomon, P. R. and Hamblen, D. G., in *Chemistry of Coal Conversion*, R. H. Schlosberg, Ed., Plenum Press, New York, 1985, Chap. 5.
35. Solomon, P. R. and King, H. H., *Fuel*, 63, 1902, 1984.
36. Solomon, P. R. and Squire, K. R., *ACS Div. Fuel Chem. Preprints*, 30(4), 347, 1985.
37. Niksa, S. and Kerstein, A. R., *Combust. Flame*, 66(2), 95, 1986.
38. Field, M. A., Gill, D. W., Morgan, B. B., and Hawksley, *Combustion of Pulverized Coal*, The British Coal Utilization Research Association, Leatherhead, 1967.
39. Smoot, L. D. and Pratt, D. T., *Pulverized Coal Combustion and Gasification*, Plenum Press, New York, 1979.

40. Smoot, L. D. and Smith, J. S., *Coal Combustion and Gasification*, Plenum Press, New York, 1985.
41. Klosek, J., Sorensen, J. C., and Wong, W., Air versus oxygen for gasification combined cycle power, paper presented at VGB Conference — Buggenum IGCC Demonstration Plant, Maastrich, The Netherlands, November 1993.
42. Smith, A. R., Klosek, J., and Woodward, D. W., Next-generation integration concepts for air separation units and gas turbines, paper presented at the International Gas Turbine and Aeroengine Congress & Exhibition, Birmingham, U.K., June 1996.
43. Sorensen, J. C., Smith, A. R., and Wong, M, Cost-effective oxygen for GCC — matching the design to the project, paper presented at EPRI 10th Annual Conference on Gasification Power Plants, San Francisco, October 1991.

Index

D

E

F

G

H

M

N

O

P

T

U

V

W

Z